The Shop Wisdom of Frank McLean

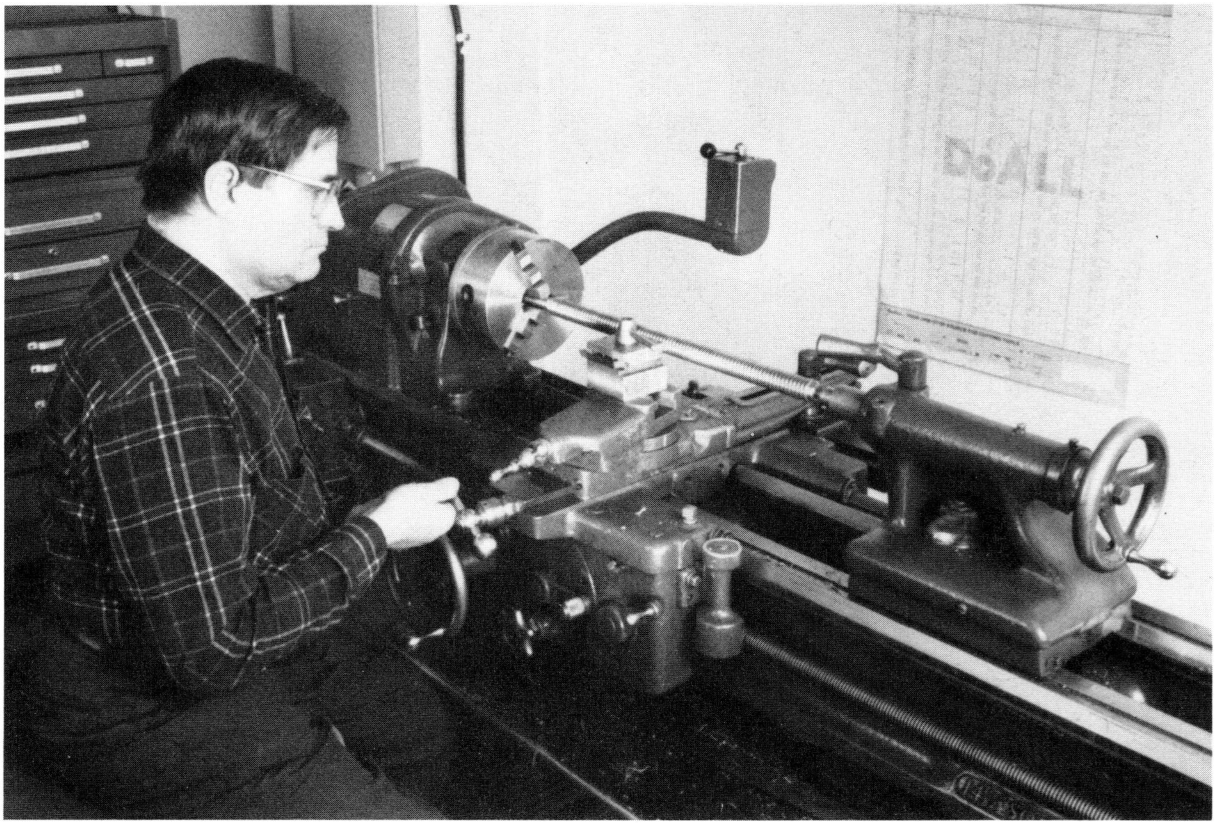

The author in his shop.

The Shop Wisdom
of Frank McLean

Edited by Joe D. Rice

Village Press, Inc.
Traverse City, Michigan

The Shop Wisdom of
Frank McLean

First printing November 1992

Village Press, Inc.
P.O. Box 1810
Traverse City, Michigan 49685-1810

International Standard Book Number 0-941653-06-4
Library of Congress Catalog Card Number 92-60089

Copy-editing by Clover McKinley
Design and Layout by Luana Dueweke

Typeset and printed at Village Press, Traverse City, Michigan, USA

INTRODUCTION

Frank A. McLean has been a well known name among thousands of readers since September 1976. His wisdom, skills, and knowledge has come from many years in industry, teaching, and now in tool design for Lee Valley Tools, one of the largest sellers of woodworking tools in the world. For years, Frank was author of "Beginner's Column" (*Live Steam*) and more recently, as contributing editor for "From the Scrapbox" (*The Home Shop Machinist*). The latter column is for both the novice and the experienced machinist, while the former was basically designed to guide the machinist with little or no experience. It did not matter to whom the columns were directed – virtually everyone read them with interest, and often found ideas, techniques, and designs that were new and unique. You'll notice a distinct difference in the drawing styles from the "Beginner's Column" to "From the Scrapbox;" but aside from that, the common sense, wisdom, and know-how he imparts does not change.

Virtually all his articles from the beginning through 1990 are published here. You'll find an enormous variety of excellent machining technique articles, but it seems to me that where Frank excels most is in the area of designing tools and accessories. They are marvelously simple, and do the job effectively for less money. His power hacksaw is one of those simple but effective designs, and anyone who has a shop without one will surely want to put that machine in action as soon as possible. Whether they are cutting tools, jigs, fixtures, or other accessories, you can't go wrong by following his guidance. The same goes for the many techniques he has developed over his many years in the trade.

Our publications have been truly blessed to have Frank on its roster of regular contributors. I hope you find the contents of this book useful.

Joe Rice

I dedicate this book to my patient and understanding wife Jacqueline (a loyal member of the workshop Widows of the World), and to my father who had the aptitude and the patience to teach a young boy.

PREFACE

Most home workshops evolve without any master plan, and mine is now exception. Fortunately, most amateur machinist are good horse traders who are always on the lookout for a bargain. The basic needs for a home shop are simple: a sturdy workbench with a proper machinist's vise with jaws 4" to 6" wide, a lathe, and then a milling machine. With these tools, almost anything is possible.

If space or funds are very limited, a lathe can do everything, but milling is difficult, so one of the new mill/drill-type machine tools will be a good investment, and can be used as a drill press.

I mention these basic tools because a young chap starting out with a family, a mortgage, car payments, and more, may become discouraged when he looks at the marvelous home workshops shown in *Live Steam* or *The Home Shop Machinist* magazines.

I was six or seven years old when my father started to take me and my brother to Bill Leggett's "Toad Swamp and Punk Hallow Railroad" at his house in St. Lanbert, Montreal, Canada. This was about 1933 or '34. As far as I can remember, we had a shop at home. My father started me on the lathe when I was nine or ten years old. He had a series of small English lathes.

I finished high school, and then completed a four-year machine shop course at the old Montreal Technical School. This was a very thorough course, and it's hard to find courses today in the technical schools that are comparable. I have worked as a machinist, toolmaker, foreman, setup man on screw machines, inspector, draftsman, and machine shop teacher. I am nearing retirement now from Lee Valley Tools, in Ottawa, Ontario, where I'm a tool designer.

I call my home shop "The Hermitage" for it is a great retreat from the cares of daily living. I'm also building a railway in 1 1/2 " scale at the back of our lot, with trestles, buildings, and whatever else seems appropriate. This will depict a small branch line during the 1890s, and will be appropriately named the "Frog Pond & Mosquito Haven Railway."

Frank Allen McLean

TABLE OF CONTENTS

Techniques

Drill Press Tools & Accessories

Lathe Accessories

Milling Machine Accessories

Shop Tools & Improvements

Techniques

Planning a Small Shop

Most small shops simply evolve without much prior planning for the shop area and the machinery acquired. I say acquired, because most shop owners seem to be born horse traders, and the shop develops in no particular order, but only as the opportunities appear to buy or swap for extra equipment.

When moving to a new house, or if much new or large equipment is being considered, then a serious evaluation of the requirements of the shop and facilities available should be considered.

Those living in the warmer climates will most likely have outdoor workshops as a separate building or part of a garage. When one half of a two-car garage is used as a machine shop, a dividing wall can be installed with the car on one side. When the car is out of the garage, the area can be used for general woodworking, patternmaking, welding, forge work or as a foundry. These dirty operations should be kept out of the machine shop so dust will not settle on the machine tool ways and, also, so the fumes from welding will not rust the tools. Woodworking always requires lots of room to pass long boards through the circular saw and jointer. The garage door can be left open when extra space is required.

For those people living in northern climates, the building will have to be very well insulated and heated to a constant temperature to avoid rusting. Gas or electric heating systems are probably most suitable.

Some of the garden workshops in Europe are quite small, simply to reduce heating costs.

The outdoor workshop will have to be well ventilated or even air-conditioned during the summer months, if you are to avoid having a "sauna bath" each time you enter your shop. A wooden, well-insulated floor laid over the concrete floor would be worthwhile to reduce heating costs and eliminate cold feet.

The shops in the southern areas will probably require a dehumidifier during some months. A cover of plastic, or thicker material such as an old blanket placed over each machine when you leave the shop, will reduce the chance of condensation settling on the machine tools, if there is a big variation in the night and day temperatures.

Most people living in the northern areas will have a basement shop simply because the basement is there and it is heated by the furnace. Any problems with humidity usually occur during the summer when a dehumidifier or good cross ventilation should cure the problem. Basement shops need to have the walls insulated to be comfortable and this is really necessary during this time of increasing heating costs. Better do this before the shop is set up or you will have to move everything away from the walls when they are insulated.

A separate outside basement entrance is really necessary for a basement shop. Many basement shops have an inside entrance from a hall or the kitchen. This means that all the machinery and long steel bars, pipes, etc., have to be carried through the house. This limits the size of the machinery in the shop and carrying oily bars through the house may not be popular with your wife. I also find that friends prefer to drop in for a visit by the basement entrance without disturbing the family upstairs.

An outside entrance with a small enclosure or shed over the entrance can have a steel I-beam and a trolley with a chain block to bring in any large machine tools or other items too heavy to lift by hand (Figure 1 and 2). If wooden steps are placed in the well, they can be moved out of the way so that heavy objects can be lowered to the floor level. The shed will have to be built to support the I-beam properly. If this shed or enclosure area is increased in size, the space could be used for welding and forge work so the fumes and the noise would not pass through the house. This work could also be done in a nearby garage or shed. I do my welding in the basement, but the fumes from brazing are a problem.

There are no fixed rules for the shop layout since so much depends upon the room available. Long, narrow rooms seem to be common. This means that the machinery and benches will have to be placed along the walls. The bench should be located so that long bars or pipes may be held in the bench vise for cutting or threading. A common arrangement is shown in Figure 3.

The lathe should be located so that a long bar may be passed through the spindle, should the

FIGURE 1
Plan View

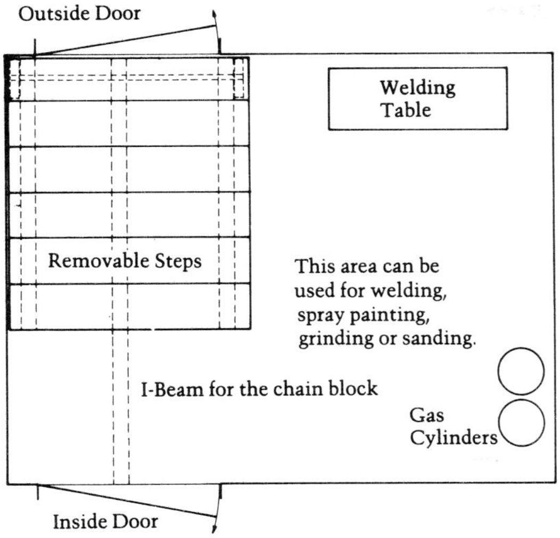

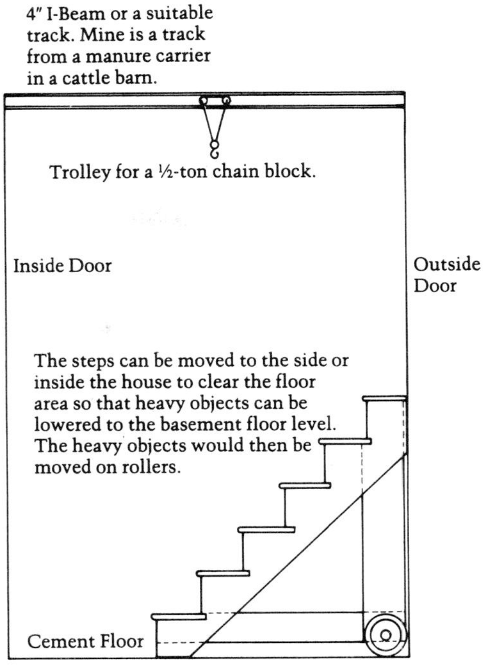

FIGURE 2
Side View

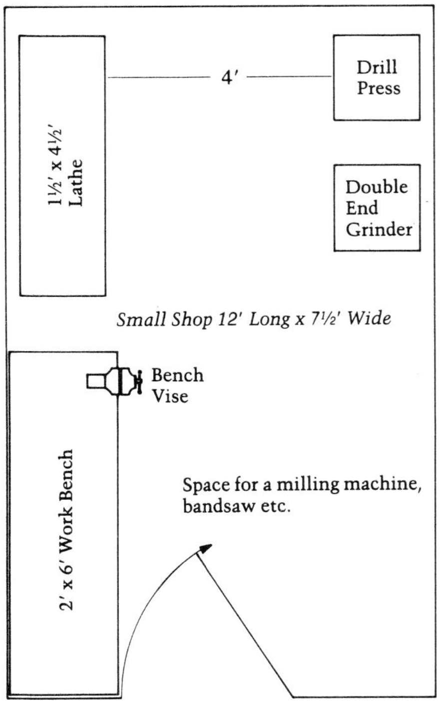

FIGURE 3

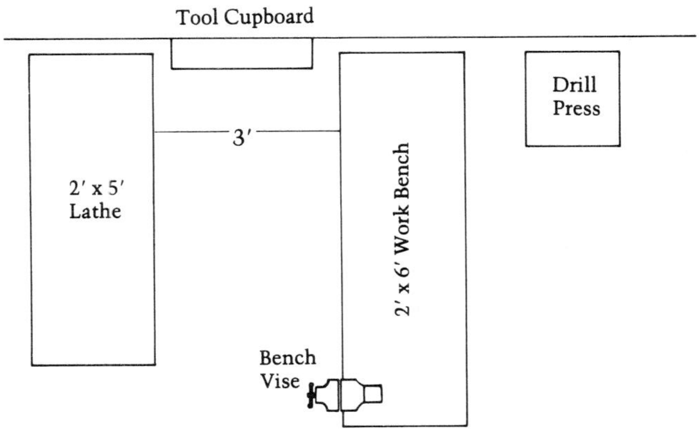

A space-saving arrangement for a lathe and a work bench.

FIGURE 4

need arise. If possible, leave enough room between the back of the lathe and the wall to be able to sweep up the shavings and to be able to get at the back of the lathe for maintenance and leveling.

Two workbenches are preferred by some; one to be used for the surface plate and for fine work, and the other for rough work, carpentry or pattern-making. Very long benches were popular in the old machine shops. Usually, they ran along one wall and had many drawers and vises at about six-foot intervals.

Try and plan for any machine tools to be purchased in the future as you set up your shop for the first time. This avoids a major rearrangement every time a new machine is brought in.

An open area, usually in the middle of the shop floor, is nice to have if some sizeable job has to be assembled. This would be called the "erecting bay" if we use the old shop terminology.

A square room gives more flexible space to place the machinery in groups, or to have the large machine tools set out in the open. This is important for large horizontal and vertical milling machines, shapers, surface and cylindrical grinders, etc., where access to all sides of the machine is necessary.

A layout I have found convenient is shown in Figure 4. As you can see, the lathe and a bench are placed so they face each other with the tailstock end of the lathe and the bench against the wall. This gives access to the back of the lathe and the bench is used to hold the tools and drawings when the lathe is in use. The tool cupboard is for drills, taps and dies, and lathe attachments. This compact arrangement may be used in a very small shop or in a large shop in order to leave space for a large milling machine, band saw or other machine requiring a lot of space.

Try to have all the tooling for each machine stored in close proximity to the machine so that you avoid unnecessary steps, just as your wife does when planning a kitchen.

Storing metal, bolts, small parts, etc., has always been a problem. Steel may be stored in another area, if room is not available in the shop itself. I store mine in a set of 18 × 36" metal shelves in a waste space area under the basement stairs. I inverted the shelves so the short lengths of rod would not roll off the shelves. The small metal cabinets with plastic drawers are so inexpensive these days that they are a very practical means of storing screws and bolts, small cutting tools and small parts. I was glad to be rid of all the glass baby food jars. A set of wooden shelves can be made up to hold all the cabinets, or they may be tucked away in any old space available.

The most suitable lighting for a shop will include the 4- or 8-foot fluorescent tubes. Aim for a good general shadow-free lighting situation. You will need a fluorescent fixture over the bench. The lathe and milling machine usually require some extra lighting, either from incandescent lights in flexible fixtures or fluorescent lights.

Cutting oils, lubricating oils, solvents, etc., are best stored outside the house to reduce the risk of fire, with only small quantities brought into the shop for daily use.

A set of steel shelves to hold the work in progress is very useful and will keep the workbench from becoming a catch-all. Small parts can be placed in bread or cake pans and placed on the shelves.

A filing cabinet is a very useful addition to the shop for all drawings, catalogs and instruction books. They do tend to pile up over the years and need to be organized and filed away.

If the shop is fairly large, a tool cabinet or "caddy" on casters, such as might be used by an auto mechanic, is very useful. This can be rolled to the machine in use and all the measuring instruments and other tools can then be right at hand, thereby avoiding unnecessary steps. If the bottom and middle sections are used, there will be a useful area on top of the chest to place wrenches and other tools when setting up the machine.

Drilling and Tapping Hints

Trying to remove a broken stud or bolt from an engine block can be a very frustrating experience. Usually the broken stud is partially hardened from the heat, so the drill bit slips off the stud into the soften cast iron, and then the block may have to be scrapped. (The alternative is a drilling and plugging operation. This usually requires a fair amount of luck to be successful.)

The stud may be removed by welding a piece of rod to the broken stud if it stands above the surface of the block. The rod is gripped by vise grip pliers or a pipe wrench. Use penetrating oil and try to unscrew the stud.

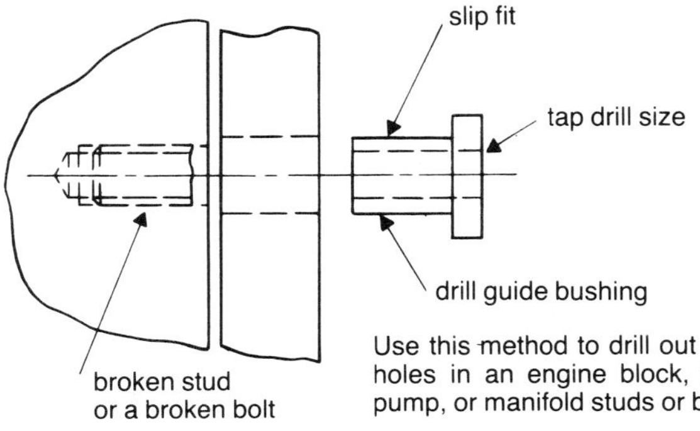

Procedure to remove a broken stud:

1. Bolt the part in place with the other studs.
2. Make a drill bushing.
3. Drill and then remove the bushing.
4. Run a tap in to clear the threads.

Use this method to drill out and re-tap the stud holes in an engine block, for a broken water pump, or manifold studs or bolts.

Figure 1

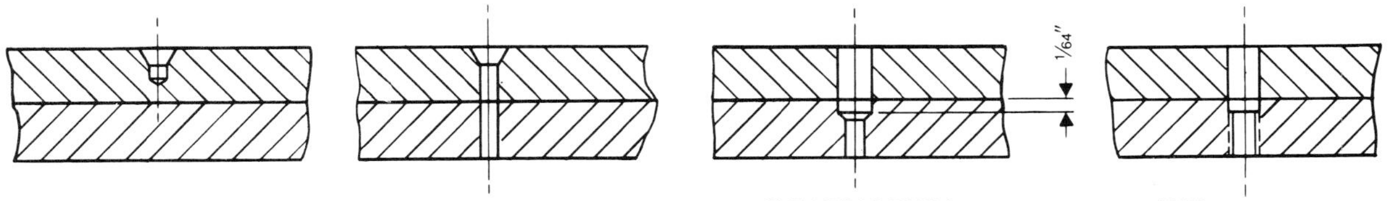

1. CENTER DRILL

2. TAP DRILL
Drill through the bottom plate or to the correct depth in a casting.

3. CLEARING DRILL
The clearing drill should penetrate the bottom plate 1/64" to 1/32".

4. TAP
Use the clearing hole to guide the tap when tapping with a tap wrench or hold the tap in the drill chuck and rotate by pulling the vee belt by hand.

THE PROCEDURE TO FOLLOW TO DRILL AND TAP TWO PLATES FOR BOLTING TOGETHER AND FOR BOLTING A PLATE TO A CASTING

Figure 2

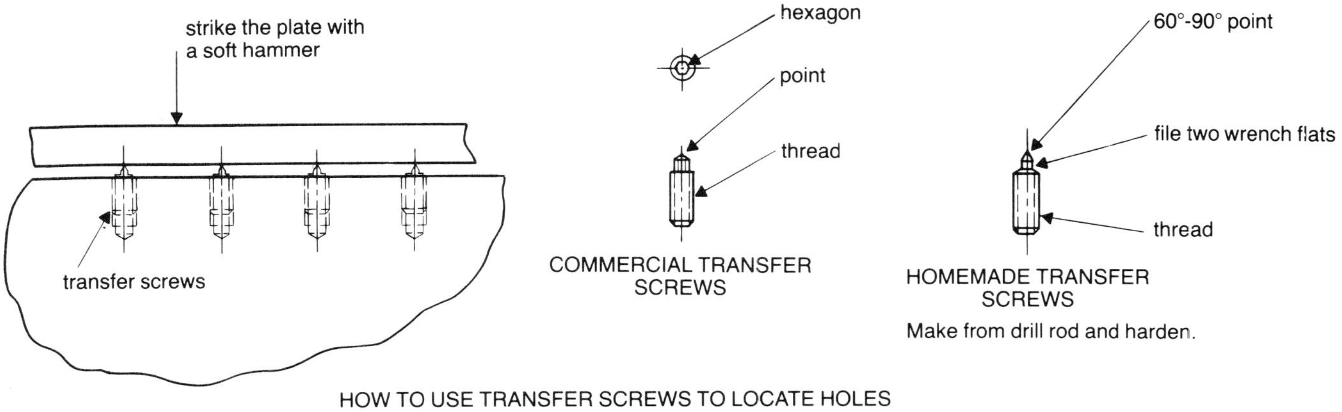

HOW TO USE TRANSFER SCREWS TO LOCATE HOLES ON A REPLACEMENT PART FROM BLIND HOLES PREVIOUSLY DRILLED AND TAPPED

Figure 3

The break usually occurs below the surface of the block. When this happens make up the bushing shown on the drawing and drill out the broken stud. If the stud is a large one, make up several bushings so you can start with a small drill, and gradually open up the hole to the tapping size. This will make the drilling easier, for you will not have to push so hard on the drills. Make sure the drills are sharp and use cutting oil. Remove and sharpen the drills again as soon as they start to dull.

Use a bent scriber to pick out the threads in the larger holes, and then pass a tap to remove any remaining threads.

Two plates should be drilled and tapped together in the sequence shown. Clamp the work in place on the drill press table, and complete each hole before shifting the work to drill and tap the next hole. This requires patience but ensures that the clearing holes and threads will be concentric. Perform a countersinking or counterboring operation at this time, if required. Be sure the clearing drill enters the bottom plate 1/32" or so. This will prevent the tapping operation from raising a burr above the surface of the plate and causing the plates to seat improperly. This is an old toolmakers' method.

To locate holes on a plate from blind holes in a block or plate is easy when you use transfer screws. If possible, buy them – but they can be easily made by threading drill rod, turning a cone point and filing two wrench flats. Part off and then harden as usual.

When the center of the holes are located by the transfer screws, use a wiggler to locate the plate on the drill press for drilling the clearance holes. Alternately re-punch the holes with a center punch, center drill, and drill to size.

To Find the Tap Drill Size

Most of us use the charts to find the correct drill (tap drill) size to use before tapping a thread. Model building often requires the use of finer-than-standard threads and for these the charts won't help.

Here is a simple method to find the tap drill size. Subtract one thread from the outside diameter of the screw.

$$\text{Tap Drill Size} = \text{Screw Diameter} - \frac{1}{\text{Number of Threads per inch}}$$

Example: a 5/16-32 screw

Tap Drill Size = 5/16"–1/32" = 10/32"–1/32" = 9/32" drill size

Often it is more convenient to change the screw size and the pitch $\frac{1}{\text{No. of Threads per in.}}$ to decimals.

Example: a 3/8-40 screw $\frac{1"}{40} = .025"$

.375 – .025 = .350"

Choose the nearest drill size – in this case, letter S.

To find the outside (major) diameter of the machine screws if they are not shown on the chart, multiply the number of the screw size by thirteen thousandths and add the base size of 60 thousandths.

Example: No. 6 screw size

6 × .013" = .078 + .060" = .138" diameter

Here is a comparison chart for the National Machine Screw and British Association Sizes:

National No.	Major Dia.	British Association No.	Major Dia.
No. 0	.060"	No. 10	.067"
No. 1	.073"	No. 9	.075"
No. 2	.086"	No. 8	.087"
No. 3	.099"	No. 7	.098"
No. 4	.112"	No. 6	.110"
No. 5	.125"	No. 5	.126"
No. 6	.138"	No. 4	.142"
No. 8	.164"	No. 3	.161"
No. 10	.190"	No. 2	.185"
No. 12	.216"	No. 1	.209"
No. 14	.242"	No. 0	.236"

Drilling Center Holes Accurately

There are situations when the center holes must be drilled exactly concentric with the circumference of a shaft or when a tool such as a machine reamer has been shortened. This is easily done by this old-fashioned method even though the three-jaw chuck is not reliable for accuracy.

Place the shaft in the three-jaw chuck and tighten the chuck. Slide the steady rest along the lathe bed until it is close to the chuck jaws (Photo 1). Advance the two lower jaws until they just touch the work and then clamp in place.

1

Loosen the chuck jaws and slide the steady rest along the lathe bed until the shaft is held by only 1/8" in the chuck jaws (Photo 2) and the shaft protrudes by 1/4" to 1/2" beyond the steady rest (Photo 3). Clamp the steady rest to the lathe bed and tighten the chuck jaws. If your steady rest has a hinged top, bring this down but do not clamp it in position.

Advance the top jaw of the steady rest until the top of the frame just starts to lift. Now clamp the frame in position and then clamp the top jaw. If necessary, adjust the top jaw until there is no clearance but the work can rotate without binding. Oil the shaft. This method is more precise than using a strip of paper between the shaft and the steady rest jaws.

The reason for holding the shaft in the three-jaw chuck by only 1/8" of the shaft is to permit the shaft to move slightly in the chuck jaws if the chuck is not accurate.

Now face the end of the shaft in the usual manner and then center drill.

For greater accuracy, set the compound slide rest at 30° to the lathe axis, then use a very small boring bar to take a light cut in the center hole.

Repeat the procedure for the other end of the shaft if required.

A variation of this method will be used to drill the center hole in the shank of a machine reamer that has been cut off or damaged.

2

How to Drill The Center Holes in a Shaft Accurately

The old ways are often best and this method is one of them.

Place the work in the three-jaw chuck temporarily to adjust the two lower jaws of the steady rest (Figure 1 and Photo 2). Slide the steady rest along the lathe bed until it nearly touches the chuck. Adjust the two lower jaws to just touch the bar and then lock them. Slide the steady rest towards the tailstock to the approximate location (Photos 3 and 4). Now loosen the chuck and pull the work out until the chuck grips the last 1/8" of the work. Set the steady rest at the other end of the bar and clamp it to the lathe bed. To adjust the top jaw of the steady rest, loosen the clamping bolt on the steady rest frame. Now adjust the top jaw until the frame just starts to lift, then tighten the frame clamping screw. Use a heavy oil to lubricate the work and the steady rest jaw.

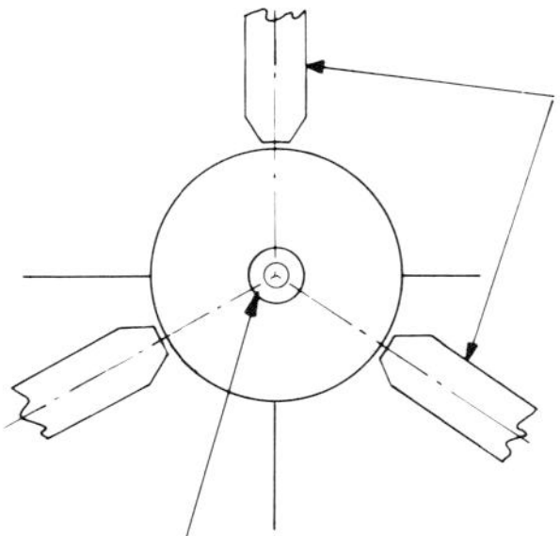

Figure 1

2

3

4

The work is placed at the extremity of the chuck jaws so that the work can move slightly if the chuck is running out of true (Figure 5).

The steady rest is placed at the end of the work to minimize the effect of the chuck jaws running out of true (see Figure 5).

Face the end at the work flat and center drill from the tailstock drill chuck. Repeat the process on the other end of the work. The accuracy obtained by this method will be suitable for most purposes.

If greater accuracy is required – say, for a center hole in a reamer shank that has been shortened – then a boring operation is required.

Attach a lathe dog on the headstock end of the work and tie the dog to the faceplate with a leather boot lace or strong string (Photo 6). Those who have a lathe with a threaded spindle should unscrew the faceplate one or two turns before tying the boot lace. Screwing the faceplate back will tighten the lace to hold the work firmly on the headstock center.

Set the compound slide rest 30° to the lathe axis. Now prepare a very small boring bar and set it at the center height. Feed the boring bar with the compound slide rest to take a slight cut to true up the center hole.

The work is turning in a bearing (the steady rest jaws) on its own circumference. Therefore, the boring operation will ensure that the center hole is concentric with the circumference to an accuracy of ±.0005".

If a test bore is being made, then repeat this operation on the other end of the work.

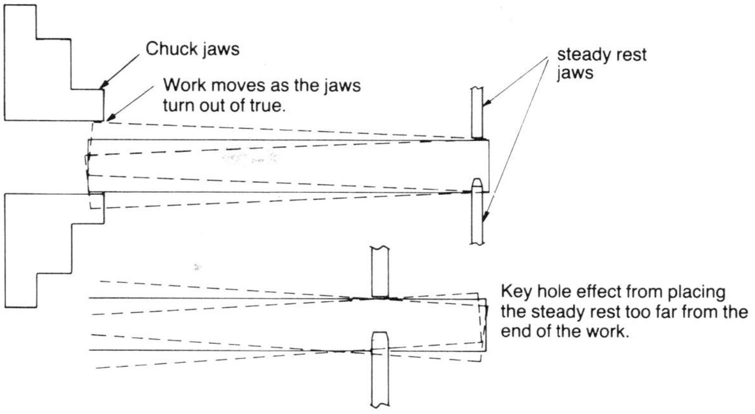

Figure 5

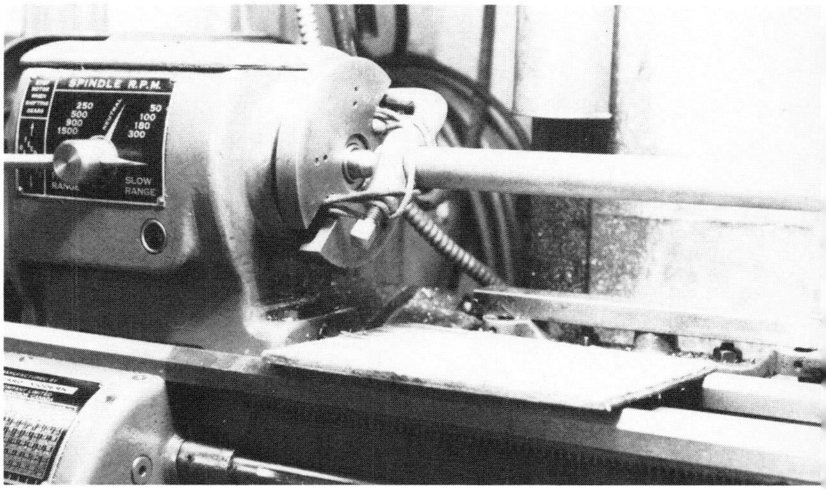

6

A Few Tips on Drilling on a Drill Press or a Vertical Milling Machine

The drawings show some easy methods of drilling a hole exactly through the center of a shaft. Either the offset method or the layout method may be used.

The finger stop bolted to the side of the vise will position the work accurately when many pieces are to be drilled. Just lay out the first piece and locate with a wiggler. This finger stop can usually be used instead of making a special drill jig.

The universal drill jig arm bolted to the top of the fixed jaw of the vise uses standard slip drill bushings. For limited use, bushings made from mild steel and not hardened will be suitable.

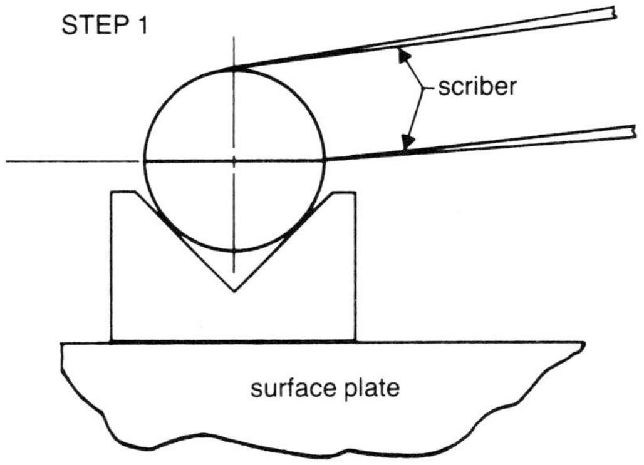

THREE STEP METHOD
TO DRILL A HOLE THROUGH A SHAFT

Bring the scriber of a surface gauge to just touch the top of the work. Move the gauge to bring the scriber to touch a rule set vertically on the surface plate. Note the reading and then adjust the scriber to ½ the work diameter. Scribe a line across the end of the work and along one side. Use a combination square to scribe a line across the center line the correct distance from the end.

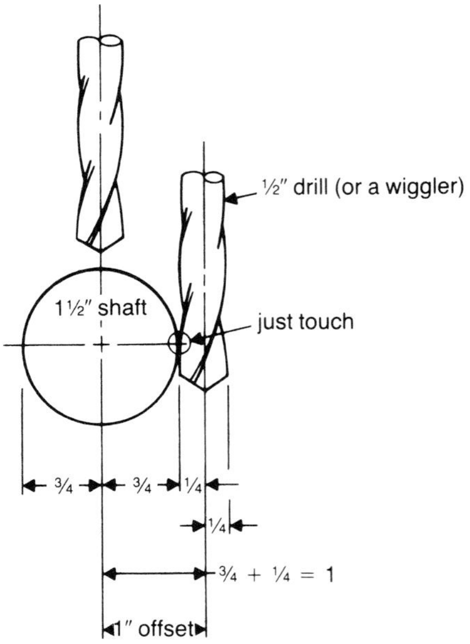

$$\text{offset} = \frac{\text{dia. of shaft} + \text{dia. of drill}}{2}$$

OFFSET METHOD FOR LOCATING A DRILL
OVER THE CENTER OF THE WORK
Use a compound table on the drill press
or a vertical milling machine.

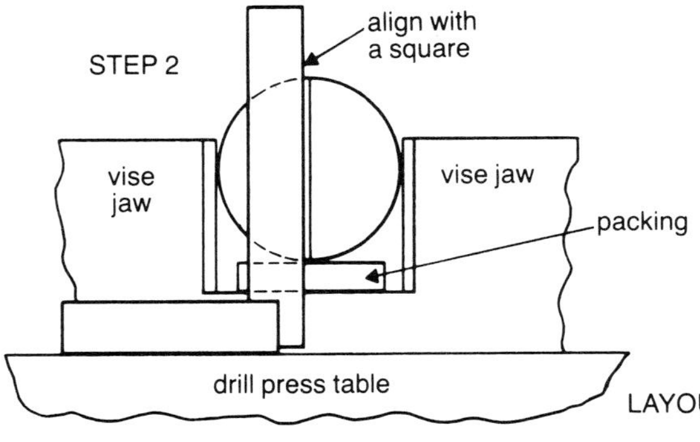

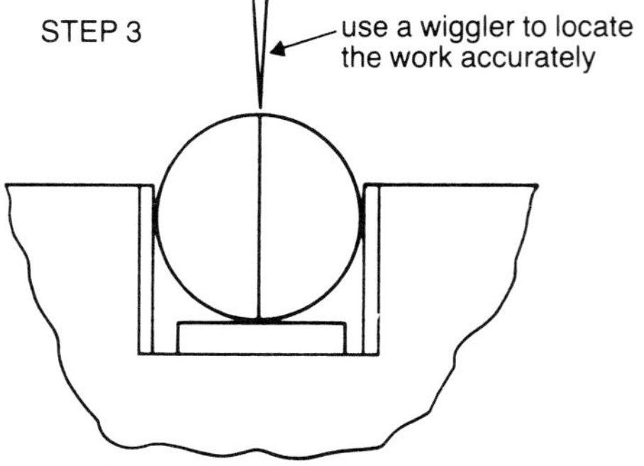

LAYOUT METHOD

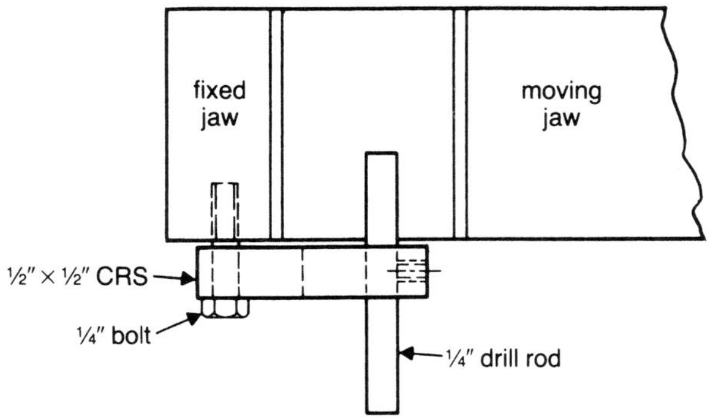

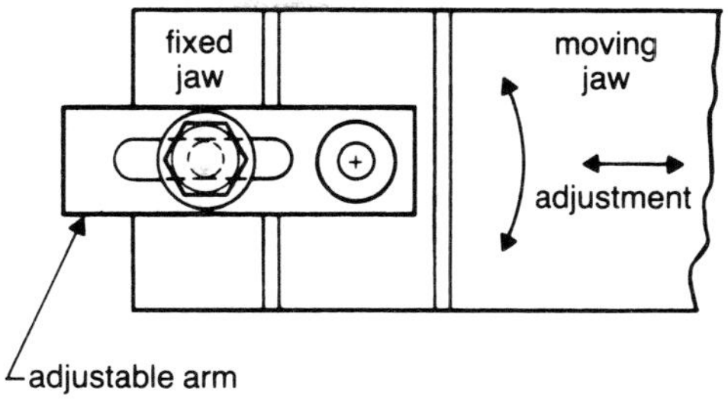

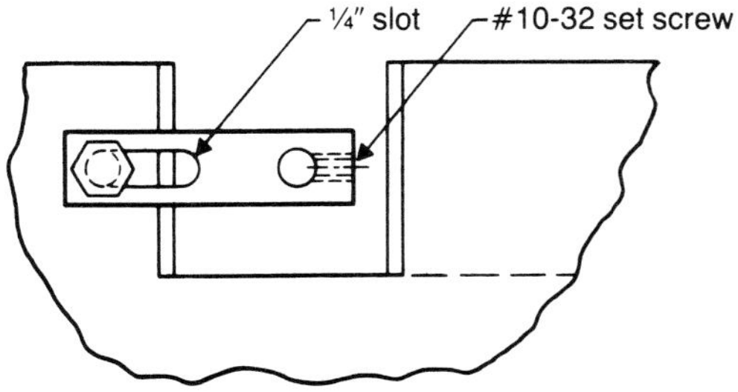

LOCATING ROD OR FINGER STOP FOR A
DRILL PRESS OR MILLING MACHINE VISE

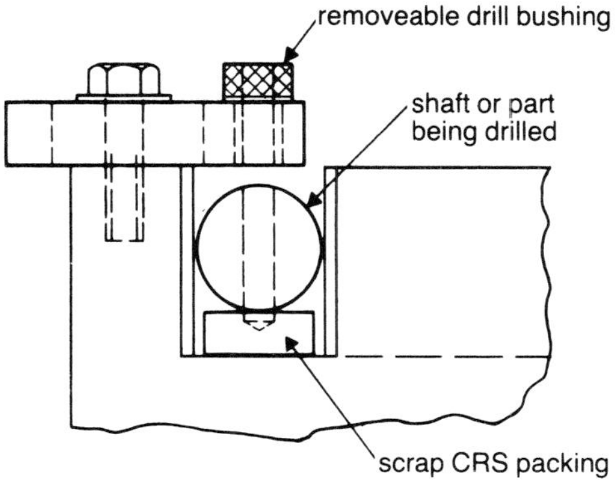

DRILL VISE CONVERTED TO A DRILL JIG

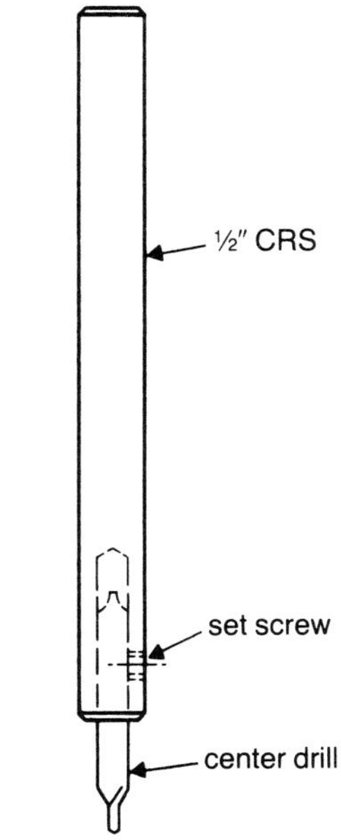

EXTENDED CENTER DRILL

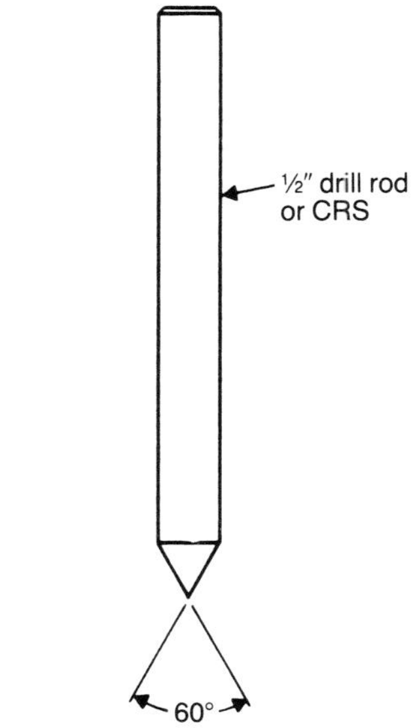

EXTENDED CENTER
For locating the work when long drills are to be used and the table must not be moved. Also to guide large taps.

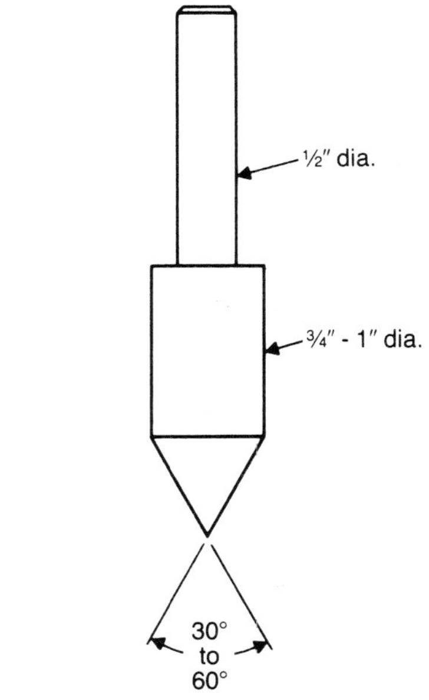

CENTER PLUG
For locating the work from a previously drilled or cored hole.

Chain Drilling Locomotive Frames

Use the chain drilling method to remove a section of metal surrounded by the frame that you do not want to cut. Toolmakers used this method to remove sections from a die block. (Nowadays, the section is sawn out on a bandsaw, after the blade has been passed through a narrow slot, and welded.)

Decide on the radius to be used on the corners. Lay out the center lines of the holes (say 1/2" diameter) and the outline of the opening. Now lay out the centers of the chain of holes along their center lines, forming the radii at the corners. Choose a common size, such as 1/4". Center punch each hole carefully. Now drill out the corner holes using several drills so the finished hole will not be oversize.

There are three methods of drilling out the small holes:

a) Drill each alternate hole 1/4" and then go back and drill the remaining holes.
b) Drill each hole with a drill 1/64" undersize so there will be a thin section between each hole. This can be easily cut through with a knock-out chisel. (I prefer this method.)
c) Use a plug (1/4" diameter – see drawing) in the hole previously drilled so the drill will not wander into the hole.

Grind a standard cold chisel to the shape shown in the drawing to make a knock-out punch (chisel).

Place the frames in the bench vise with the section to be removed just above the vise jaw for maximum support. Knock out the sections between the holes and shift the frames in the vise so the section being cut is close to the vise jaws.

A hacksaw or a band saw could also be used to saw through the sections between the holes. A

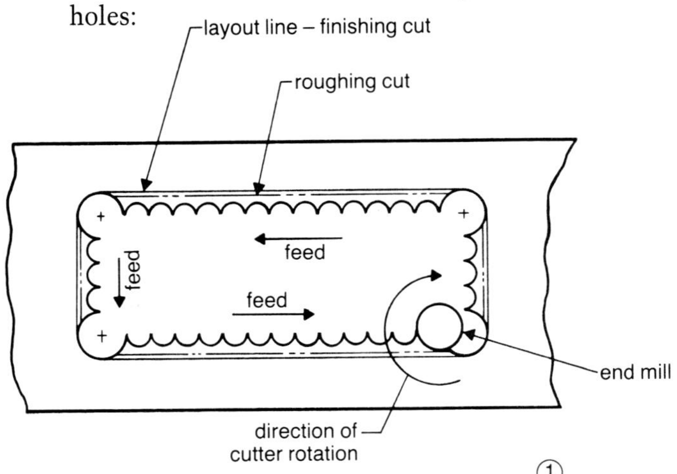

FRAME AFTER SECTION HAS BEEN REMOVED. READY FOR MILLING ①

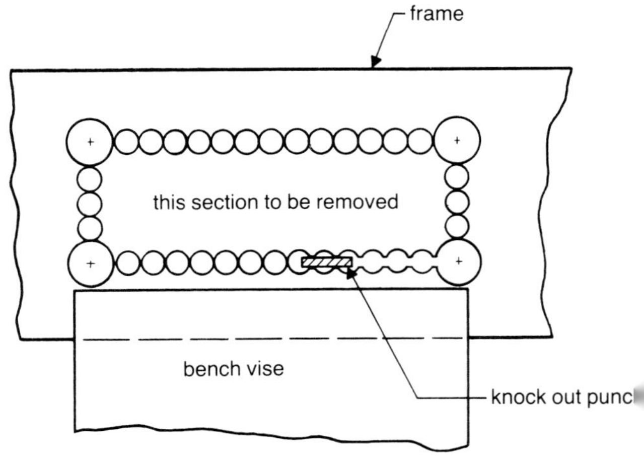

HOW TO USE THE KNOCK OUT PUNCH ②

Do not use a standard cold chisel the wedging action may bend the frame.

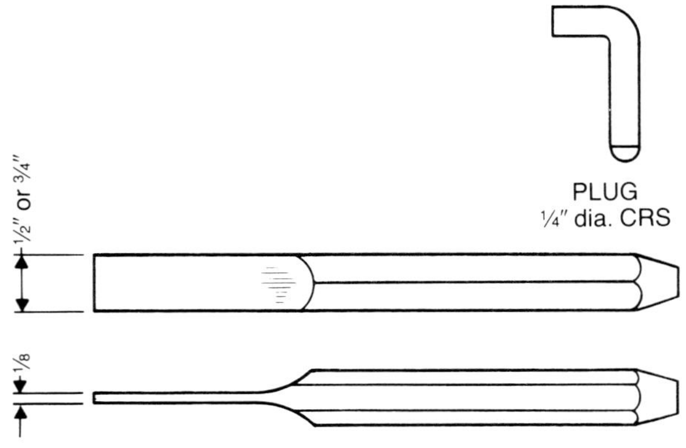

KNOCK OUT PUNCH
make this from a standard cold chisel by grinding

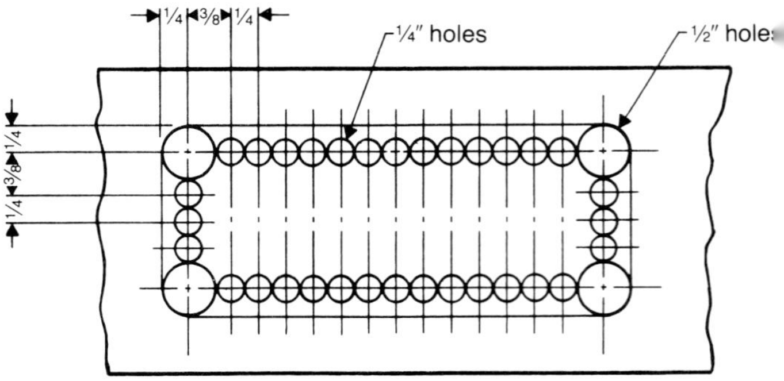

FRAME-LAYOUT FOR CHAIN DRILLING

1/2" wide hand hacksaw blade will just pass through a 1/2" hole in one of the corners. The bandsaw blade will have to be broken to pass through the hole and then welded for each section. For those with a DoAll or Powermatic band saw with a blade welding attachment, this will be easy and could eliminate drilling the chain of holes. A cutting torch with a small tip could also be used without too much distortion.

The rough edges inside the opening can be finished by filing or milling (on a vertical mill).

For filing, use a double-cut file about 12" long. Place a 1/4 × 1" strip of metal at the corners to protect the radii in case the file slips.

For milling, use an end mill the same diameter (or slightly smaller) as the holes in the corners. Align the frames from the front edge of the table with a combination square and clamp the frames down on parallels so the frames will clear the table. Use a round rod in the spindle collet the same diameter as the corner holes to locate the table limit stops to limit the table travel lengthwise (X axis) and crosswise (Y axis).

This is a difficult milling operation because of the alternate light and heavy cuts along the scalloped edges.

Set the table .010-.015" inside the limit stops to make the roughing cuts. *Do not* attempt to mill the full depth of the frames at one pass, but make a series of cuts along one side about .050" deep each cut. The cutting lips will be doing the work and these can be easily resharpened if you are using a two-lip end mill. A full depth cut can cause the cutter to deflect into or away from the work (see drawing). Rough the mill one side to the full depth before relocating the table to mill another side.

To make the finishing cut, place the cutter in one of the corner holes at the full depth and move the table up to the limit stops. Mill around the opening for the finishing cut. Don't forget to feed the table against the direction of the cutter rotation. A final cut with the table moving in the direction of the cutter rotation with the table against the limit stops will take a very light cut from the "spring" in the cutter. This will give a fine finish and is called a "feather cut."

The two frames can be milled together. How-ever, if the frames are very thick, they can be drilled together and then separated for milling. If the layout is carefully done, the openings should be identical.

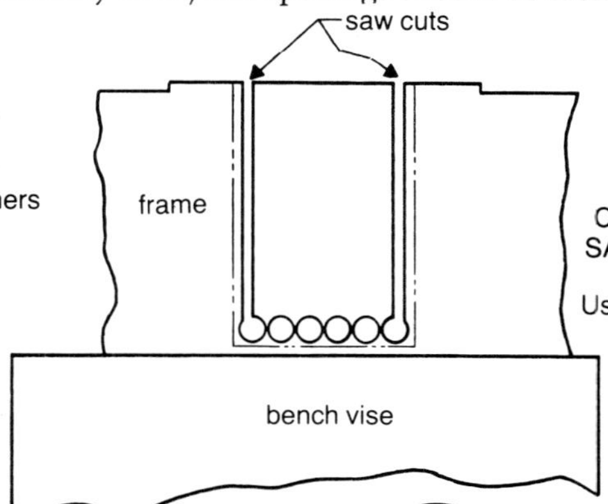

CHAIN DRILLING AND SAWING OUT AN AXLE BOX OPENING
Use an ordinary cold chisel. The wedging action will remove the section.

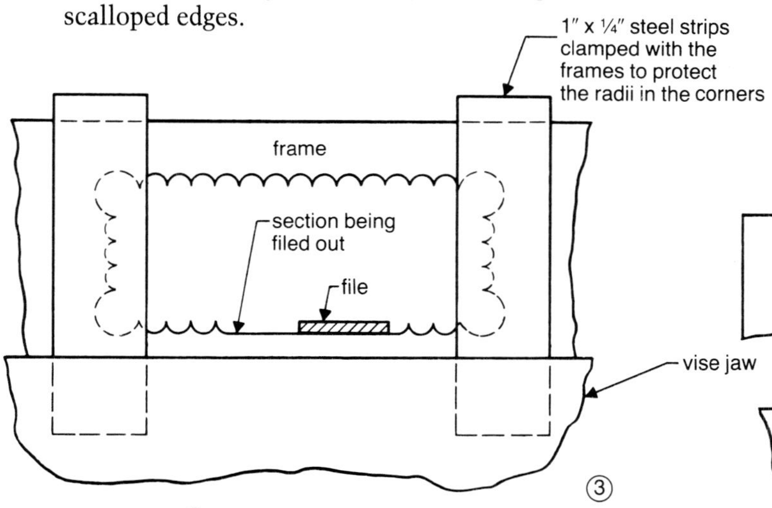

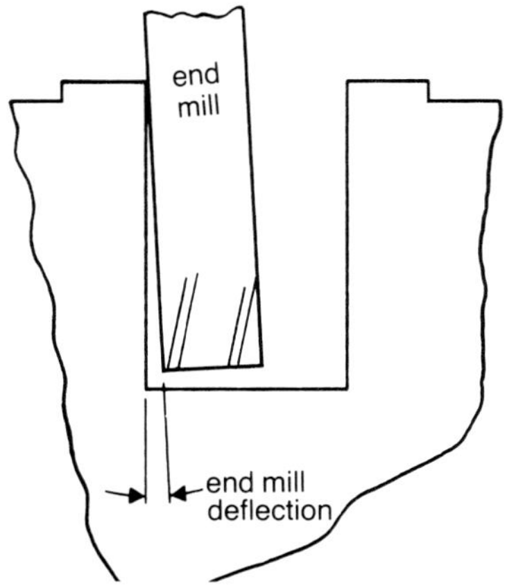

When a full depth cut is taken with the side of the cutter (for roughing) use a full depth cut for a light (.005"-.010") finishing cut.

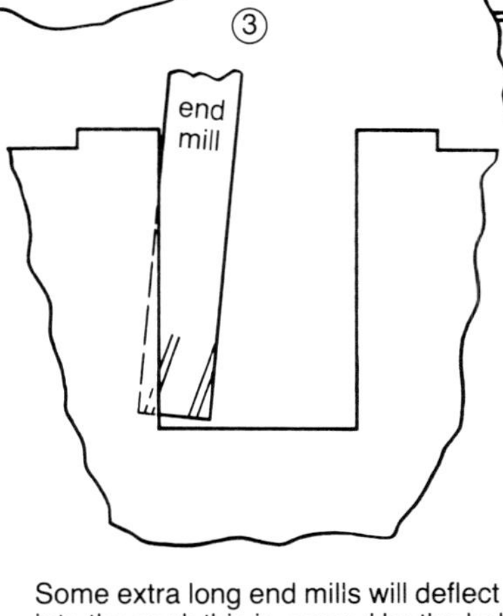

Some extra long end mills will deflect into the work this is caused by the helix angle of the flutes.

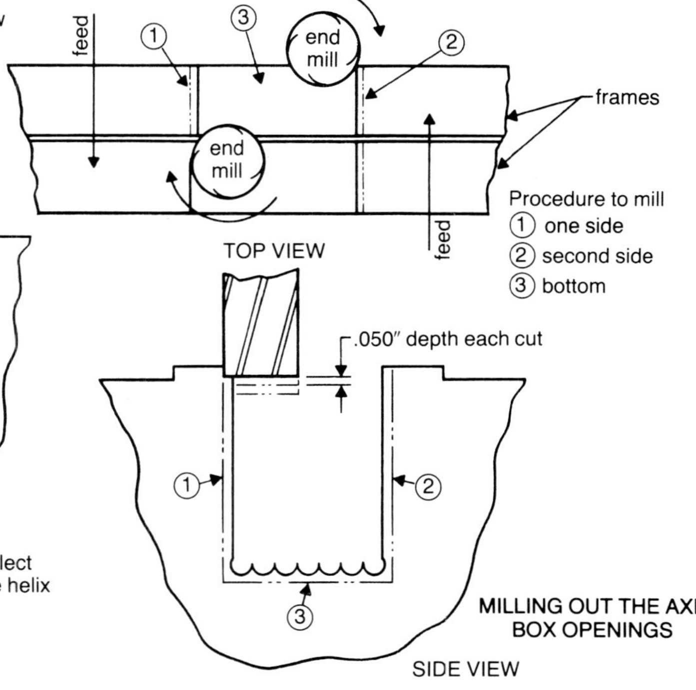

MILLING OUT THE AXLE BOX OPENINGS

Using Your Drill Grinding Attachment

Most beginners have difficulty correctly sharpening twist drills. Using a drill gage to check the 59° angle and the length of the cutting edges helps, but the drill still has to be ground freehand, and the 12° relief angle varies considerably.

A good drill sharpener such as the Darex will grind drills accurately, but the cost of such a tool may be a bit too high for some people. The alternative may be to buy a drill sharpening attachment such as one made by General Hardware and sold by Sears and other stores.

Many people have difficulty obtaining satisfactory results with this grinding attachment because the screw holding the base to the bench has to be loosened to permit the attachment to be moved closer or farther away from the grinding wheel to suit the diameter of the drills being ground. This is critical since the relief angle of the drill is determined by the distance of the attachment from the grinding wheel.

Douglass McIntosh and Grant Wood, both members of the Ottawa Valley Live Steamers and Model Engineers (my club), have published a description and drawings for a suitable adjustable base for this drill grinding attachment in our club newsletter, "The Link."

The following description is by Doug and the original drawings are by Grant. I recommend this adjustable base because it works and can be made in two or three evenings.

How to Successfully Use Your Drill Grinding Attachment

by J. D. McIntosh

Photo courtesy of Author
Drawings by Grant Wood

On numerous occasions, I have heard comments from fellow model engineers indicating that they have a problem using their grinding jig. These problems have also been the subject of letters to the editor of *Model Engineer* (e.g., Volume 146, pages 373 and 764).

While some of the unsatisfactory results obtained may be due to inadequate or nonexistent information supplied by the manufacturer, my experience has been that the jig must be mounted on an adjustable base if consistently satisfactory results are to be achieved. Incidentally, the information supplied with my Sears Craftsman attachment is excellent in all respects except for this one point.

Most drill sharpening jigs, whether sold under the Sears, General, Reliance or another trade

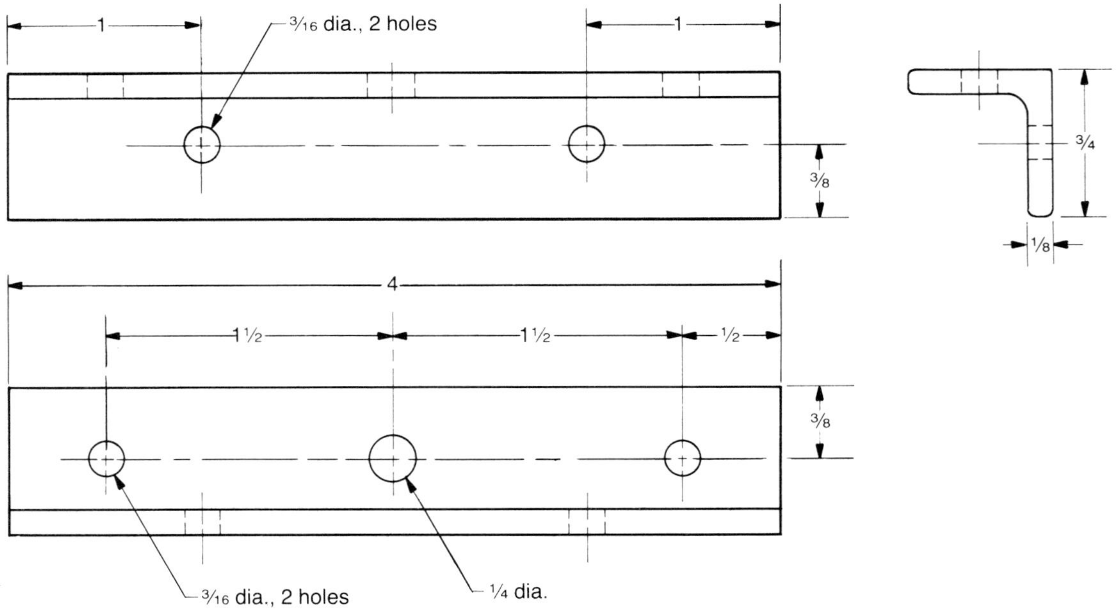

END FRAME
steel, 2 req.

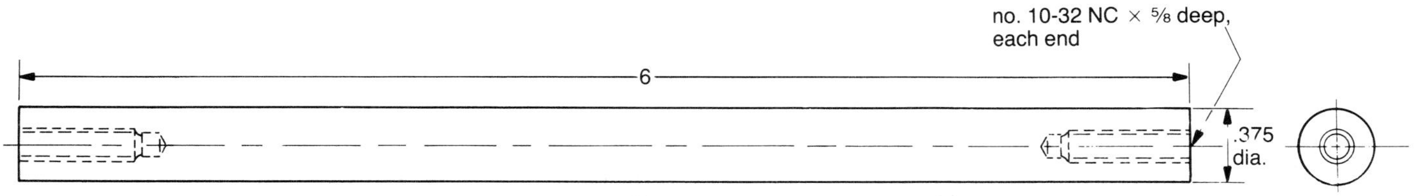

SPACER BAR
drill rod, 2 req.

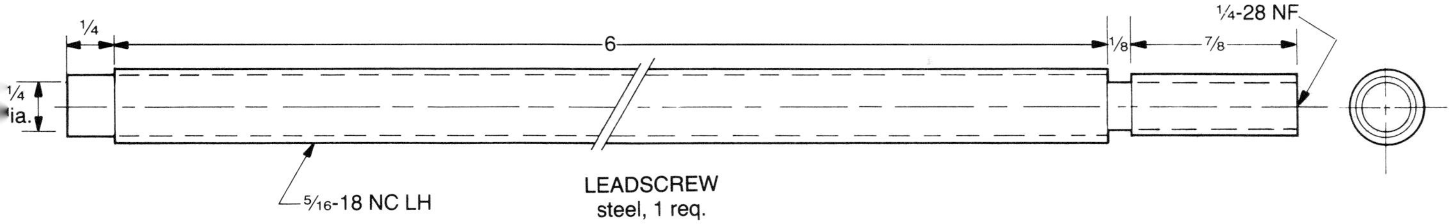

LEADSCREW
steel, 1 req.

name, appear to operate in essentially the same manner:

a) An adjustment is provided for various drill point angles, such as 88, 68, 59 (for general use), 49 and a CSK setting of about 41°. The included angles in each case are double these figures.

b) The drill is held in a vee-shaped trough and positioned by an adjustable lip rest which ensures that both lips are presented equally and evenly to the grinding wheel.

c) The whole assembly rotates upon a vertical spindle sufficiently offset from the center of the Vee to produce the required chisel edge and lip relief angles.

d) The drill may be moved toward or away from the grinding wheel by means of an adjustable slide bracket for coarse adjustments and a feed screw and lock-nut assembly for fine adjustments.

Note: Either of these secondary adjustments cause the drill to be rotated in its trough, and except when small adjustments are made, may require repositioning the base to maintain correct chisel edge and lip relief angles.

The sequence of loosening the retaining bolt, moving the attachment toward or away from the wheel as required and retightening the bolt is not always understood nor easily accomplished. The purpose of this article is to

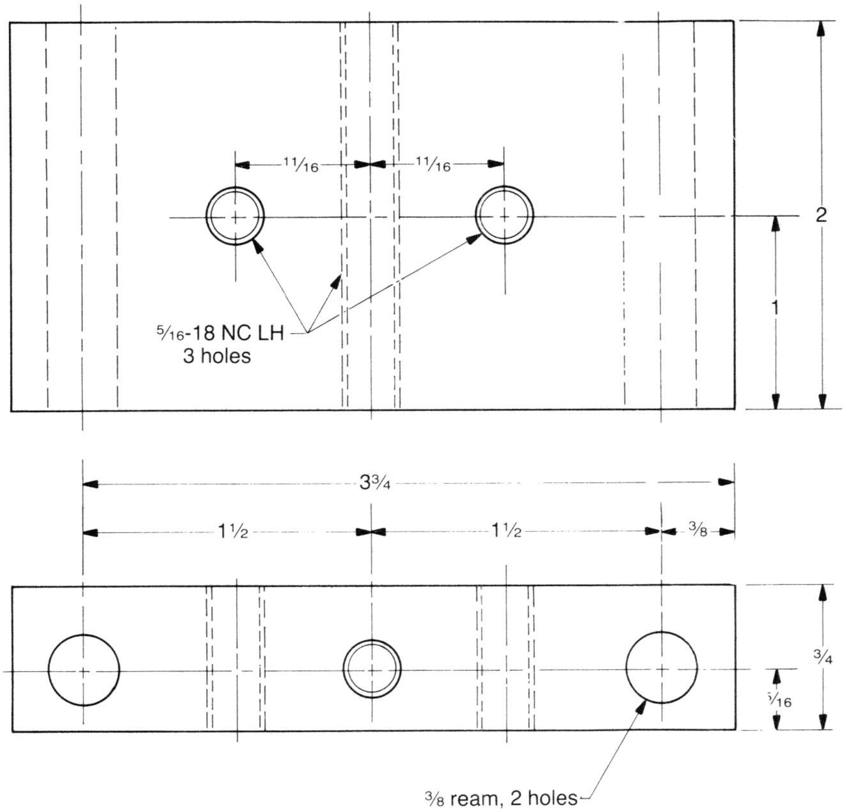

SLIDE
aluminum, 1 req.

provide details for making an adjustable base for your drill grinding attachment and to describe how the primary adjustment can be more easily accomplished.

Make up an adjustable base similar to that described in the drawings. With the drill jig mounting on the base, the jig may be moved toward or away from the grinding wheel in a systematic manner. Sufficient travel is built into the base to accommodate the various lengths of drills and the differing spirals likely to be encountered. The drill is secured in the jig as recommended by the manufacturer and adjusted on the lip rest by the adjustable slide bracket and/or the feed screw and nut so that the cutting edges are approximately vertical. Using the leadscrew knob, move the entire assembly toward the grinding wheel until a light cut can be taken. Unlock, rotate the drill, relock using the same settings and grind the other lip. If the chisel edge and lip relief angles appear correct, a further cut may be taken on each lip following a small adjustment using the feed screw and locknut.

If sufficient lip relief angle is found, move the entire assembly away from the wheel with the leadscrew and readjust the drill toward the wheel using the feed screw and nut. If excessive lip angle is found, back off the feed screw and nut adjustment, then move the whole assembly toward the wheel using the leadscrew. When you have obtained the correct setting, note the angle that the cutting edges are displaced from the vertical and use this same setting whenever other drills are being sharpened later.

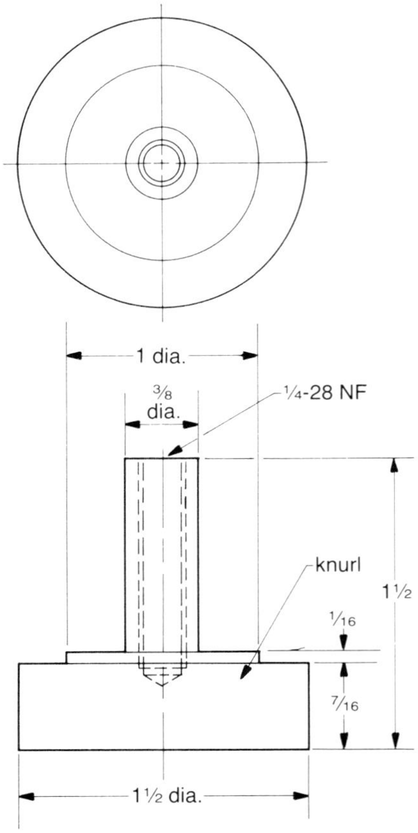

KNOB
aluminum, 1 req.

A Few Thoughts on Drill Presses and Tool Grinders

Most shop work seems to be drilling holes. Therefore, I think the purchase of a good drill press and a bench grinder (double-end-type) should precede the purchase of a lathe if funds are limited.

The use of a lathe as the only means of drilling holes is inconvenient and dangerous. Anyone who has used the old fashioned T-type drill holder on a lathe, drilled a round bar with a crotch center, or drilled a flat bar or a plate with a drill pad in the tailstock spindle will agree with me.

Neither a drill press nor a lathe are of much use without some means for sharpening drills and tool bits. The old-fashioned hand grinder is difficult to use to sharpen tool bits, and almost impossible to use for sharpening drills. Therefore, a good double-end bench grinder is necessary. Use a coarse wheel on one side and a medium to fine wheel on the other side. Either a self-contained motor type or one that is belt-driven from a separate motor will do.

There are many excellent drill presses being imported from the Orient and their cost is reasonable. Most of the drills of 1/2" capacity or larger come with a geared table elevating mechanism (Photo 1) and this is a worthwhile feature. Trying to loosen the table clamp with one hand and lift the table loaded with a vise and heavy workpiece with the other is difficult and dangerous. The drill presses of 5/8", 3/4" and 1" capacity have Morse taper spindle sockets. This means that standard drills and machine reamers with Morse taper shanks can be used instead of Prentice (or Blocksmith) drills with 1/2" diameter shanks. The Morse taper drills and reamers are often obtainable from surplus sales at a fraction of their new cost. Larger drills need to turn more slowly, so drill presses are supplied with an extra vee pulley mounted between the spindle and the motor. These pulleys, called speed reducers, are easily made up if you want to add one to a drill press. This enables you to use countersinks and counterbores without chattering (Figure 1).

If many holes smaller than 1/4" are to be drilled, then you should consider the purchase of a small drill press as well as a larger one. I have three: one with 1/4" capacity, another with 1/2" capacity, and a third with 5/8" capacity and a No-2 Morse taper spindle. I often use more than one when several size holes are to be drilled. I use my vertical milling machine for drilling and reaming holes between 1/2" and 1" diameter (Photo 2). This machine has a No. 3 Morse taper spindle socket and the speed range and the power for these drills. The machine also has a lever and a handwheel for rapid or fine feeding.

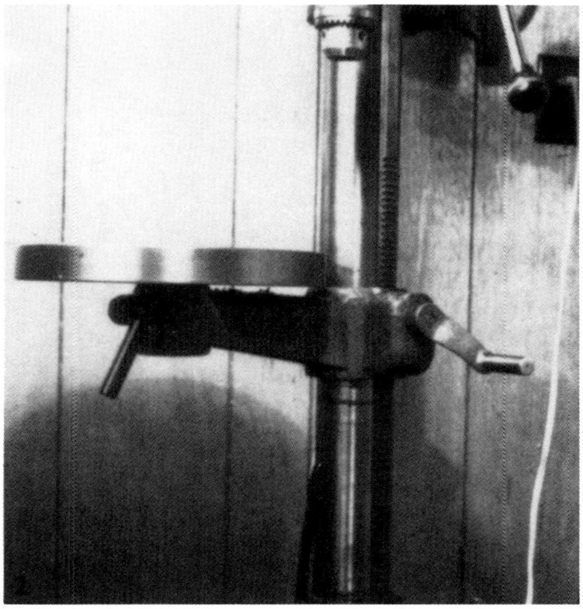

1

Sometimes a good used machine may be found, but you must check the machine very carefully. Make a general inspection to see if the machine has been abused or is worn out. Are there any pieces knocked out of the vee pulleys? Is the table full of holes? (I have seen a heart and initials chiseled in some.) Are the clamping handles bent, broken or missing?

Does the quill move easily up and down in the head casting when you rotate the feed lever? The pinion gear or the rack in the back of the quill may be damaged. If they are damaged do not consider purchasing the machine, because someone probably tried to use it as an arbor press. Grasp the chuck or spindle to see if there is any shake or looseness (a) between the quill and the head casting, or (b) in the spindle bearings. The bearings can be replaced, but if the quill clamp has to be tightened a lot to remove the shake, then the machine is worn out or was poorly manufactured.

2

To test the spindle for straightness, place a piece of 1/2" drill rod in the drill chuck and bring a dial indicator to touch the rod near the chuck jaws (Photo 3). Rotate the spindle by hand. The total indicator reading (TIR) should be less than .002". For those machines with a Morse taper spindle socket, use a new Jacobs

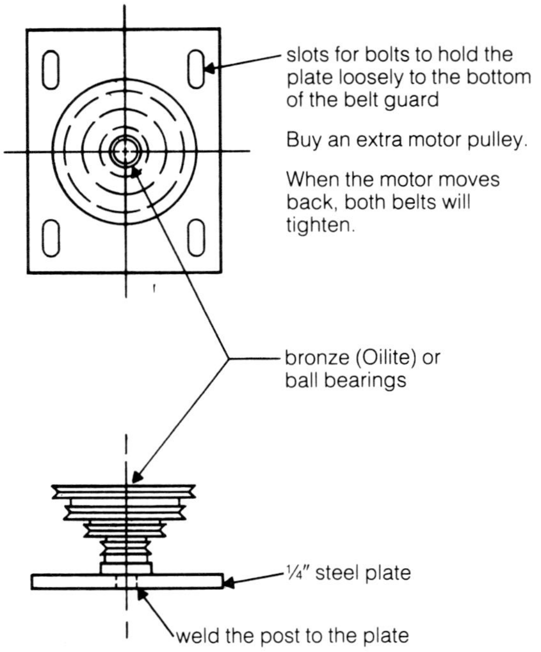

SPEED REDUCER PULLEY

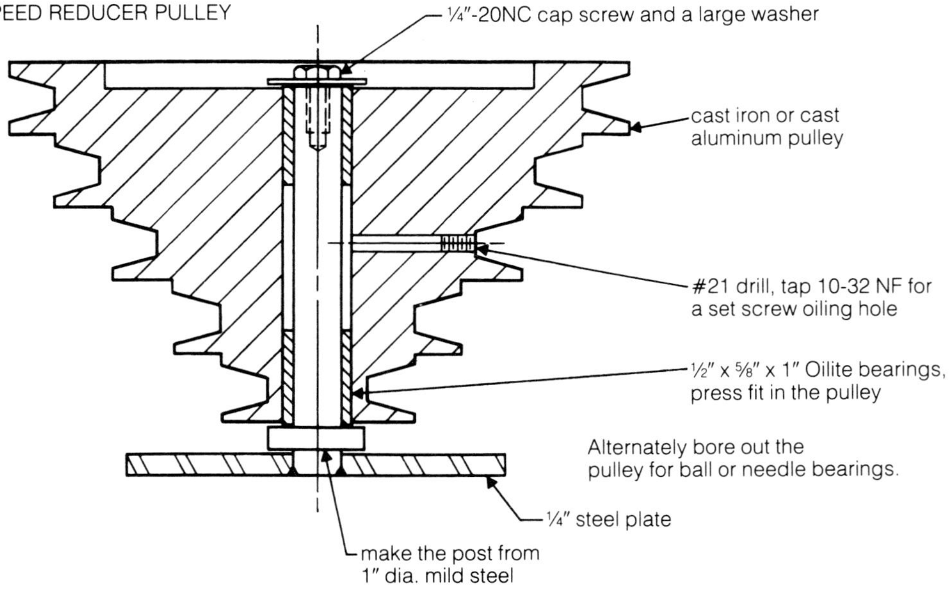

SPEED REDUCER

Figure 1

chuck arbor and place the indicator contact on the taper that goes into the chuck.

If the machine seems to be in good condition but the chuck is a cheap one or badly worn, this can be replaced. The drill chucks are usually fitted on a short taper on the end of the spindle. Some spindles have a nut above the chuck that can be used to force the chuck off the taper. Others will require a special forked wedge to remove them (Figure 2).

A broken quill return spring may be replaced if parts are available. Sometimes the end of the spring can be softened and bent or a hole punched as required.

On my vertical milling machine, I replaced the broken return spring with a counterbalance system. I placed a 3" vee pulley on the pinion shaft and drilled a small hole through one side at the bottom of the vee. The end of a bicycle brake cable passes through the hole. The ferrule or rod at the end of the bicycle cable should be on the outside of the pulley. The cable is wrapped around the pulley several times, and then passes over another pulley fastened to a mounting bracket at the back of the machine. The cable hangs down the back with the cable attached to an eye bolt screwed into a piece of 2" round bar stock for a counterbalance weight. The advantage of this system is a constant force to counterbalance the weight of the quill. A schematic of this arrangement is shown in Figure 3.

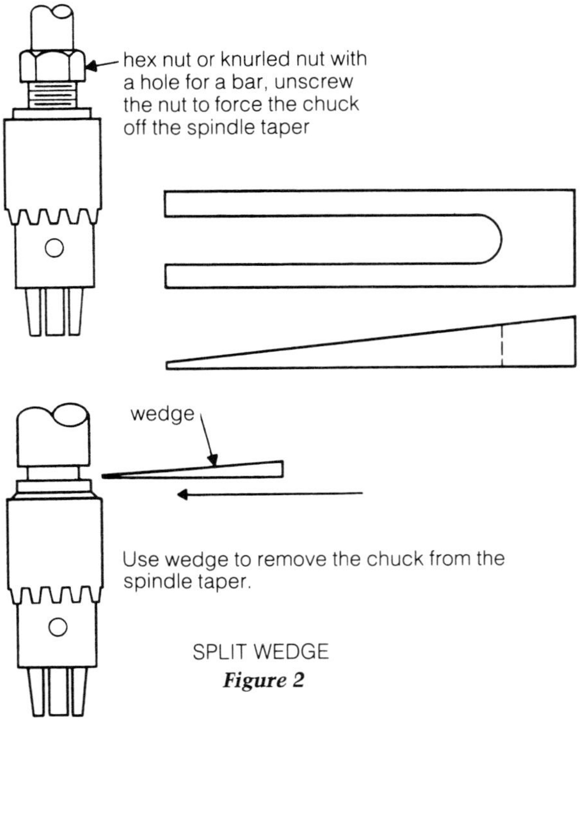

SPLIT WEDGE
Figure 2

3

4

Some of the floor model drill presses have a piece of thin wall steel tubing for a column instead of a cast iron column that has been turned and ground. The cast iron column is much more rigid and, of course, more expensive. The drill presses with a thin wall column often vibrate when the drill is running. This can be prevented by fastening the column to the wall in back of the machine (Figure 4). Hardwood blocks will be satisfactory and the price is right. To make the blocks, start by drilling the bolt holes and then bolt the blocks together with a piece of cardboard between them. Hold the blocks in the 4-jaw independent chuck and bore them out the same diameter as the column. Remove the cardboard, clamp the blocks around the column just under the head casting, and

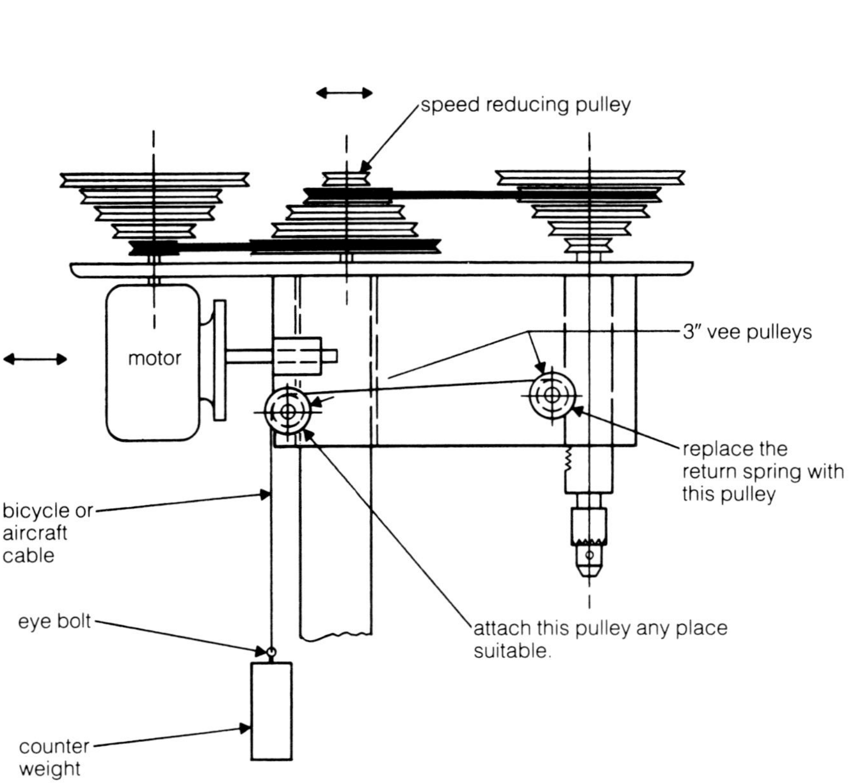

SPEED REDUCER AND COUNTER BALANCE WEIGHT

Figure 3

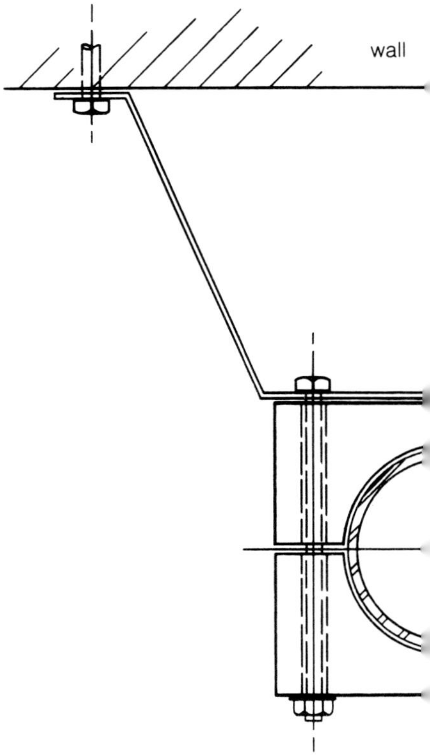

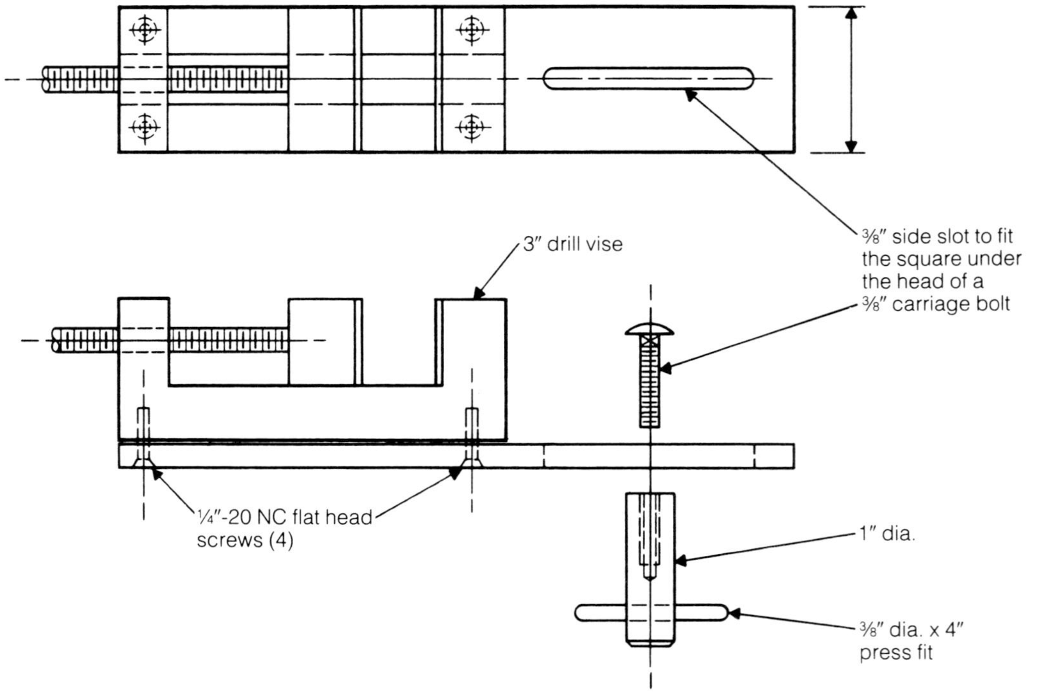

DRILL PRESS VISE CLAMPING PLATE

Figure 6

attach the metal legs as shown.

There are several accessories worth considering. I have a foot switch on my 1/2" drill press so I can hold small work with my left hand and operate the feed lever with my right. A tapping head is a good investment if many holes are to be threaded. So is a rotary table with a cross and longitudinal traverse for locating holes accurately. For long work, fasten a board to the table with flathead screws (Figure 5). Attach a clamping plate to the bottom of the drill vise

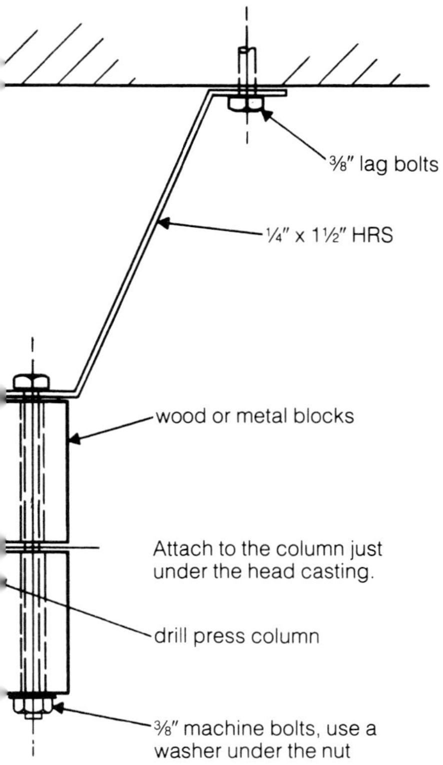

DRILL PRESS SUPPORT

Figure 4

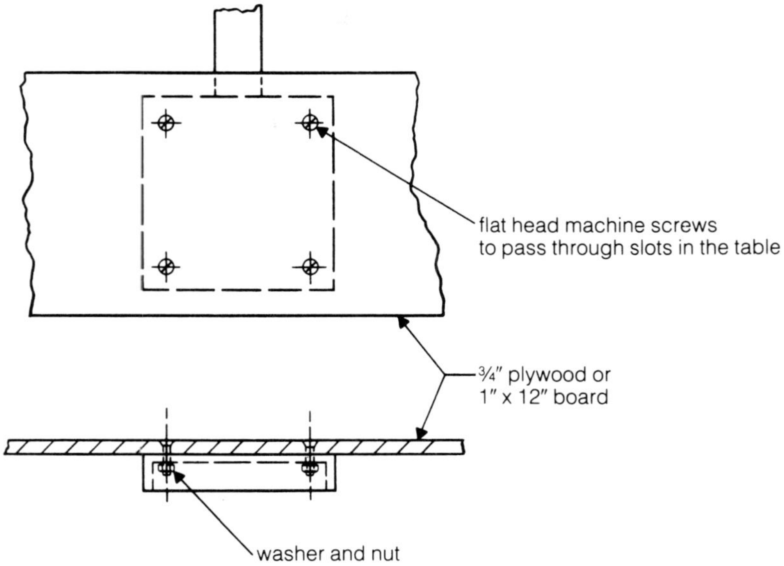

EXTENDING THE DRILL PRESS TABLE FOR LONG WORK

Figure 5

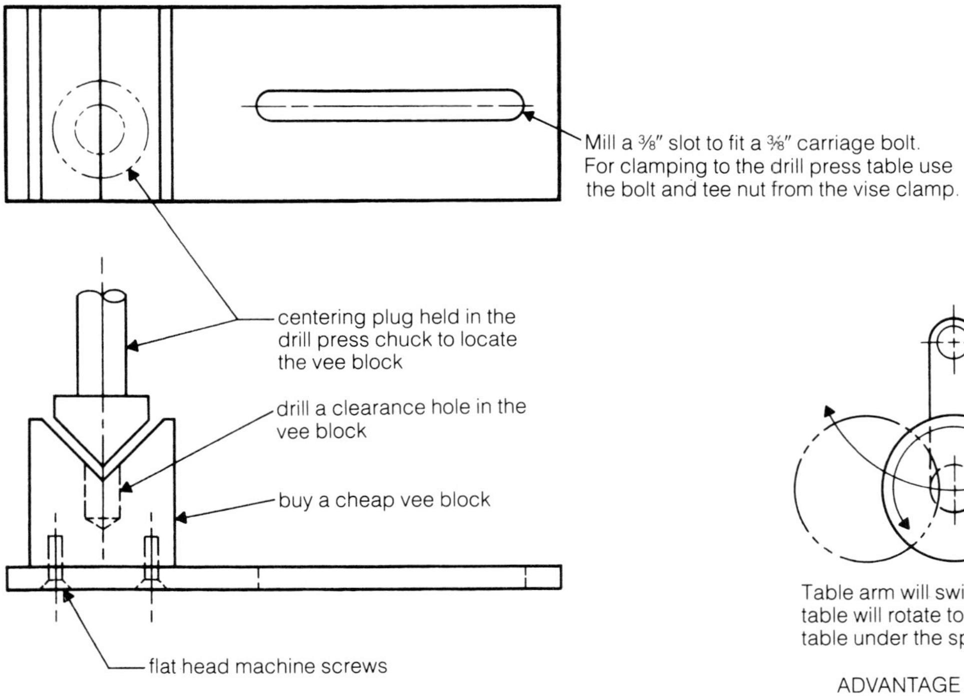

DRILL PRESS VEE BLOCK

Figure 7

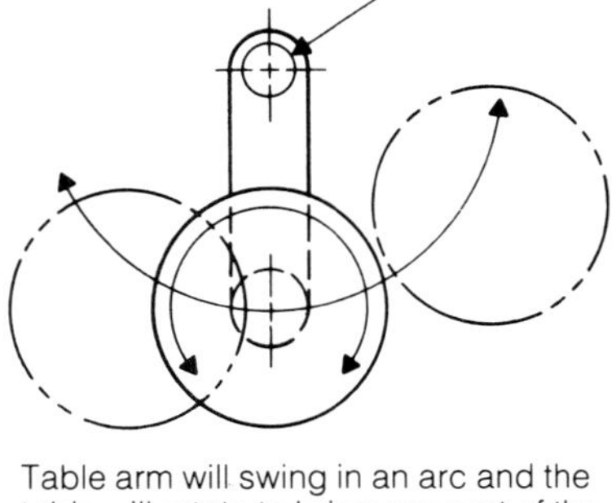

Table arm will swing in an arc and the table will rotate to bring any part of the table under the spindle axis.

ADVANTAGE OF A ROUND DRILL PRESS TABLE

Figure 8

for easier clamping (Figure 6 and Photo 4). Make up a simple vee block with a clamping plate and a centering plug (Figure 7). Consider the advantages of a round table instead of a square table when purchasing a new machine (Figure 8).

One of the more important accessories is a wiggler to locate the intersection of the layout lines exactly under the spindle axis.

Lay out the center lines, but do not center punch. Set the wiggler to run true with your

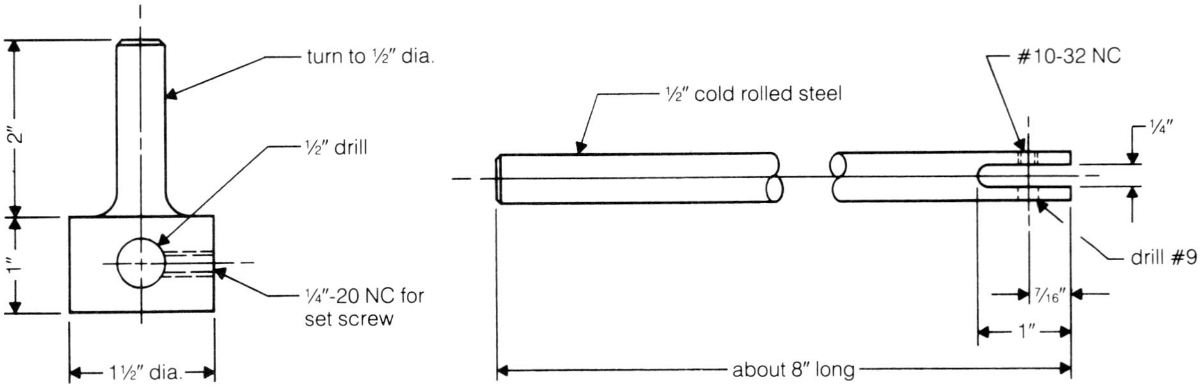

ADAPTER FOR STANDARD LUG MOUNTING FOR DIAL INDICATORS

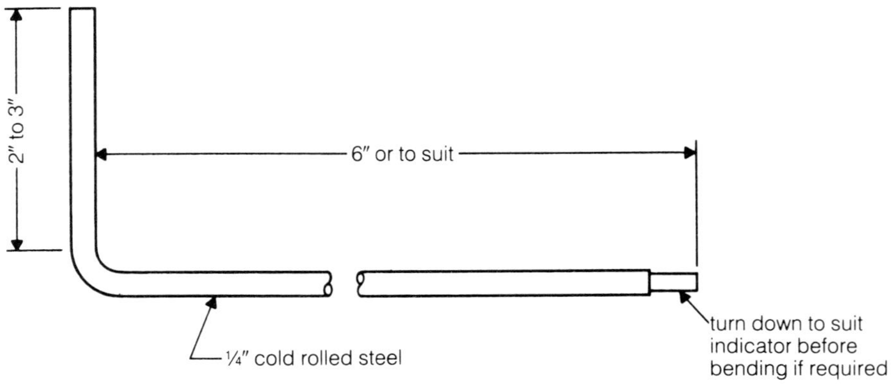

ADAPTER FOR A SMALL DIAL INDICATOR
STARRETT "LAST WORD" TYPE

Figure 9

thumbnail when the drill is running at a moderate speed (1500-2000 RPM). Lower the spindle until the point of the wiggler just clears the work. Sight down each line in turn and adjust the work until the point is over the intersection. Clamp the work in place and start drilling with a center drill. Be sure there is enough clearance between the chuck and the work so you can use a larger drill without moving the work.

THE TURN-AROUND TEST

You cannot do accurate work on the drill press if the table is not exactly at 90° to the spindle axis. If the table is out of square by only 2°, the hole at the bottom of a 4" thick block will be out of location by more than 1/8".

Make up a holder to suit your dial indicator (examples shown in Figure 9) and align the tables as in Figure 10. Loosen the clamping bolt between the table and the column bracket, then tap the table with a plastic or rawhide hammer or a block of wood until you have the same dial indicator reading on both sides of the table. The front of the table may be .001" higher to allow for deflection or spring from pressure when drilling. Use cigarette paper shims between the table and bracket, if required.

Try pressing down on the front of the table when the dial indicator is contacting it to see how easily the table is deflected. This demonstrates why you should use several drills in progressive steps for accurate work instead of a single step with the large drill.

A sturdy, solid drill press stand is absolutely essential. Figure 11 offers some suggestions on that subject.

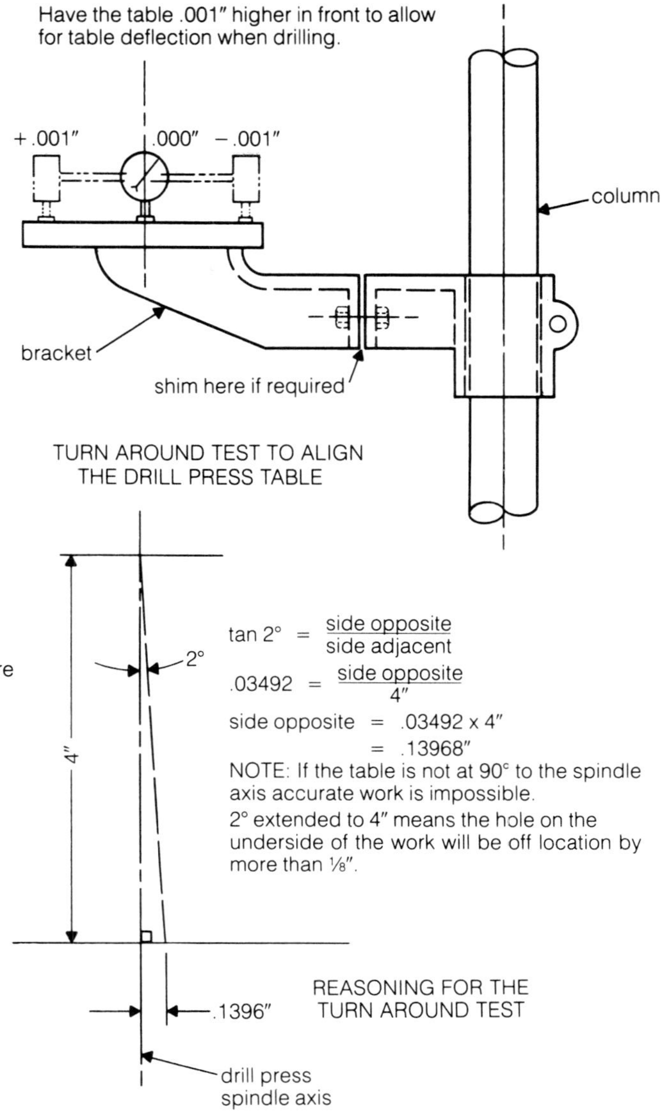

Figure 10

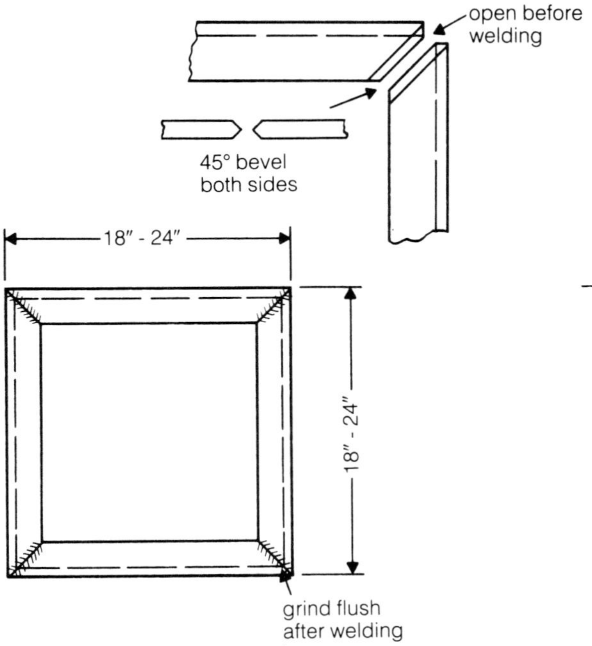

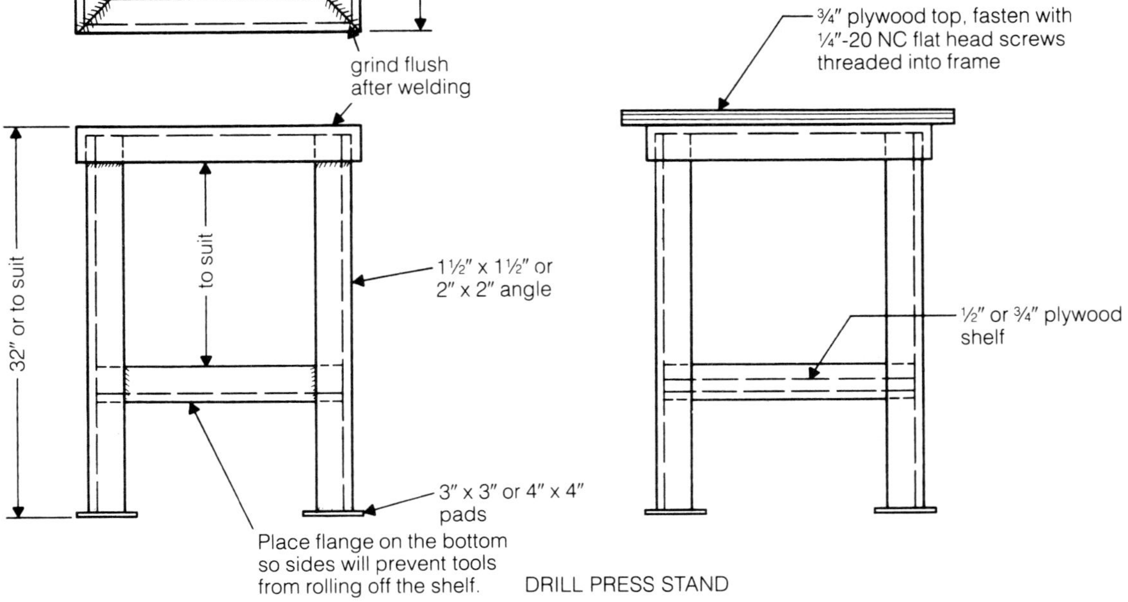

Figure 11

A Fireside Chat About Lathes

There is no doubt in my mind that the most important machine tool in the home shop is the lathe, for this is the one tool that is absolutely essential.

The lathe must be in good condition and accurate to do the great variety of work undertaken in a small home shop. An old, worn-out lathe with strained or worn-out chucks is not capable of good work except in the hands of an experienced, skilled machinist, and it requires the patience of Job.

Most home shops have a South Bend, Logan, Atlas or similar lathe, or one of the smaller lathes, such as Myford, Covel or Economat. There are some quite good lathes coming from India and other eastern countries and these should be considered instead of buying a used lathe of doubtful condition, if funds are limited. Some people waste years looking for a bargain in a used lathe, and when the lathe is finally purchased, properly rebuilt and equipped, it is not such a bargain after all. I have been down this path.

There is a trend toward larger lathes in many home shops. However, I recommend keeping the smaller lathe with all its chucks, collets, and taper attachments, steady and follower rests, etc., and buying a used machine of 13" to 16" swing. This will preserve the accuracy of the small lathe for precision work. The large lathe can be used for rough work, such as large castings and large diameter work. Factories and government workshops will gradually be changing over to the metric system, and you may be able to find a really good lathe, such as a Monarch, Dean-Smith-Grace, LeBlond, or another lathe of this quality, for a reasonable price. A 16" lathe with a gap bed would be ideal for anyone building a steamboat or a large traction engine. There was a 15-1/2" swing Colchester lathe in the school shop where I used to teach, with a 24" swing when the gap piece was removed. A large faceplate was supplied so that this lathe could turn or bore large flywheels, propellers, brake drums, etc. The carriage had tee slots in the carriage wings so that large cylinders and similar work could be bolted to the carriage for boring.

Anyone contemplating rebuilding a used lathe should buy the book *Machine Tool Reconditioning and Applications of Hand Scraping*, by Edward F. Connelly, from Machine Tool Publications, 811-S Commerce Building, 4th Street, St. Paul, Minnesota 55101.

Any amateur starting to rescrape a lathe bed will have a better understanding of the magnitude of the job, and can then decide if having the bed resurfaced by grinding at the factory or at one of the machine tool rebuilders is a more practical solution. If the bed is resurfaced by grinding or scraping, the screw mounting holes in the rack, the lead screw brackets, and the quick change gearbox will probably have to be elongated into slots by filing the tops of the holes. The rack and lead screw will have to be lowered to compensate for the material removed from the lathe bed.

To test for wear in sleeve-type spindle bearings, place a flat bar under the spindle nose with one end of the bar resting on a block of wood at the rear of the lathe bed. Place a dial indicator on top of the spindle and note the deflection when the bar is lifted. The bearing caps usually have some shims in the joint that can be peeled off with a knife if more than .001" play is present.

Some lathes with thin wall spindles have a bent spindle from a previous "dig in" during a parting off operation or boring a bronze casting. To test the spindle for straightness, a test bar will be required. This can be made if you have the use of a good lathe or a cylindrical grinder. A bar of 1" to 2" diameter, depending upon the size of the spindle socket, and about 14" to 18" long is set up for turning between centers. The bar is turned parallel and then one end is tapered to fit the headstock spindle socket. (Do not use the sleeve.)

The tapered part is then placed in the spindle of the lathe being tested with a dial indicator at the outer end. Rotate the spindle slowly by hand and note the reading; ±.002" will be acceptable. This bar can be used to check the alignment of the headstock with the lathe bed by running a dial indicator along the side and the top of the bar with the carriage, when the dial indicator is mounted in the tool post. You can run a similar test to check the tailstock alignment with a suitable test bar. Any reputable machinery dealer will let you make these tests on his machines and will often supply test bars and a dial indicator. If the dealer objects to your making these tests, leave and try another dealer. The alignment of the cross slide can be tested by mounting a dial indicator in the tool post and running the dial indicator across the faceplate. The reading should be so that the lathe faces flat or slightly concave (.002"), never convex. This alignment is usually corrected by scraping the wings of the carriage where they run on the bed rather than on the dovetail of the cross slide, if this seems to be in good condition.

Check the chucks to see if they fit snuggly on the spindle register and if the jaws are loose in their slots. If the jaws are loose or the scroll is worn or strained, you will have to buy new chucks, for they cannot be reconditioned properly in a home shop.

A lathe should be mounted on a heavy steel or cast iron stand, not on a wooden bench. A good lathe stand can be made at home. This will be covered in the next article, and the method of leveling the lathe will also be described.

My own lathe stand is made from 2 × 2 × 1/4" angle iron and all joints are welded. There are drawers for accessories and shelves underneath for chucks, etc. As you can see in Photo 1 there is a shelf on top to hold drills and taps, and a tray to hold drills in order, so I can use them in sequence almost as quickly as a turret lathe. The shelf also has a lamp on a flexible mounting for peering into holes being bored and to help me see the graduations when swiveling the compound slide rest. The shelf has holes in front to hold all the chuck keys, and nails to hold the other wrenches so they can be returned to their proper place immediately after use instead of being thrown into the chip pan. There is a splash guard mounted under the shelf to keep cutting oil off the wall. I also have a paper towel bracket under the shelf for wiping the lathe bed when turning cast iron or brass. I place a clean rag or a piece of towel on the shelf to hold micrometers, vernier calipers, telescoping gages and rules. A piece of 1/2" insul-board is mounted on the wall at the end of the lathe, so drawings may be pinned up where they can easily be seen and will not become dirty or torn.

You cannot realize the full potential of the lathe unless it is well equipped. Three- and four-jaw chucks (Photo 2) are necessary and you should attempt to have several sizes, if possible. I have a four-jaw self-centering chuck, as well, for square bar stock. If you are buying a new lathe, try to buy all the accessories you want with the lathe, for somehow you never buy them afterwards. I didn't do this because I had to lick the spoon to buy the lathe and, soon after, had to buy a micrometer carriage stop and a taper attachment. I was doing some commercial work and didn't have time to make them.

1

2

A good vertical milling slide is very useful even if you have a shaper or a milling machine. The Myford milling slides are among the best for small lathes, for the tee slotted table will hold a vise, a vee block or an angle plate. The work may also be clamped directly on the table. The milling slide can also be used to hold a milling or drilling spindle driven by a small motor or an overhead driving system. In the first shop I worked in, there was an old Hendy tool room lathe with a relieving attachment for making taps, hobs and form cutters. This lathe had a long cylinder mounted on hangers above the lathe which was driven by a stepped pulley from the countershaft so that the speeds could be varied. This cylinder was, in effect, a long pulley enabling the belt to travel along the cylinder as the carriage moved along the bed. The overhead drive was used to drive internal or external grinding spindles and a milling spindle for cutting key seats in a shaft mounted in the lathe.

Centering Small Square Stock

Small square stock can be centered easily in a three-jaw chuck. A little setup work is needed, but it can be done quickly and is accurate, if carefully done.

Measure the distance across the corners of the square bar with a micrometer. Choose a bar of cold-rolled mild steel about 1/8" to 3/16" larger in diameter than the diagonal of the square stock. Cut off a 2" length, face both ends in the lathe and center punch one end, on the periphery, just in front of the number 1 chuck jaw. This is a relocating mark, to insure later trueness.

Next, still in the lathe, drill the length of the bar with a drill of the same diameter as the diagonal measurement of the square bar.

The sleeve will now have to be removed from the chuck and split with a hacksaw. Hold the sleeve in a bench vise, squeezed on the ends to avoid squashing things, and make the split on the side opposite the center punch mark.

Replace the sleeve in the 3-jaw with the square bar inside it and the punch mark back in front of the number 1 jaw and your square bar will always be centered.

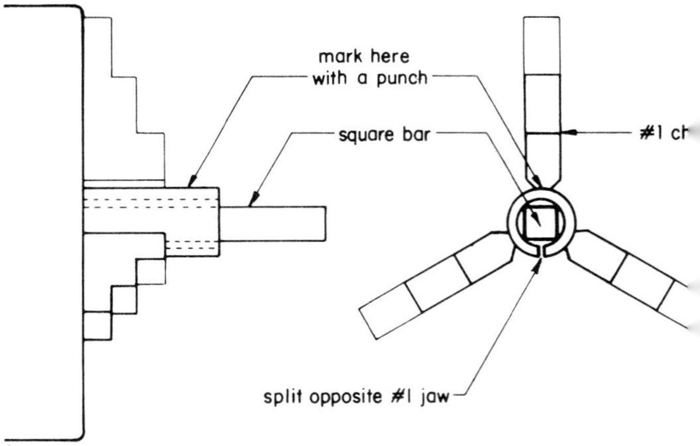

A Fireside Chat About Lathe Chucks

When you buy a new lathe chuck, or receive several chucks with a used lathe, you should check the fit of the back plate on the register of the lathe spindle. I have found a difference between the diameter of the spindle register and the recess of the back plate up to .040". The diameter of the recess of the back plate should be about .001" larger than the spindle register so the chuck is a "wringing fit" on the spindle. The threads hold the chuck on the register, so they can be a loose fit. However, the fit on the register is very critical.

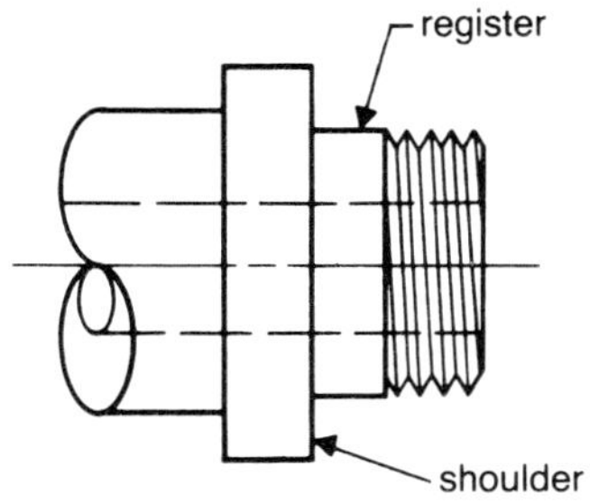

If the diameter of the recess is a good fit on the spindle register, then check the fit against the shoulder. Put a thin film of Prussian Blue on the spindle shoulder and screw the chuck in position. Remove the chuck and see if there is a uniform contact all over the back surface of the back plate. If there is Prussian Blue showing on one or two spots, scrape these with a hand scraper. Replace the blue on the spindle and replace the chuck to test for contact and then scrape the high spots. Continue until there is a uniform contact of many small spots all around the shoulder. If the chuck does not fit the spindle properly, then the chuck cannot be removed and replaced in exactly the same position.

Sometimes the back plate is a "green" casting and has warped slightly, making the chuck "wobble." To check this, remove the chuck from the back plate and screw the back plate on the spindle. Use a dial indicator to test if the back plate casting has warped. A light facing cut on the back plate will correct this.

The scroll on a universal or 3-jaw chuck is a true spiral with a constantly changing radius.

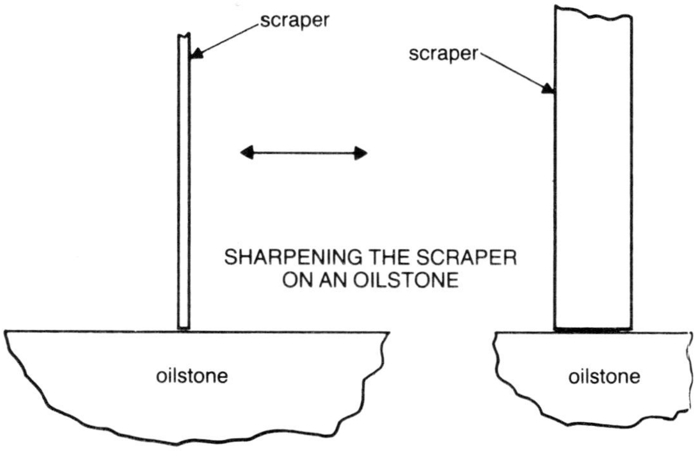

SHARPENING THE SCRAPER ON AN OILSTONE

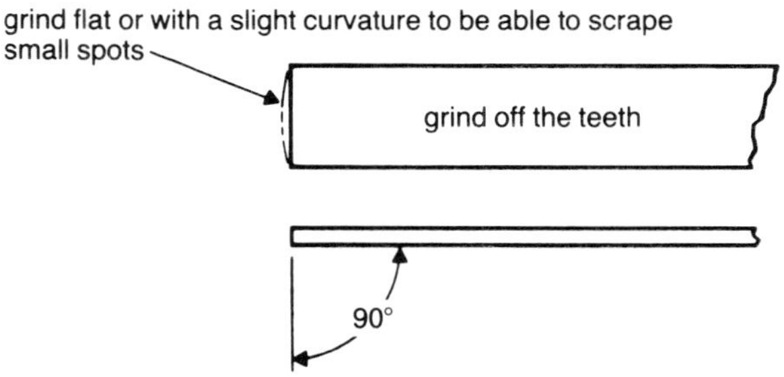

HAND (FLAT) SCRAPER MADE FROM AN OLD FILE

The curvature of the teeth on the back of the jaws must match the smallest radius near the center of the scroll so there is only a very limited contact near the outside diameter of the scroll. The scroll can be easily damaged by over-tightening the jaws or by holding irregularly shaped work that places an extra strain on one jaw. The most common causes are trying to hold a square piece of stock in the 3-jaw chuck instead of taking the trouble to change to a 4-jaw chuck, and using a piece of pipe on the chuck key to gain extra leverage. If heavy cuts are required even on round stock, then place the work in a 4-jaw independent chuck, for this chuck is stronger and holds the work more securely than a universal chuck.

When a 3-jaw chuck has been strained or has received a lot of use, the chuck will no longer hold the work accurately on "run true." Some people try to correct this by placing a ring on the steps of the jaws and opening the chuck against the ring. They then use a toolpost grinder to grind the inside of the jaws. The mistake made here is that the jaws are being opened out against the ring, but when the chuck is being tightened on a bar after grinding, the scroll is contacting the outside of the jaw teeth instead of the inside, as was the case when the ring was in place. Industrial 3-jaw chucks have separate jaws bolted on to the jaw base with teeth. These jaws can be removed and soft jaws bolted in place. A ring or a disk is placed inside the chuck at the back of the jaws which are usually recessed here. Then the soft jaws are bored to suit the work. This is a reliable method of rechucking work so that both ends and any bores will be concentric.

We can hold round work accurately in an old or strained 3-jaw chuck if we modify the back plate by reducing the register between the back plate and the chuck by twenty or thirty thousandths. The bolt holes in the back plate will also have to be enlarged by 1/64" to 1/32".

The setup procedure is as follows: Place the bar or work in the chuck and tighten the jaws. Tighten the chuck equally in all the key holes and, if there is an arrow or mark by one of the key holes, tighten this one last. Now loosen the screws clamping the chuck to the back plate slightly. Place the contact point of a dial indicator on the work and rotate the chuck by hand. Note the high spot on the dial indicator reading and tap the chuck with a lead or plastic hammer. Repeat until the dial indicator gives a "0" reading. Tighten the bolts on the back plate and then check the work with the dial indicator again just in case the chuck has moved during the bolt tightening. Now any work of the same diameter can be chucked and it will run true. A chuck similar to this with adjusting screws is available from Bernard-Pratt. The trade name is a *Grip-True* chuck and one of these would be a welcome addition to any shop. (Some day I would like to buy one.) Some of the *Grip-True* chucks have six jaws in place of the usual three so they can hold rings and thin wall work without distortion. I have one chuck that is a bit unusual. This is a 6" 4-jaw self-centering chuck. This is great for square bar stock and plates.

If the recess in the back plate is ten or twenty thousandths larger than the diameter of the spindle register, you have a problem. Turn up a steel ring with a wall thickness of about 1/8" and the bore should be about .020" less than the spindle register. Round off or chamfer the outer corner with a file and then part off so the width is a little greater (say 1/32") than the depth of the parallel section in the chuck back plate. Now put this ring in the freezer.

Remove the jaws from the chuck being repaired and hold the chuck in a large 4-jaw

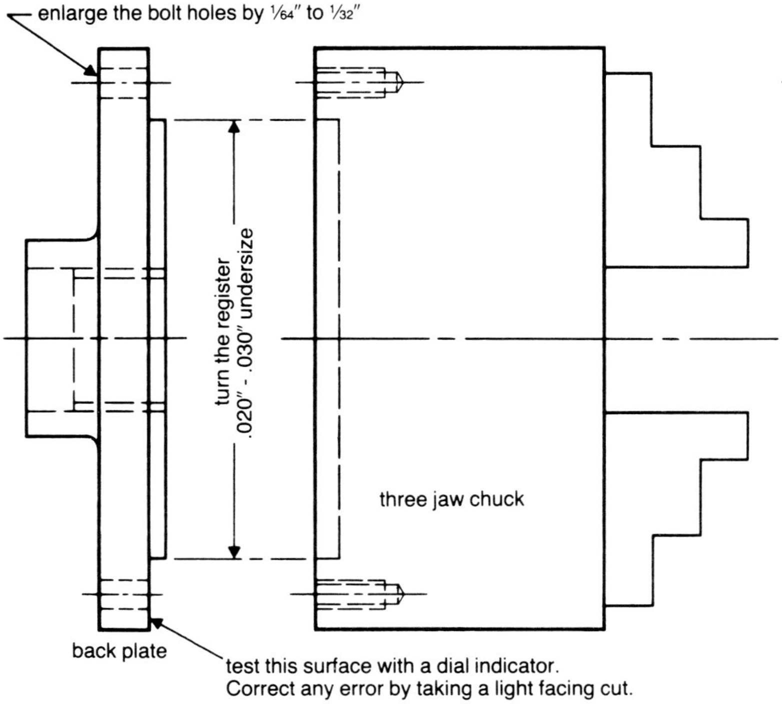

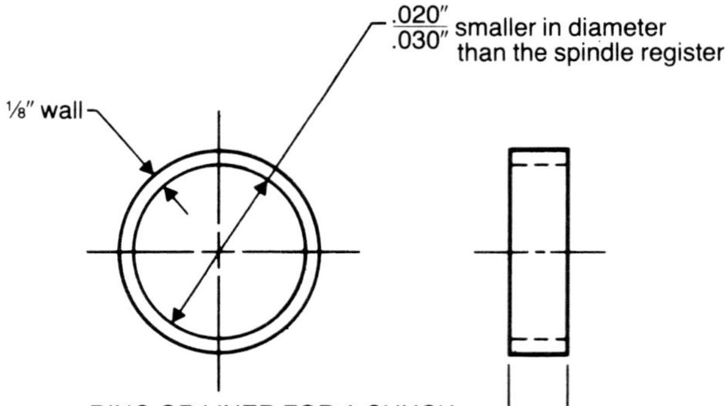

RING OR LINER FOR A CHUCK BACK PLATE

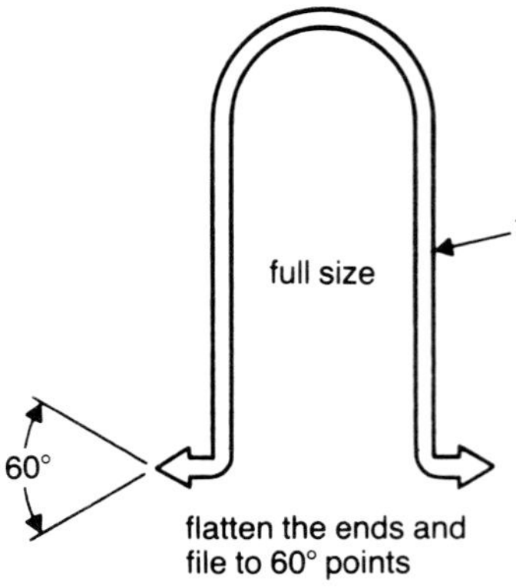

CHUCK THREAD CLEANER

chuck with the back plate facing the tailstock. Place the contact point of a dial indicator on the parallel portion inside the back plate and set the chuck to run true. Now bore out this parallel portion until the bore is about .002" less than the outside diameter of the ring. Use a scraper to scrape a slight lead in the bore. Now bring the ring from the freezer and tap in place with a brass or plastic hammer. Do this quickly so the ring will not heat up and expand before it is in place. When the ring has warmed up to room temperature, finish the bore to .001" more than the spindle register and then face until you take a light cut off the back plate where it butts against the shoulder on the spindle.

A small air compressor is a must for any home shop, for there are many uses, but a blow gun will be very useful for blowing the shavings out of the mounting threads and the scroll and the tightening screws of the 4-jaw chucks. If the chuck has been placed on the lathe spindle when the threads were dirty, there may be a shaving lodged in the back plate threads. To remove these shavings, make up the screw thread cleaner shown in the drawing. Rotate the spring cleaner in the back plate threads and any shaving lodged there will be removed.

Some old-time turners would place a ball of waste or a rag inside a 3-jaw chuck when short pieces were being chucked. This prevented the shavings from entering the scroll behind the jaws. A piece of brass rod filed up to look like a screw driver but made to just fit in the scroll will be useful to clean by revolving it when the jaws have been removed. Always check the jaws and scroll for shavings when you start to work.

If there is difficulty turning the chuck key on a universal chuck, and the scroll seems to be clean, some shavings may be behind the scroll in the bevel gear. To take the chuck apart, make two orientation marks on the split line where the chuck key enters the pinion gears. Use an old knife and drive this into the space between the two halves of the chuck. Open the chuck evenly all around. Note where the pinion gears are so they can be replaced in the same position. Remove the scroll and file off any burrs, clean and replace. Repack with grease and then reassemble the chuck.

Four-jaw independents often have the inside of the jaws worn "bell mouth" so the inside edge of the jaw is no longer at 90° to the chuck face. This can be corrected by grinding the inside edge or face on a surface grinder by holding the jaw on an angle plate or in a vise on the magnetic chuck. Good chucks are essential for accurate lathe work, so take good care of them.

For small diameter bar work, collets are par excellence and should be purchased when funds permit.

Jacobs used to make a drill-type chuck that could be screwed directly on the lathe spindle for 6" Atlas lathes and the standard 1-1/2-8 spindles of the 9" to 12" lathes. I have one and this is almost as accurate as a collet.

A Fireside Chat About Lathe Tools

Most small shops use high speed steel tool bits for most turning operations for the simple reason that these are still the best for most work. Although the price of tool bits has risen a great deal over the past few years, if you shop carefully, they are not too expensive.

Tungsten carbide and ceramic-tipped tools are excellent for production when used on the heavier types of industrial lathes designed to use them. There are many carbide-tipped tools offered for sale by tool suppliers at discount prices, and their purchase is worthwhile for turning or boring cast iron and for turning alloy and tool steels. However, for best results, one should buy the right type of tip for each application, such as turning mild steel, alloy steel, cast iron or for heavy interrupted cuts.

The main problem arises when these tools have to be resharpened. A silicon carbide (green) wheel should be used for grinding, but to finish the cutting edge properly, a diamond wheel should be used. The cost of these wheels is prohibitive for occasional use in a small shop, so a diamond slip, like a small oilstone, can be used to remove the rough surface left from the green grinding wheel. When grinding these tipped tools, do not dip them in water to cool them, for the thermal shock may crack the carbide tip. Cooling naturally in air is the best way, but if you are too impatient, then you can try cooling the butt end of the shank in water, or in a stream of compressed air, and let the shank conduct the heat away from the tip.

For a small shop, a nice luxury item is a carbide tool holder with clamp-on tips that can be indexed as each corner becomes dull. Mine is a General Electric for triangular tips. These have three cutting edges on one side, and then I turn the tip over for three more cutting edges. The .093" radius on the tips is just right for machining the radius at the flange for 1-1/2" scale railroad car wheels. The tips are available with a rake for steel, or flat (no rake) for cast iron.

Carbide tools should be held in an open side-type tool block, a four-tool turret or an Alina-type tool holder for maximum rigidity. If the tool holder is not strong enough, or the lathe is very light in construction or badly worn, the cutting edge will probably chip in use.

Because tungsten carbide is usually more brittle than high-speed steel tool bits, the clearance and rake angles are minimal. This means that more force is required to drive these tools into the work. For close dimensions on light lathes, the high-speed tool bits, with their sharper cutting angles, will give more accurate results on the finishing cuts.

An alternative to carbide tools are the Stellite tips. These are harder and more heat resistant than high-speed steel tools, but they can be sharpened with the ordinary aluminum oxide grinding wheels.

If you look at one of the books written about eighty years ago, you will see many odd shapes for lathe turning tools. Most of these shapes are unnecessary and are redundant today. The small lathes in use at that time were driven with a foot treadle or were driven with a flat belt from a line shaft. Very little power was available, so the diameter of the spindle bearings was small and the tools had to cut as freely as possible.

When high-speed steel tools were introduced about the time of the First World War, the lathes had to be redesigned with larger pulleys and heavier spindles with larger and better spindle bearings, to take advantage of the new cutting tools. The same thing occurred when tungsten carbide tools became common in industry. Most small lathes are intended to use high-speed steel tools most of the time, and carbide tools only for harder steels and cast iron.

The old idea of cutting action of lathe tools was that of an axe splitting wood. This is no longer valid, except for some free-cutting materials, for most cutting action is similar to plowing wet snow; the shavings are short and tend to pile up on top of the tool bit. Often, production tools have a curved side rake, ground in to help the tool to cut more freely and to make the shavings curl and break into short pieces instead of having a long ribbon-type shaving. These long shavings are hot and very sharp and must be handled with a hook. (Make one from a coat hanger.) Some tools have a chip breaker ground in the top of the tool. The carbide tools sometimes have a carbide chip breaker clamped in place on top of the tip.

The Armstrong and Williams-type tool holders usually have 15°-20° back rake built in. Now that lathes are more robust and powerful, this back rake is usually ignored in four-tool turrets and open side tool holders. The clearance angles are absolutely necessary for the tool to cut; the rake angles help the tool to cut more easily. To be sure of the clearance (8°-10°), make up a simple gage of tin. This does not have to be very accurate. Simply lay out the angle from a protractor, cut out with tin snips and bend a tab at the back to 90° so the gage will stand by itself on a surface plate. Use the drill press table if you don't have a surface plate. When possible, use the tool holder to hold the tool bit for grinding. This is safer than holding the tool bit in your fingers, and more pressure can be applied when grinding. Some grinders have the tool rests set low, so the tool bit will be near the center height of the grinding wheels when a tool holder is used. The side and end clearance angles change as the tool holder is elevated or depressed on the rocker to bring the cutting edge to the center height of the lathe (or slightly above). This becomes a real problem with bent tool holders and especially the bent or offset parting tool holders.

Small four-tool turrets can be removed from the lathe to grind the tool bits while still in the turret to save time, or the tool bits can be removed and held in your fingers. Use a gage to check the angles until you have enough experience to judge them.

Before grinding, make sure the wheels are clean, sharp and straight by dressing the wheels with a star wheel dresser. This is a form of crush dressing and gives a cooler, faster cutting action than when a wheel is dressed with a diamond. For new tools, start by grinding the end- and side-clearance angles and the side rake, if required, on the rough wheel. Cool frequently and grind up to 1/32" to 1/16" from the cutting edge. Use a fine wheel for finish grinding. To start finish grinding or resharpening, lower the end of the tool holder in your hands so the wheel starts cutting at the bottom of the tool bit. Then elevate the end of the tool holder until the whole face of the tool is being ground. This method will avoid accidentally grinding too much off the cutting edge and then extensive regrinding. Move the tool bit slowly in a straight line across the grinding wheel so that each face being ground is a flat surface and not multi-faceted like a diamond. (This will require some practice to achieve.) Watch for sparks to appear at the top of the tool along the cutting edge, for this indicates that the grinding wheel has reached the cutting edge. Finish with an oilstone slip by three or four strokes flat on each face, and then round off the vertical edge at the tip very slightly. For a finishing tool, the radius can be increased to .010-.015", or a small flat (1/32" to 1/16") can be stoned on the end.

These small oilstones are also called abrasive sticks or slips, and are three to four inches long. Aluminum oxide is most common; silicon carbide will give a faster cutting action but will wear away more quickly. Several stones of various grits are a worthwhile investment. When the tool bit starts to dull the finish on the work, there may be flecks on it or it may be developing a rough finish. If an oilstone is used immediately, then the tool bit can usually be resharpened several times before a major regrinding is necessary. (My Scottish ancestry cannot be denied.)

Tool holders are very expensive nowadays, so I have included the drawings for a simple tool holder I saw when I first started working in a tool room over thirty years ago. The dimensions are for a nine- or ten-inch lathe and can be altered to suit your lathe. The multi-purpose holder is very versatile and can be used close to a shoulder. I have included drawings showing all the uses. I have also included a design for a tool holder with back rake. One or two evenings spent making each of these holders will not be wasted.

For best results, do not exceed the recommended cutting speed for each material. Use this formula to find the revolutions per minute at which to set your lathe or drill press. For milling, reduce by 1/4 to 1/3, if heavy cuts are to be taken.

$$\text{RPM} = \frac{\text{CS} \times 4}{\text{Diameter}}$$

Example: A 6" diameter cast iron pulley. The cutting speed (CS) is 60.

$$\text{RPM} = \frac{\text{CS} \times 4}{\text{Dia.}} = \frac{60 \times 4}{6} = \frac{10 \times 4}{1} = 40$$

This is a simplified formula, but perfectly suitable for finding the nearest RPM setting on the machine. These cutting speeds are for light machines and can be increased or decreased when you see how the tool is cutting and

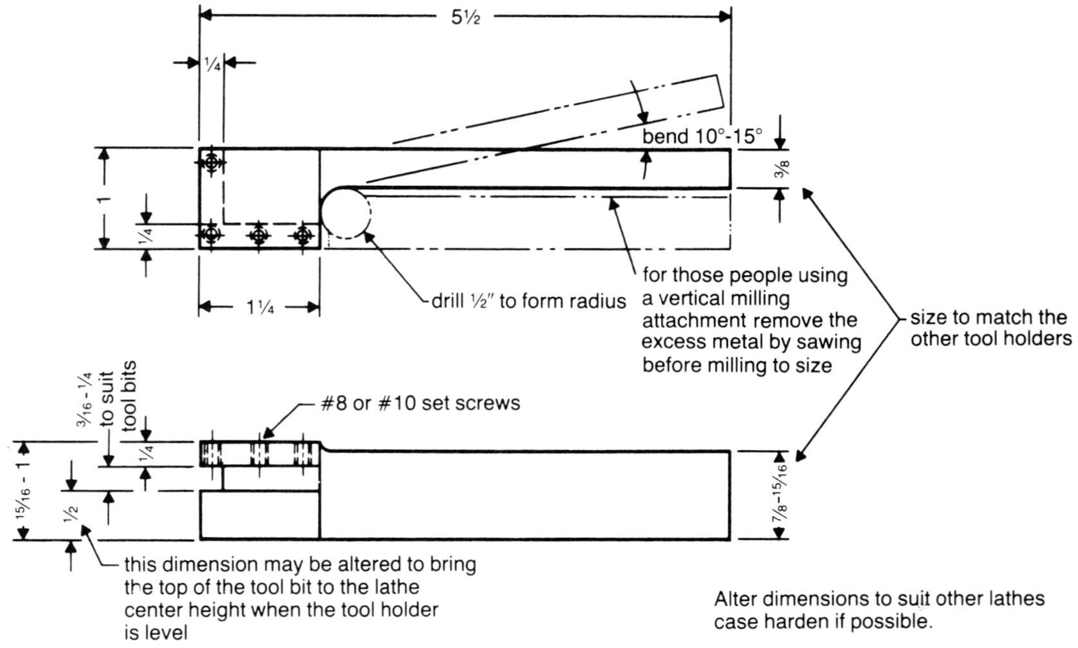

TOOL HOLDER FOR 9"-10" LATHES

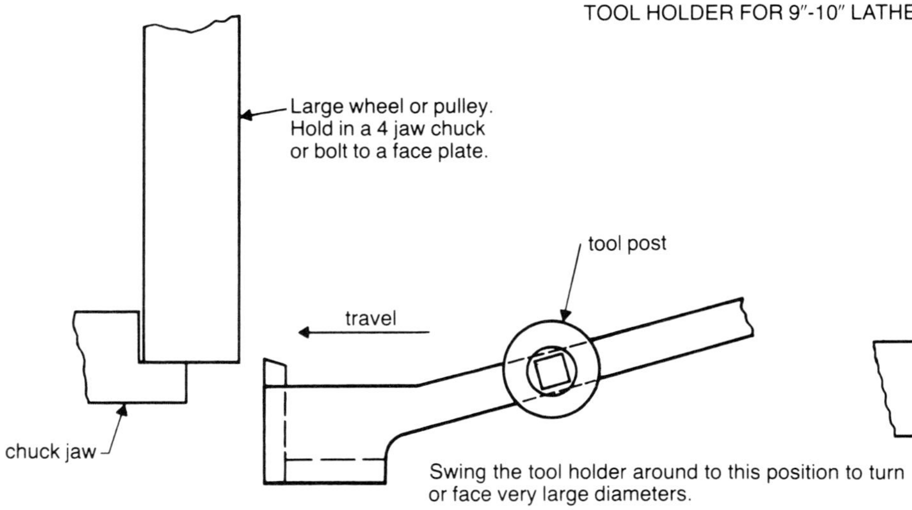

TURNING THE OUTSIDE DIAMETER

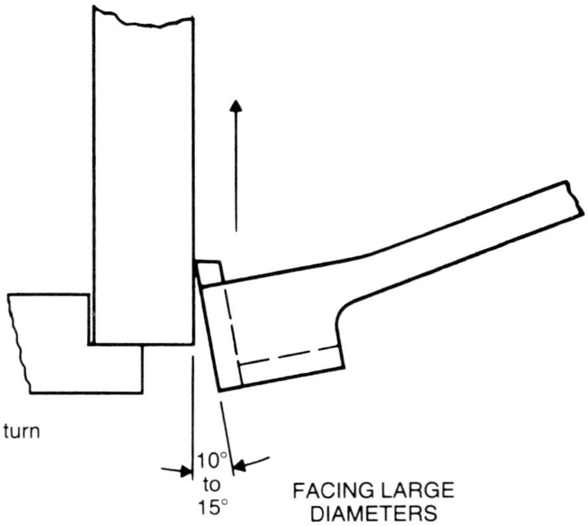

FACING LARGE DIAMETERS

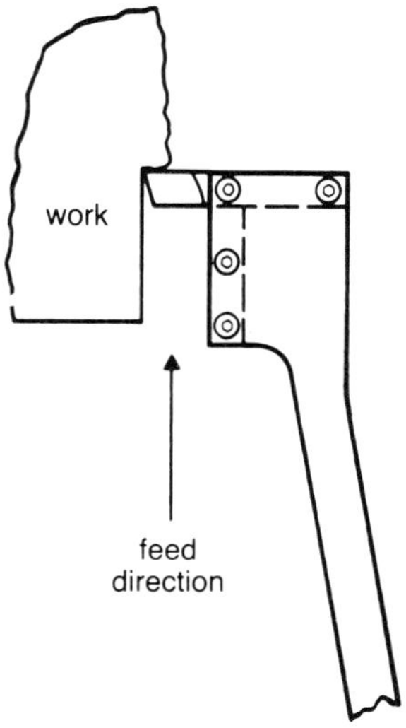

USING A LEFT HAND TOOL BIT FOR ROUGH FACING

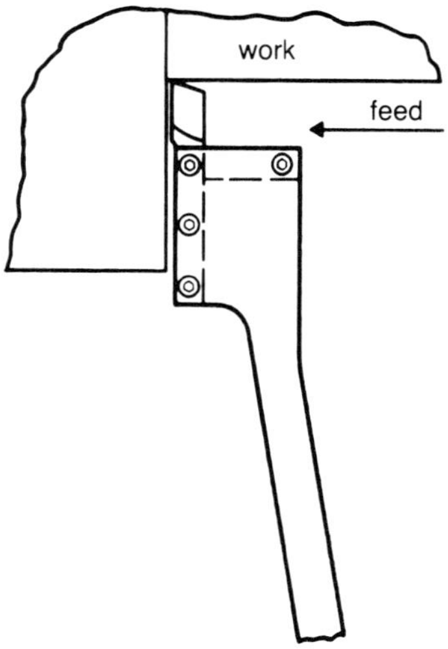

TURNING UP TO A SHOULDER WITH A RIGHT HAND TOOL BIT

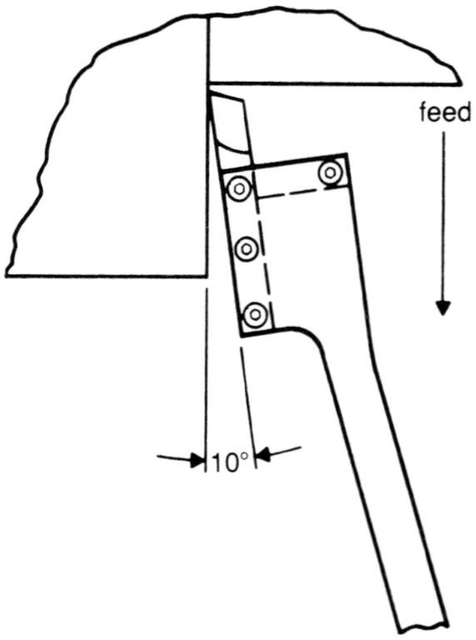

FACING A SHOULDER WITH A RIGHT HAND TOOL BIT

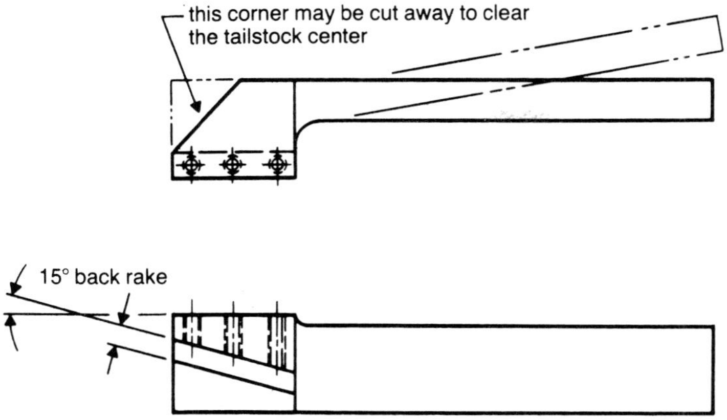

ALTERNATE DESIGN TO PROVIDE
A 15° BACK RAKE

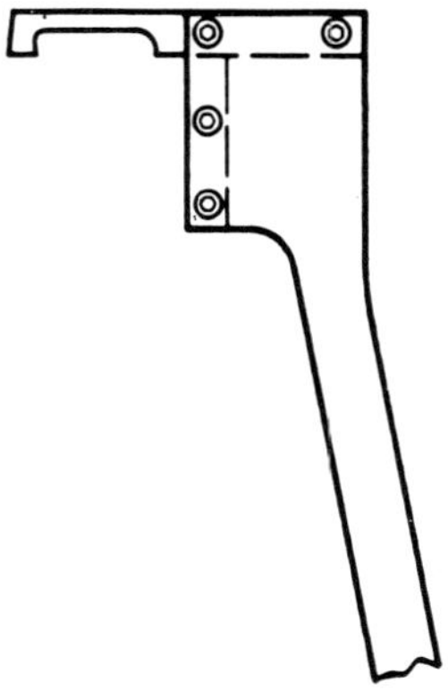

RECESSING OR
UNDER CUTTING
BEFORE THREADING

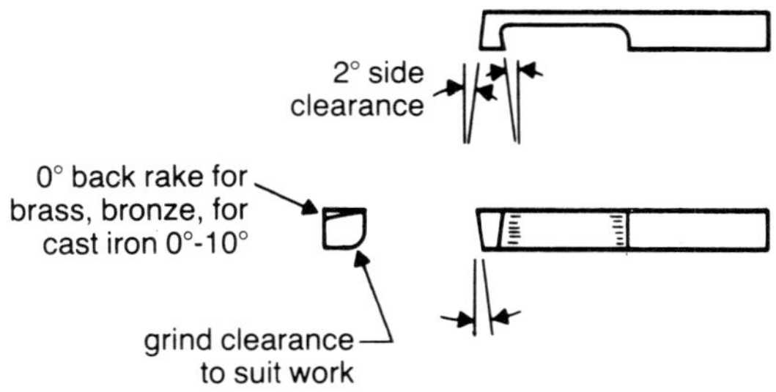

RECESSING TOOL

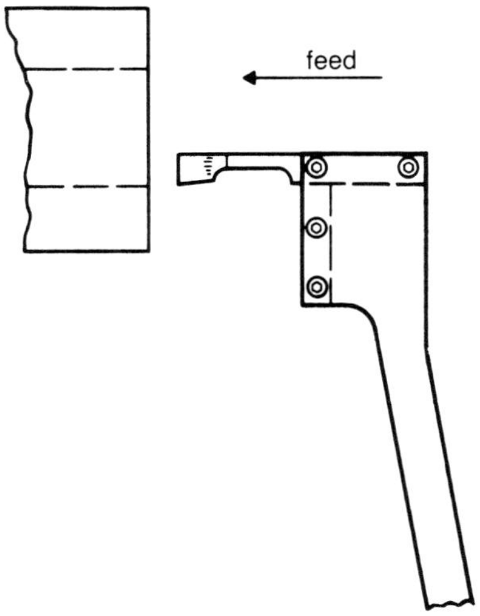

SET UP FOR BORING

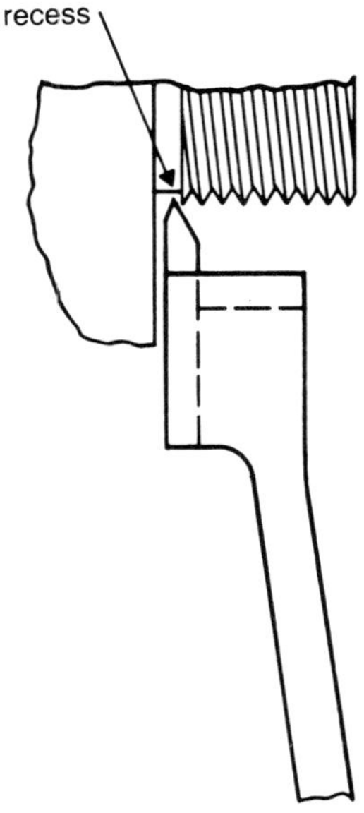

THREADING
TO A SHOULDER

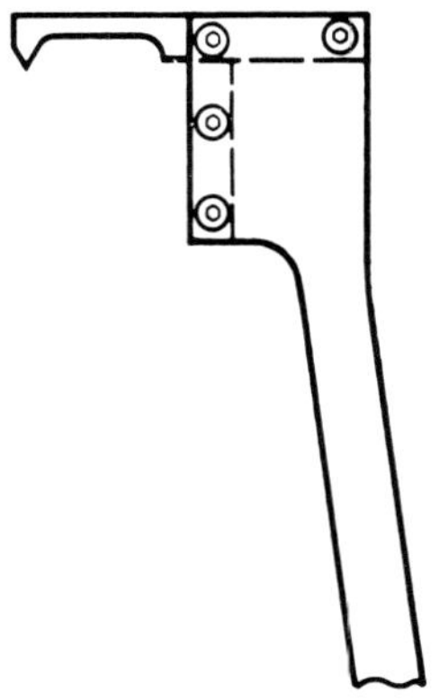

INTERNAL
THREADING

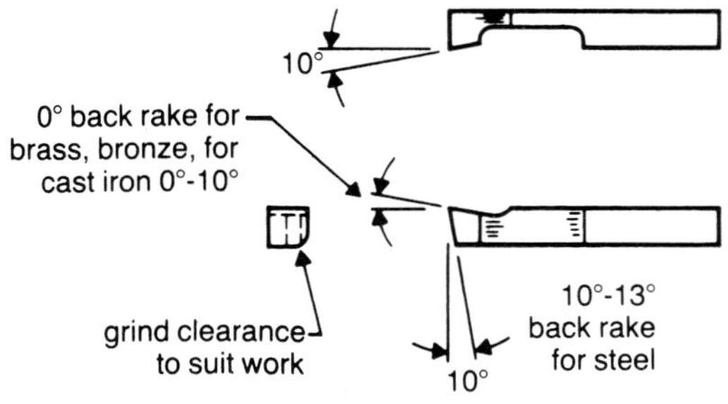

BORING TOOL

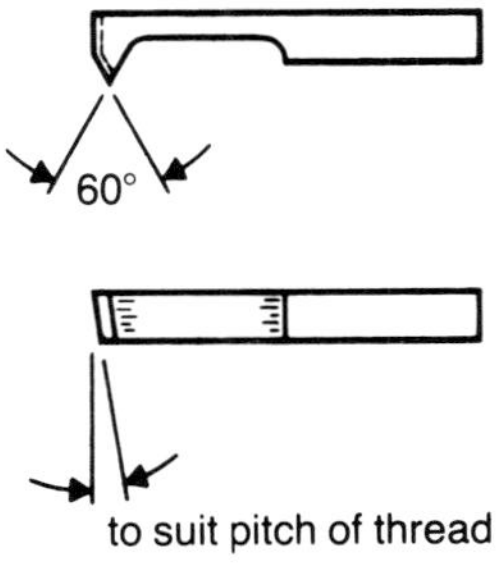

THREADING TOOL

whether you are taking a heavy cut or a light finishing cut.

Mild Steel and
Leaded Screw Stock ......80 to 100 ft. per min.

Tool Steel and
Alloy Steels .... 20 to 60 fpm

Cast Iron....................... try 60 to 100 fpm

Aluminum .... 200 fpm

Brass .... 300 fpm

Bronze .... varies

The machinability of bronze varies as there are many types. Some machine easily, but then some are so tough that machining them is similar to cutting a cheap steak with a dull knife. Try 100 to 150 fpm to start.

For tungsten carbides, increase the cutting speeds by 200% to 400%.

When calculating the setting for fractional size bars, drills or end mills, invert the fraction and multiply or use the decimal equivalent and divide. Example: A 1/4" drill in mild steel. CS = 80.

$$RPM = \frac{CS \times 4}{Dia.} = \frac{80 \times 4}{1/4"} = 80 \times 4 \times \frac{4}{1} = 1280$$

I hope this discussion has been helpful.

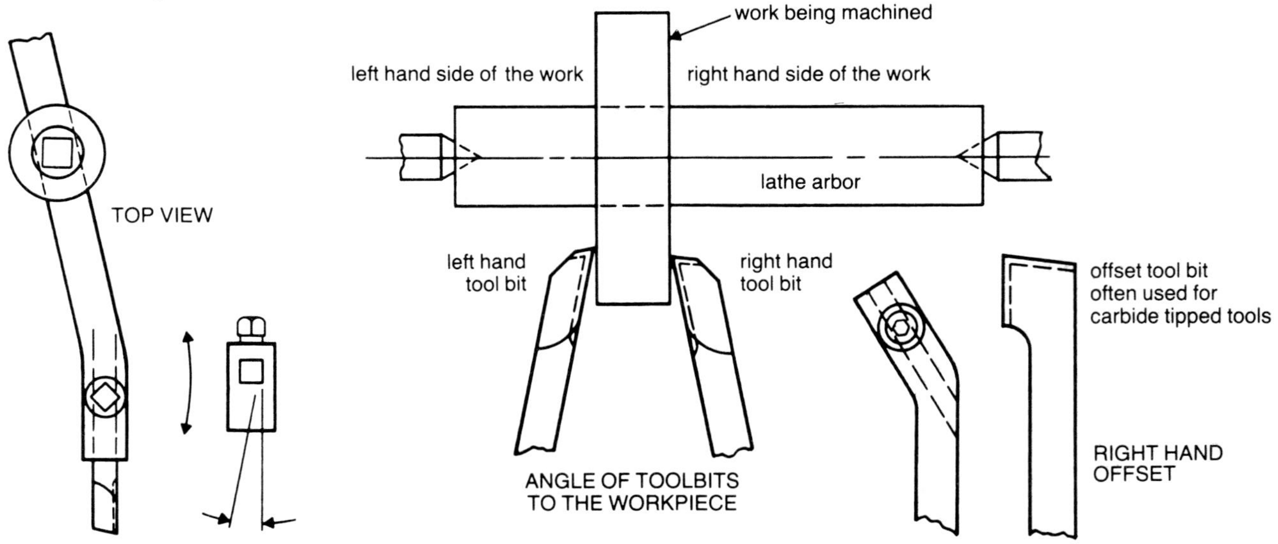

LEFT HAND BENT TOOL HOLDER

This clearance angle changes as the tool holder is raised or depressed.

RIGHT HAND BENT TOOL HOLDER

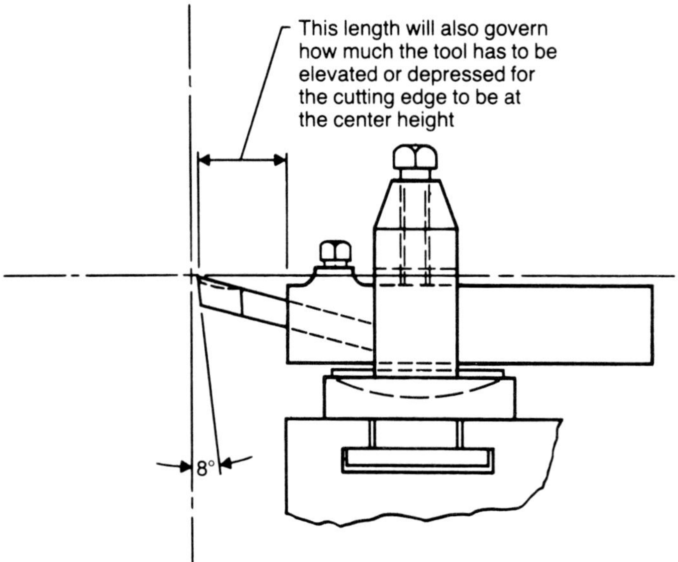

CORRECT SETTING FOR TOOLBIT AND TOOLHOLDER TO THE WORKPIECE

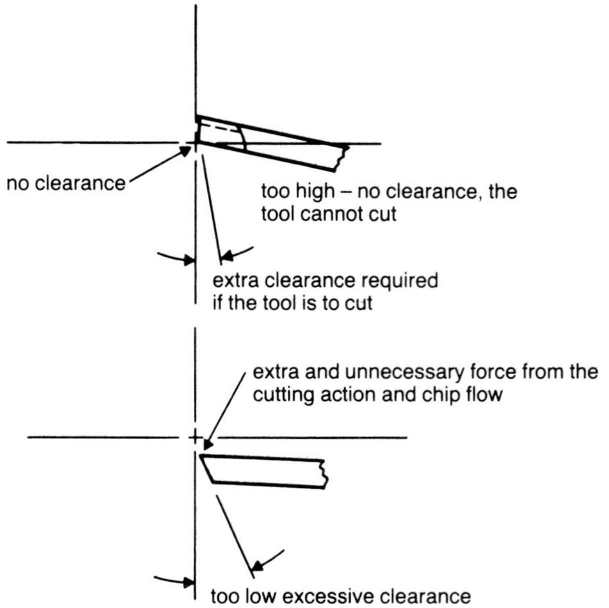

INCORRECT SETTINGS
The tip of the tool is weak.

MATERIAL REQUIRED

You will need one piece of 1" square hot- or cold-rolled mild steel 5-1/2" long, and five 10-32 × 1/4" Allen setscrews.

OPERATIONS

Face both ends of the bar smooth and square with the sides. Hold the work in a four-jaw chuck, turn fairly fast (250 to 500 RPM) and take light cuts from the center towards the outside with a right-hand tool set exactly at the center height and at an angle of 10° to 15° to the surface being faced.

If hot-rolled mild steel is being used, then the sides will have to be machined.

Those who own a milling machine or a shaper will have no problem, provided they remember a cardinal rule when machining a block square: the fixed jaw of the machine vise must be exactly vertical if square work is to be had. Test as shown in the sketch. After machining one side, this becomes the datum surface and is used against the fixed jaw of the vise to machine the two adjacent sides and is placed on parallels on the bottom when machining the fourth side. (Again, see the drawing.)

For those who must do all the work on a lathe, the solution is very simple and just as accurate. Hold the work in a four-jaw chuck as shown. If parallels are required to hold the work clear at the chuck jaws, get someone to help you hold the parallels in place as you tighten the jaws on the work. Tap the work back against the parallels with a soft hammer. Use a brass, copper, plastic, wooden or rawhide hammer. Do not use a regular ball-peen hammer as the work will "bounce" off the parallels and it will be impossible to make the work seat properly. If no help is available, the parallels may be held in place with scotch tape. Remove the parallels before machining by means of a brass bar and a hammer. *Do not switch on the lathe with the parallels in place* for you could be seriously injured if one of the parallels should fly out and hit you in the face.

Use a slow speed (50 to 100 rpm) and start facing from the outside toward the center using a left-hand tool. Set the power cross-feed to on or about .005" per revolution if available. Take a finish facing cut from the center toward the outside with a right-hand tool in the usual way.

Repeat for each of the other sides. Be sure the finished side is held firmly against the jaws if square work is to result.

Be sure to always remove the burrs caused in machining after each machining operation. Hold

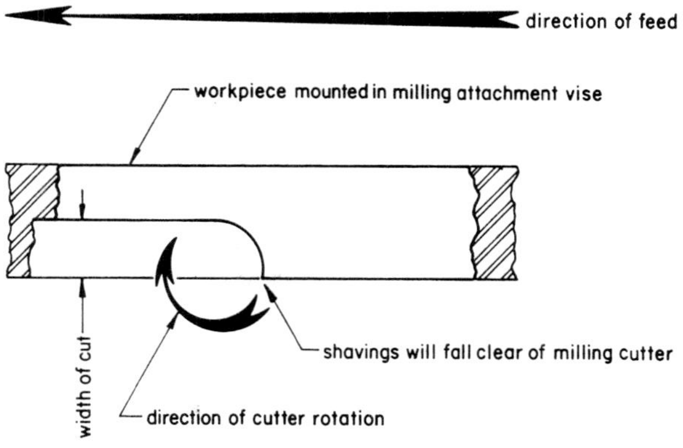

Repeat until entire surface has been machined.

MILLING WITH A MILLING ATTACHMENT IN THE LATHE

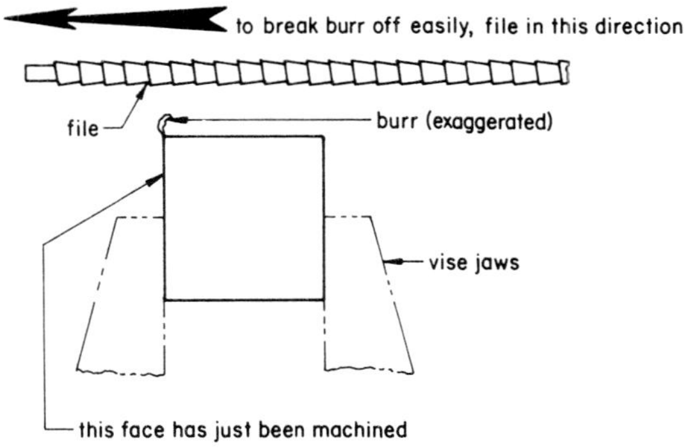

DE-BURRING PROCEEDURE

F. A. McLEAN/grb

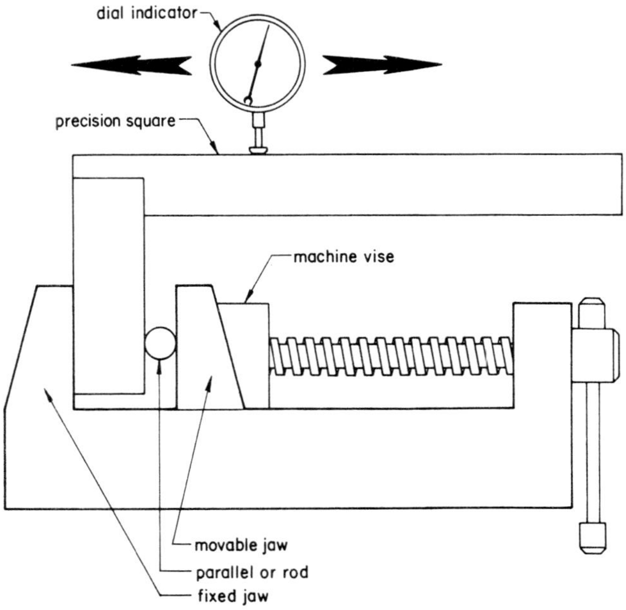

To test fixed jaw of machine vise for vertical accuracy, mount indicator in mill spindle and pass along blade of square and compare readings.

TESTING THE FIXED VISE JAW FOR VERTICAL ACCURACY

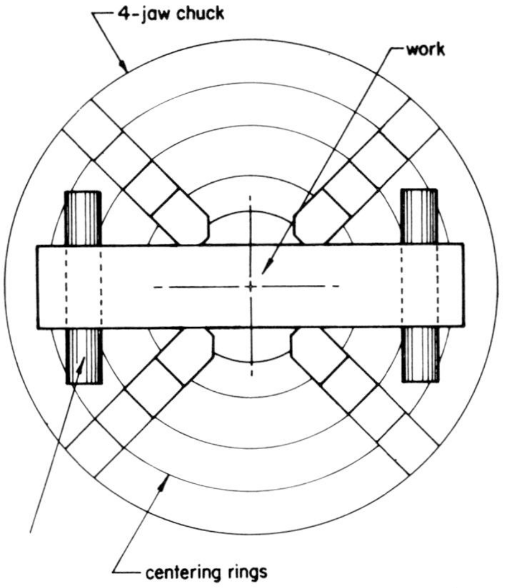

Use parallels to set the work out beyond the chuck jaws, if necessary. Tap the work back into the chuck, against the parallels, with a soft hammer—brass, wood, rawhide or plastic. Remove the parallels with a hammer and a brass bar before machining the work.

WORK SET IN A 4-JAW CHUCK FOR FACING SQUARE

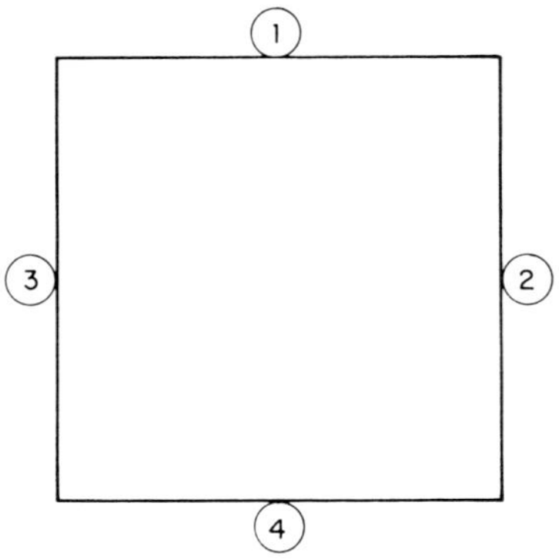

ORDER OF MACHINING SIDES FOR SQUARE BLOCKS

the work in the bench vise and file against the burr as shown in the sketch. Finish by holding the work in the left hand and pass a single cut file flat across each of the adjacent sides. Finish by filing a narrow (1/64") chamfer along the edge (45°) with a single cut (mill) file.

Paint the work with layout blue. If you don't have any, never mind. Warm the bar with a blow torch until it turns a brown to blue color and allow to cool slowly.

Lay out the design on the bar. The top, side and end views are just as shown on the drawing except that you are drawing in three dimensions. Use a small adjustable square and a scribe. Set the square to each of the measurements and set the stock against the work. Slide the stock along the work and scribe the line at the end of the rule with the scriber. Be sure to tip the scriber at a slight angle to have the lines where you want them.

Use a small center punch and mark the location of the screw holes.

Drill a 1/2" hole to form a radius at the back of the head or else just saw at an angle. Saw along the shank 1/16" from the line to remove the excess metal.

The shank can be finished on a milling machine or shaper. For those with only a lathe, this is where a milling attachment can be used to save a lot of filing.

To set a milling attachment square with the lathe axis, place the faceplate on the spindle and bring the attachment up against the plate and then tighten the clamping screws. To set the milling attachment parallel to the lathe axis, chuck a piece of cold-rolled steel in the three-jaw chuck. Bring the attachment up against the bar and tighten the clamping screws. Simple isn't it? It is also much quicker and nearly as accurate as using a dial indicator.

Hold the work in the milling attachment vise with parallels behind the work so that the surface to be machined just clears the chuck jaw. Use a 1/2" to 3/4" diameter mill or a fly cutter held in a collet or in the three-jaw chuck. Set the speed to about 300 RPM and advance the carriage along the bed for a very shallow depth of cut, approximately 1/64" to 1/32". Feed against the direction of the cutter rotation, use cutting oil and be careful. Milling in the lathe is an operation for which the lathe has not been designed, so very light cuts are in order. Lock all the slides that are not being used. That is, lock the milling attachment and the carriage before starting the cut with the cross slide.

Use a 1/4" diameter end mill to machine the slots. Set the lathe to 700 to 800 rpm and use a shallow depth of cut (1/64" to 1/32"). Clean the cuttings away with a small brush or use a jet of compressed air, if you have a compressor. Most small end mills are broken by the shavings jamming the cutter in the slot. Use cutting oil applied with a small brush.

Set the work in the drill press vise for drilling and tapping the holes for the setscrews. Use a No. 21 drill at about the third fastest speed on a standard 1/2" drill press. Do not move the work after drilling but replace the drill in the chuck with a 10-32 tap. Pull down on the feed handle with your right hand and pull the V-belt around with your left hand. As the tap is starting exactly in line with the hole, there is little chance of the tap breaking. Back the tap out after it breaks through into the slot. Repeat for each hole. The vise should be clamped to the drill press table and cutting oil should be used. Deburr by running a file along the slots.

Bend the shank by holding the head in a bench vise. One or two blows with a heavy hammer will bend the shank easily.

For those who want to case harden the holder, use a case hardening compound. Heat to bright red and sprinkle the compound on with an old salt shaker. Reheat and quench in water. Be sure to wash your hands after this operation. Most hardening compounds are very poisonous and the fumes are dangerous, too.

Leveling the Bench Lathe

Many bench lathes have the bed twisted when bolted to the bench after leveling. This is one main reason we hear so many complaints about lathes boring taper. Leveling the bench lathe is a most important function. Here are tips on how to do it.

MATERIALS REQUIRED

The best machine level you can buy or borrow.

A dial indicator and some type of stand.

A 3/4" to 1" diameter bar of drill rod or cold rolled steel about 3 feet long (to reach the end of the bed).

A pair of parallel bars for those whose lathes have the raised V-type beds. (Buy an 18" length of ground flat stock and saw in half. Use 1/2 × 1" or a similar size.)

OPERATIONS

The lathe must be mounted upon a sturdy bench that will not shake when the lathe is in use. Remove the tailstock from the lathe if yours is a short bed lathe. Remove any burrs from the flatways of the bed with a very fine oilstone. Clean the bed thoroughly and wipe dry. Loosen the bolts holding the lathe to the bench.

Test the level of the bed along the length by placing the level along one of the flatways of the bed. While the bed should be level lengthwise, it is not critical.

The leveling of the lathe across the lathe axis is *very critical* if the bed is not to be twisted. Put the parallels in place on the flatways of the bed. Slide the parallels slightly to make sure there is no grit and to break through any oil film. Place the level on top of the parallels, again testing to make sure there is no grit or any oil. *Do not turn the level end for end.* Make this test in front of the headstock and at the end of the bed. Look directly down upon the graduations of the dial to have as accurate a reading as possible.

When the lathe is level, place the bar in the three-jaw chuck. Set the dial indicator on the bed or on the carriage with the contact point touching the side of the bar. (Do not take this reading from the cross slide.)

Gradually tighten the bolts holding the lathe to the bench. Any deflection in the bed will be seen on the dial indicator. Shim the feet as necessary and retighten. On my lathe I used studs with nuts below and above the feet so that a minute adjustment is easily made.

With your lathe properly mounted, it will give you its best service for years to come and eliminate many frustrations.

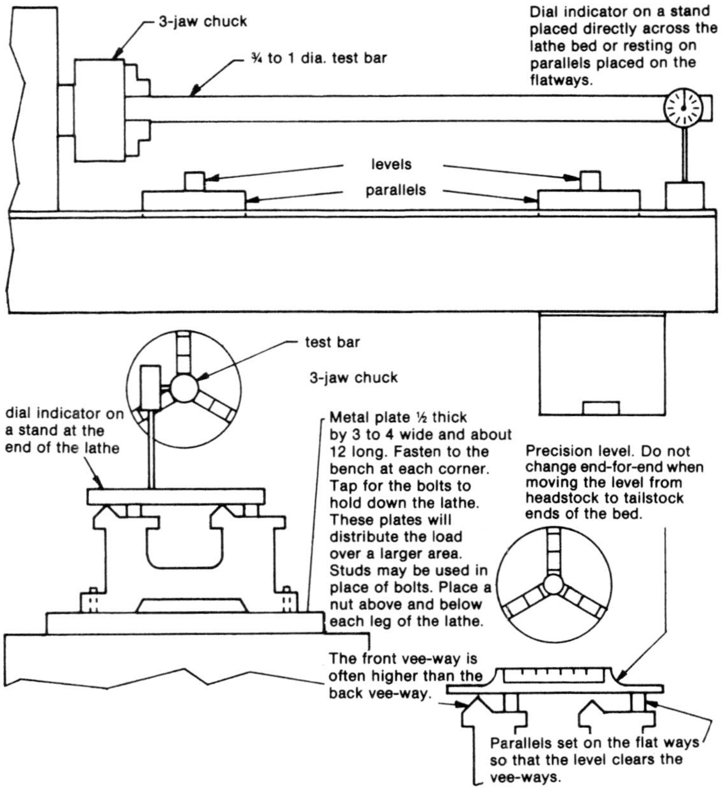

Tips on Turning Between Centers

The standard equipment with a 9" or 10" lathe includes two hardened centers, usually No. 2 Morse taper and a sleeve to fit into the No. 3 Morse taper in the headstock spindle. The Myford and similar size lathes omit the sleeve so the center fits directly into the spindle.

To have work that has been turned between centers exactly concentric with the center holes in the ends of the work, the headstock center *must run true.* Clean the spindle taper, sleeve and the center very carefully and wipe them dry with a clean rag or a soft paper towel. *Note:* Never try to clean the spindle taper with the spindle turning under power. Your finger may "catch" in the taper with the rag and your finger could be twisted off. (I know of a case where this happened.)

You can tell from the sound when the center and sleeve are "popped" into the spindle taper and there is a perfect fit without any oil or dirt between the tapers.

Now place a dial indicator in contact with the 60° cone surface and rotate the spindle slowly by hand. If the center is not "running true," check to see if there is any dirt between the tapers and look into the taper of the spindle to be sure that the surface has not been scored (scratched).

A scored taper can be repaired with a Morse taper reamer. Support the back end of the reamer on the tailstock center. Put a lathe dog on the reamer shank (or squared end) with the tail of the dog resting on top of the compound slide. Turn the lathe spindle by pulling on the belt by hand and feed the reamer slowly into the spindle by the tailstock handwheel. A light "scraping" cut is enough to restore the accuracy of the spindle taper.

If there is some run-out or eccentricity when testing with a dial indicator and the surfaces are clean, try removing the center and sleeve and replacing them in a different position to compensate for any inaccuracies in the sleeve and center.

If this does not work, then the 60° cone surface will have to be ground in place with a tool post grinder. Set the compound slide rest at 30° to the lathe axis and grind the center until it runs true. Usually the lathe spindle runs in reverse at a moderate speed. Be sure to cover all parts of the lathe with rags or plastic covers and clean the lathe with *Varsol* (solvent) very thoroughly after grinding.

When the center is running true, make small marks or nicks in line on the spindle nose, sleeve and center so the center can be replaced in exactly the same position each time.

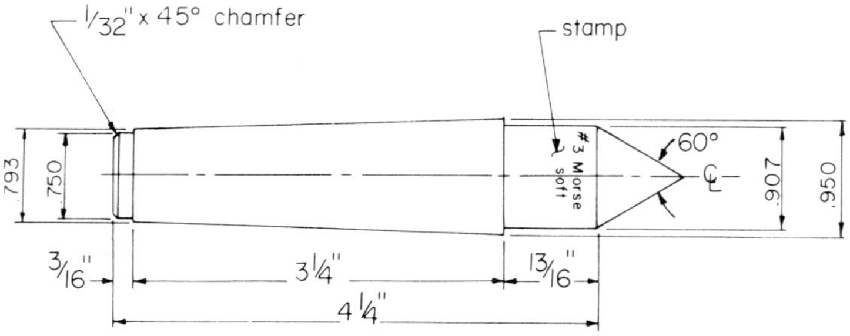

Notes: stock size 1 1/8" dia. x 5 1/2" H.R.S.
taper /ft. is .602
1 only required

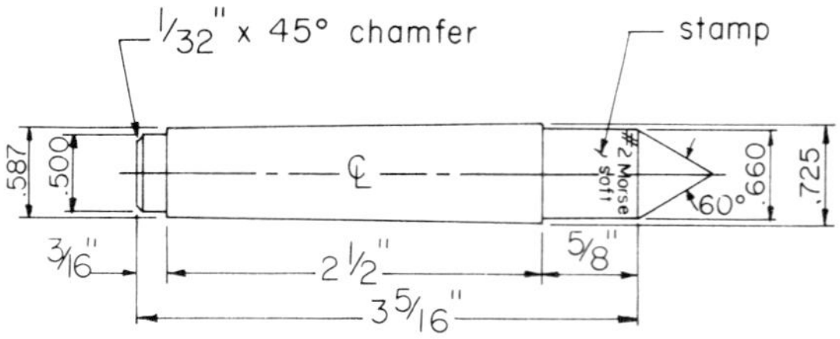

Notes: stock size 7/8" dia. x 4 1/2" H.R.S.
taper /ft. is .599
1 only required

Old-time turners avoided grinding on the lathe by using a soft center in the lathe headstock spindle. They usually omitted the sleeve and placed the center directly into the spindle taper for greater accuracy.

When setting up the lathe for turning between centers, they would take a very light cut on the center with the compound slide rest set at 30° to the lathe axis to make sure that the center was running true.

To make a soft center, cut off a piece of steel (hot or cold rolled) and chuck in the universal chuck with sufficient length projecting beyond the chuck jaws to complete the Morse taper. Face and center drill the end. Place the tailstock center in position to support the work. Now set the compound slide rest at an angle of about 1-1/2° to the lathe axis and turn the taper as described in the article on the Tailstock Die Holders. When the taper is complete, remove the lathe chuck from the spindle and set the soft center in place in the spindle taper. Now set the compound slide rest at 30° to the spindle axis and turn a 60° cone on the center. *Note:* The tool bit must be sharp and set exactly at the center height.

When turning between centers, the lathe centers and the center holes in the work must be in perfect condition. If you have a ball bearing center for the tailstock, use it. For those using a plain center, be sure to use a good grease such as *Lubriplate* No. 630AA or No. 930AA (Fiske). Make sure there are no shavings hidden in the grease in the center hole.

Locate the tailstock along the bed so that the tailstock spindle projects just enough for the tailstock casting to clear the cross slide or carriage when the work is placed between the centers. When viewed from the tailstock end, the knob on the tailstock handwheel should be in the 3 o'clock position. The weight of the handwheel knob will apply sufficient force on the work without squeezing out the grease. Lock the tailstock spindle clamp with moderate force and use the same amount of force each time.

The bent tail of the lathe dog must fit freely in the faceplate slot.

When the tailstock pressure is correctly set, the lathe dog will just rotate from its own weight in the faceplate slot when the dog is in a horizontal position.

When taking a heavy cut, stop the lathe frequently to make sure the work has not expanded too much from the heat generated from the cutting and the grease has been squeezed out of the center hole. *Note:* Because the headstock center is turning with the work there is no need to use grease there.

When taking an interrupted cut, the lathe dog may be wedged into the faceplate slot with a piece of leather or a little wedge made of wood. Another method is to tie the lathe dog tail in place with a leather boot lace or a piece of string.

Remember! Turning the work between centers is the only way that the work can be returned to the lathe for re-turning easily and accurately.

Turning Tapers

Tapers are usually expressed as the taper per foot. This means the amount of the taper if the length of the taper was one foot (see drawing). Refer to the *Machinery's Handbook* or similar books for the dimensions of the common tapers such as Morse, Brown & Sharpe, Taper Pin, etc.

There are three common methods of turning tapers. (a) By setting the compound slide rest to the correct angle and turning the taper with the compound slide rest when the carriage is clamped to the lathe bed; (b) with a plain or telescoping taper attachment using the lathe power feed; (c) by offsetting the tailstock and turning the work between centers. I shall describe the procedure to turn a taper with the compound slide rest in detail, as this is the most common method used in home workshops.

(a) Set the compound slide rest to the correct angle (see drawing). If you are turning a Morse taper this will be about 1½° to the lathe axis. If possible, place a drill with a Morse taper shank in the lathe 3-jaw universal chuck with the taper shank projecting beyond the chuck jaws. Place a dial indicator in the tool post. Traverse the dial indicator along the taper with the compound slide rest and adjust the compound slide rest setting until the dial indicator has the same reading all along the taper. Some slight alteration of this setting may be necessary when cutting the taper because of some movement in the slides. The tool bit should be set exactly at the lathe center height.

If the length of the taper to be turned is longer than the travel length of the compound slide rest, there will not be a problem. Simply return

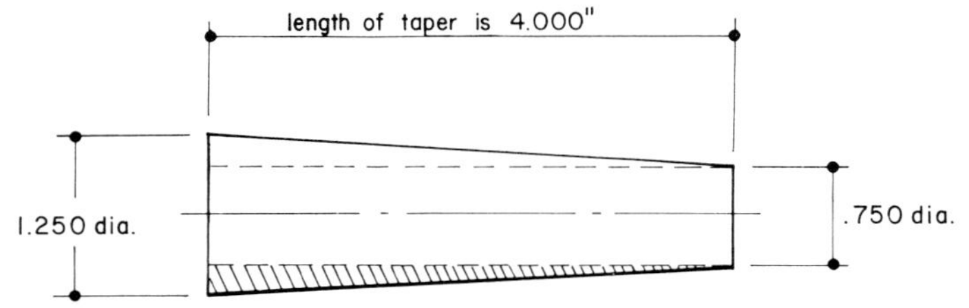

$$\text{TAPER PER FOOT} = \frac{(\text{large dia.} - \text{small dia.}) \times 12}{\text{length of taper}} = \frac{(1.250 - .750) \times 12}{4.000} = \frac{.500 \times 12}{4} = 1.500"$$

set the taper attachment to 1½" taper per foot

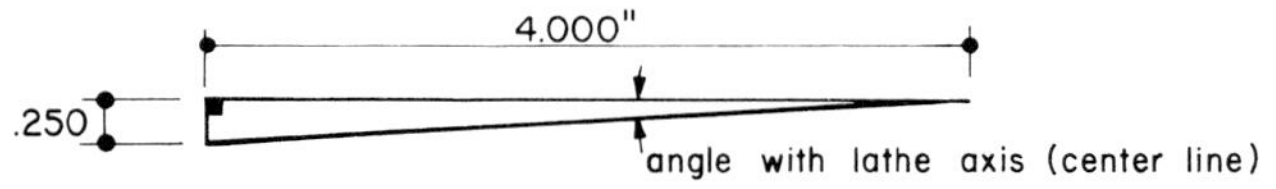

$$\text{TAN. OF THE ANGLE} = \frac{\text{side opposite the angle}}{\text{side adjacent the angle}} = \frac{.250"}{4.000} = .0625 \text{ or } 3°34'$$

(= about 3½°)

Calculate this angle for setting the compound slide rest.

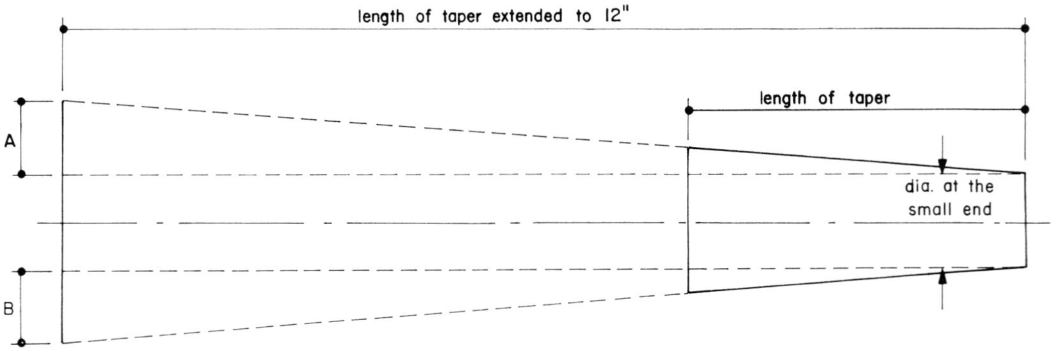

The taper per foot is A + B

the slide rest to the starting position and advance the carriage along the bed. Bring the tool bit to "just touch" the turned portion of the taper and continue turning. Any slight mark left where the second cut starts can be easily removed with a single cut "mill" or "lathe" file when the taper has been turned to the correct size.

The angle of the taper is checked with a sleeve. Use the sleeve from the headstock spindle or buy a Morse taper sleeve. The actual turning and testing is done by trial and error, but a little experience will make the job seem easy. After the first few cuts, try the sleeve on the taper and rock the sleeve to determine which end of the taper is oversize. Adjust the compound slide rest setting and take another cut. When the fit between the taper and the sleeve seems close, make three or four equally spaced chalk lines along the taper. Place the sleeve on the taper and rotate about one-half a turn. The sleeve will rub off the chalk on the tight end. Correct the slide rest settings and take light cuts until an even bearing is shown by the chalk lines. Now turn the taper down to size by measuring the diameter at the small end of the taper with a micrometer. If the small end of the taper is to be drilled and tapped for a draw rod, do so before removing the work from the lathe chuck. If a tang is required, turn the end of the taper parallel and face the end to the correct angle for the drift key. Check with the chart in *Machinery's Handbook* or other publications for the dimensions. The sides of the tang can be machined on a vertical milling machine (see drawing) or brought to size on the shaper or by sawing and filing.

(b) The taper attachment is an attachment that should be on every lathe if tapers are to be turned frequently.

There are two basic types of taper attachments. Plain taper attachments are the most common types on home workshop lathes. To use these, the compound slide rest is set parallel with the cross slide and the depth of cut is made with the compound slide rest. The nut on the cross-feed screw is disconnected by removing the bolt in the cross slide. This permits the cross slide to move freely and follow the taper angle as the shoe or saddle slides along the taper attachment bar.

The telescoping taper attachments have the cross slide feedscrew made in two pieces and these can "telescope" to follow the taper attachment. With these, the depth of cut may be set with the cross slide or the compound slide rest may be set parallel with the cross slide and the depth of cut will be set with the compound slide.

Place the work in the lathe chuck or between centers. Locate the taper attachment along the lathe bed so that the shoe can travel along the taper attachment 1" to 2" before starting to cut the taper. The bracket extending from the cross slide over the taper attachment will be clamped to the shoe.

For Morse tapers, set the taper attachment to about 5/8" taper per foot or 3° included angle or 1½° with the center line (check the graduations).

Take a few light cuts and test and adjust the taper attachment until the chalk lines are rubbed off evenly all along the taper. Be sure the tool bit is set exactly at the center height.

Using the tailstock offset method, calculate the amount to offset the tailstock. To move the tailstock over accurately is easy. Reverse the toolholder in the tool post and bring the back end up to touch the tailstock spindle. Pinch a piece of paper between the toolholder and the tailstock spindle until the paper can be just barely pulled out. Note the reading on the cross slide dial and back the cross slide out a bit more than the offset to compensate for the backlash and then place the dial to the required setting.

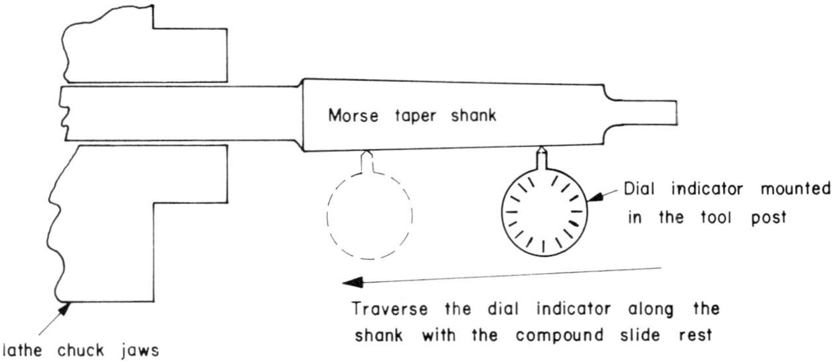

One method of setting the compound slide rest to the correct angle

Offset the tailstock until the paper is pinched and the force to pull out the paper is the same as before. Note: the bolt clamping the tailstock sole plate to the bed will have to be loosened before the tailstock body can be offset on the sole plate by the two screws at the sides of the tailstock body. To offset the tailstock to the front, loosen the back screw and tighten the front screw. Do not forget to tighten the clamping bolt to the lathe bed before starting to turn the taper.

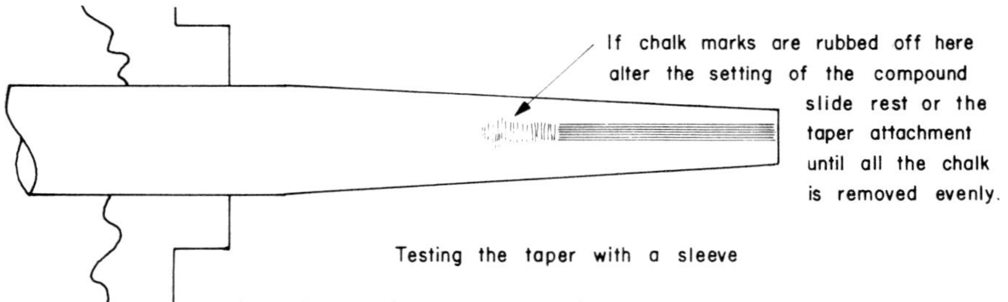

Testing the taper with a sleeve

Make three or four equally spaced chalk lines along the taper. Rotate the sleeve about one half a revolution.

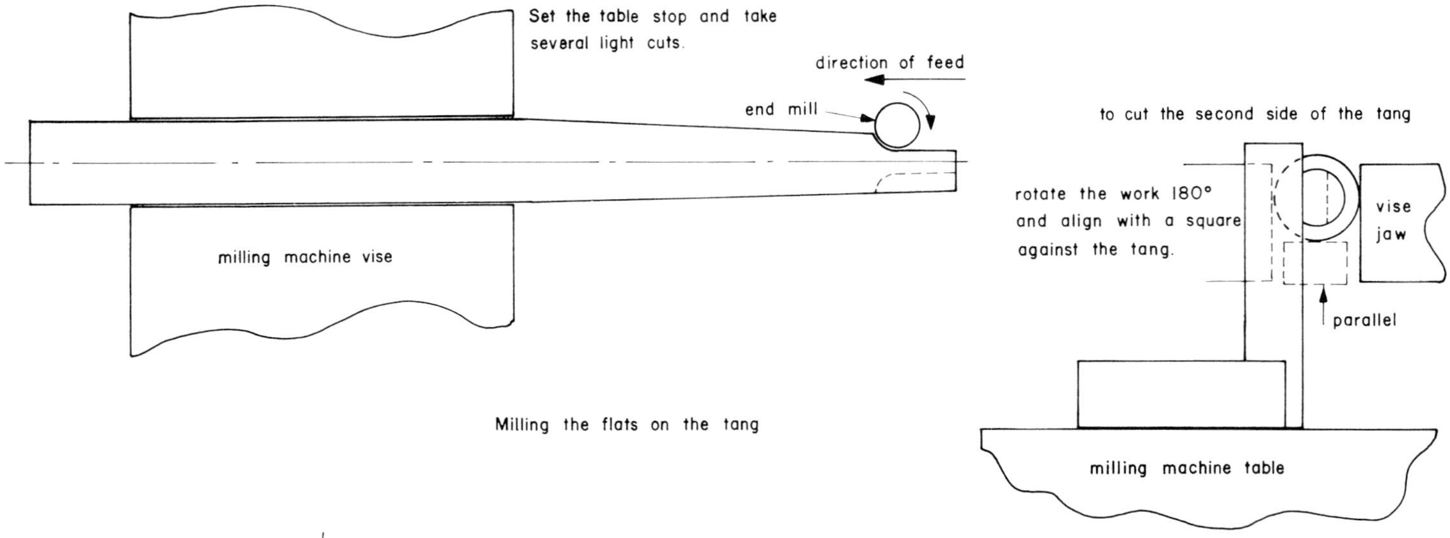

Milling the flats on the tang

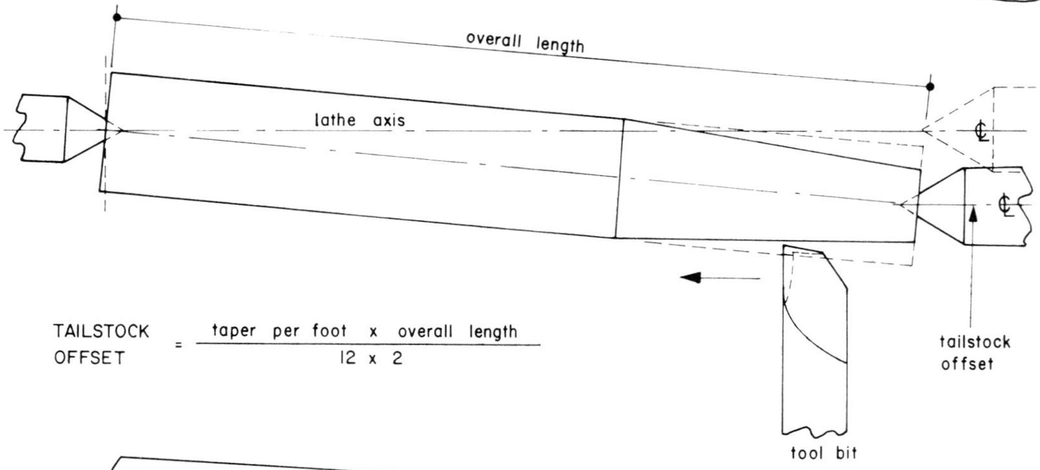

$$\text{TAILSTOCK OFFSET} = \frac{\text{taper per foot} \times \text{overall length}}{12 \times 2}$$

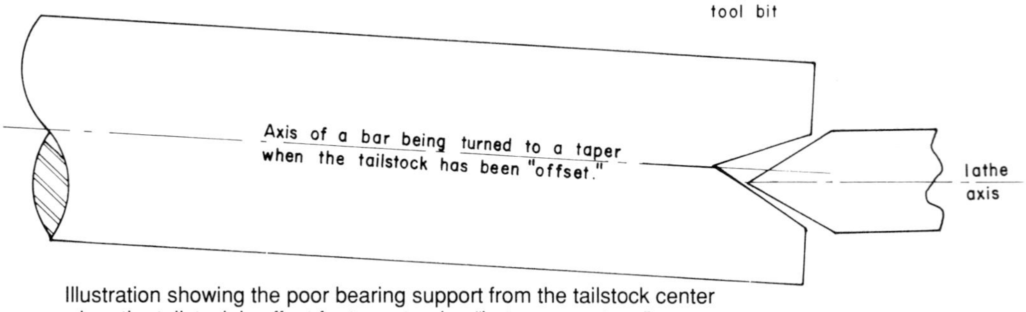

Illustration showing the poor bearing support from the tailstock center when the tailstock is offset for taper turning "between centers."

How to Align the Tailstock Easily

Place both of the lathe centers in the lathe. Test the concentricity of the headstock center by placing the contact point of the dial indicator on the conical section of the center and revolve the spindle by hand. If the spindle is not running true, take corrective action.

Now place the test bar between the centers (below left) with the tailstock spindle extended the usual distance for turning between centers. Clamp the tailstock to the bed and tighten the tailstock spindle clamp moderately with the same force each time.

Place the contact point (lever) on the test bar at the center height and run the carriage to traverse the dial indicator along the bar.

If the tailstock center is not in line with the headstock center, the dial indicator will show the direction and the amount of the offset. To adjust the tailstock, loosen the clamping nut to the bed. Adjust the two setscrews at the base to bring the tailstock center in line with the headstock. Tighten the clamping nut to the bed and then test the alignment with the dial indicator. Repeat these operations if the tailstock is not quite in line with the headstock (below right).

With some practice, the whole operation should take less than 5 minutes.

Equipment required: A test bar (see drawing), and a dial indicator.

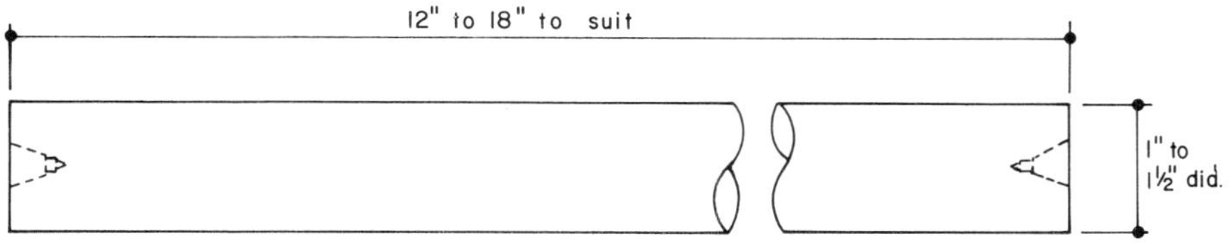

Center drill and then bore center holes using a steady rest so the center holes will be concentric with the circumference.

TEST BAR FOR ALIGNING THE TAILSTOCK

Repairing an Old Lathe

A friend of mine has a lathe that was made before the turn of the century. The tool post sits on a tee slot on the cross slide. There is no compound slide rest. The cross slide lead screw was very loose in the nut, and the intermediate gear in the apron for the hand feed had many broken teeth. The owner is on a long-term disability pension, so I offered, at no charge, to repair the lathe so he could use it again.

The cross slide screw was a vee thread, 12 threads per inch and .406″ diameter. I have no idea what thread this is. The cross slide never had a dial as fitted on lathes today, so the pitch of the screw was not important.

The cross slide nut was a bronze casting with a round plug that fitted into a hole in the underside of the cross slide. However, the tapped hole seemed to be off-center. I wanted to retain the same location so the nut was held in the milling machine vise and centered under the spindle by adjusting the table and cross slide handles until a drill would pass down through the nut. The slides were locked and the nut was drilled out to 29/64″. This was the maximum the nut casting would allow.

I turned up a bushing on the lathe for a slip fit in the nut casting (.002″-.003″ undersize) and then tapped 3/8-16NC left-hand. I have several left-hand taps and dies bought from Kitts Industrial Tools. They are moderate in price and the quality is excellent.

I could have drilled a hole in the nut casting, but I was afraid the drill might "catch up" and damage the thin wall of the casting, so I used a

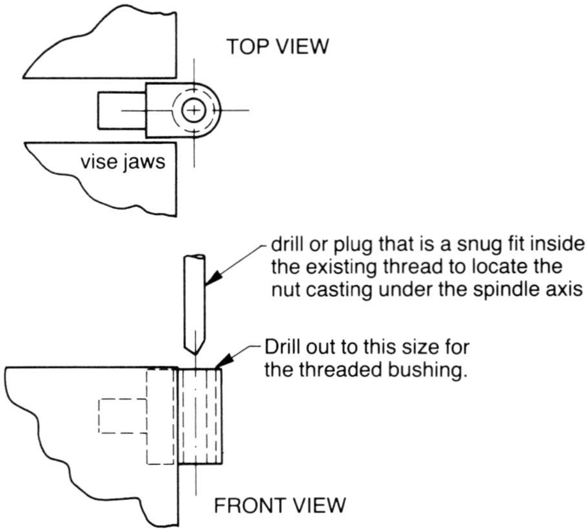

round file to file a slot through the wall for soldering the bushing in place.

The bushing was coated with soldering flux and placed in the casting. The nut was placed on a firebrick and heated with a propane torch. When the correct temperature had been reached, the filed notch was filled with solder (Photo 1) and capillary action carried the solder all around the bushing (Photo 2). I used 63-37 solder; this is the solder used most often in electronic factories. It melts at the lowest melting point, and is called "Eutectic Solder." A 60-40 or 50-50 mixture solder can, of course, be used.

The gear to be replaced was cast iron, with a diametral pitch of 20, and 39 teeth. We could have purchased a change gear from Boston

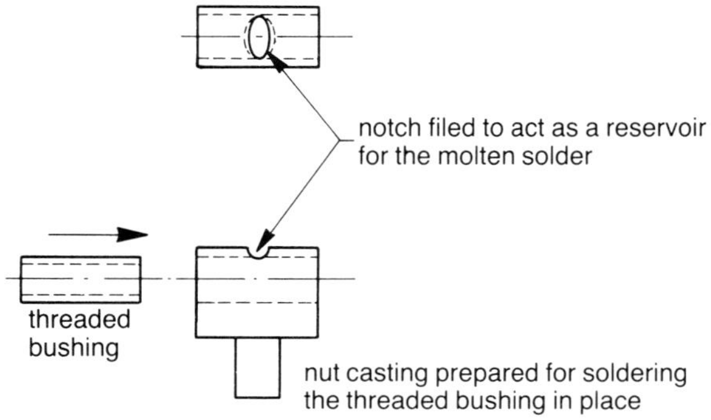

1

2

3

Gears; however, I decided to make one instead. The gears are only turned by hand, and do not have to be perfect to function.

I had a fly cutter arbor with the cutter slot at 90° to the spindle axis, and an assortment of tool bits ground up as gear cutters. However, I couldn't find it so I used an arbor on hand and ground up a tool bit on the edge of the grinding wheel until the tool bit fitted into the good teeth remaining on the gear.

For a job like this, a dividing head is really a necessity (Photo 3). Follow the procedure shown in Figure 1 and the whole operation is really very simple.

The turning and threading operations were simple. Remember, if you perform this task, to swing the compound slide rest to the left when cutting left-hand threads. The threading tool will cut on the other side so extra clearance is required on the right-hand side of the tool bit. Start each cut in the recess. When the tip of the tool bit touches the bottom of the recess, the thread should be on size.

I have shown the outside diameter to be .370″-.375″ diameter. The .375″ is, of course, ⅜″, the nominal diameter. However, for threading, if the diameter is .004″ to .005″ undersize, there is less chance of a burr on the

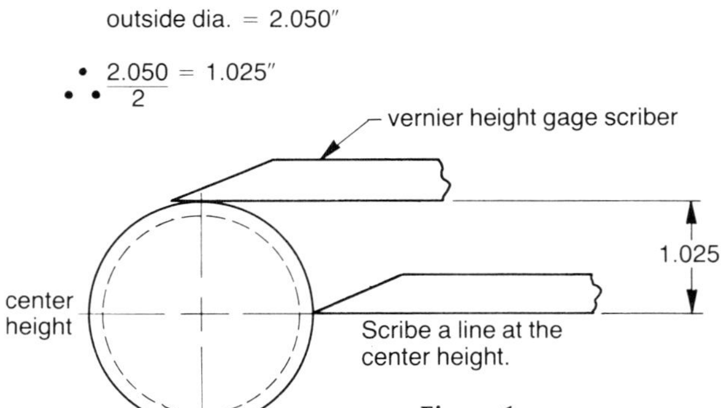

Figure 1

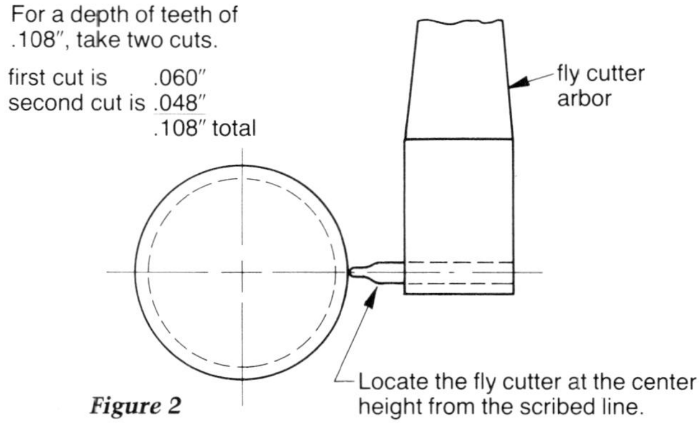

Figure 2

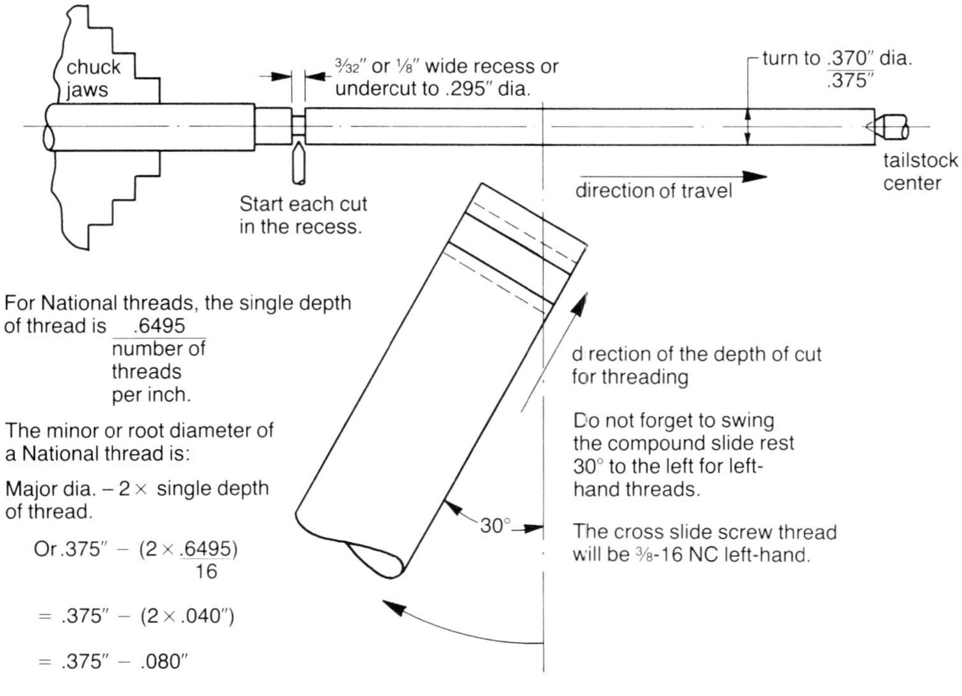

Figure 3

threads that might give a false reading when you try the nut after threading.

Sometimes a worn machine can be made serviceable again by a few simple repairs. Frequently, the machine is so old that an extensive and expensive rebuild is not justified, but the machine can be used again if its limitations are taken into consideration.

39-tooth gear **20 diametral pitch**

$$\text{Outside Diameter} = \frac{\text{Number of teeth} + 2}{\text{Diametral Pitch}} = \frac{39+2}{20} = \frac{41}{20} = 2.050 \text{ diameter}$$

$$\text{Whole Depth} = \frac{2.157}{\text{Diametral Pitch}} = \frac{2.157}{20} = .10785'' \text{ or } .108''$$

$$\text{Indexing is } T = \frac{40}{N} = \frac{40}{\text{Number of Teeth}} = \frac{40}{39} = 1\tfrac{1}{39}$$

(One full turn of the crank plus one hole in the 39 hole circle.)

The Whole Depth of Teeth is the sum of the Addendum and the Dedendum.

$$\text{The Addendum} = \frac{1}{\text{Diametral Pitch}}$$

$$\text{The Dedendum is } \frac{1.157}{\text{Diametral Pitch}}$$

(.157 is for clearance at the bottom of the teeth.)

The Pitch Diameter of a Gear is

$$\frac{\text{Number of Teeth}}{\text{Diametral Pitch}}$$

Therefore, the addition of two addendums will give the formula of:

$$\frac{\text{Number of Teeth} + 2}{\text{Diametral Pitch}} = \text{Outside Diameter}$$

How to Replace the Damaged Thread on a Drawbar

The thread on the end of the drawbar of a vertical milling machine is often damaged or worn out from frequent use. A complete new drawbar can be made, but a better method is to replace the threaded portion with a long setscrew with an Allen or hexagon socket. These setscrews are made from very good quality steel and have been heat-treated to make them fairly hard and very strong and long wearing.

Place the drawbar in the three-jaw lathe chuck or lathe collet with about ½" of the bar beyond the chuck jaws. Turn off the damaged thread. Now face, center drill, pilot drill and drill with the correct size drill before tapping the thread. Use 5/16" or 21/64" for 3/8-16NC threads and a W drill for 7/16-20NF threads. The thread should be about ¾" to 1" deep. Place the shank of the tap in the drill chuck and rotate the chuck by hand to be sure the tap starts exactly in line with the hole. Transfer the work to the bench vise to finish the tapping operation.

An Allen-type setscrew used to replace a damaged thread on a drawbar.

Blow the shavings out with compressed air and screw the setscrew in place. Hold the drawbar in the bench vise so the setscrew may be tightened properly. If the setscrew loosens and stays in the arbor, you will probably damage the thread inside the drawbar. The setscrew can be easily replaced if it becomes worn or damaged.

Boring Internal Threads

If the work to be threaded has a hole right through, threading will be easier, for then the half nuts can be disengaged when the threading tool has passed through to the other side of the work. When the hole to be threaded is "blind" – that is, the hole stops part way through the work – a recess or undercut will have to be formed at the bottom of the hole. This recess is necessary, for we want the threading tool to be free from the thread before we disengage the half-nuts from the leadscrew and then withdraw the threading tool to clear the thread when we return the carriage to the starting position.

The first operation is to face the end of the work flat. This gives a surface to measure from to determine the depth of the hole and

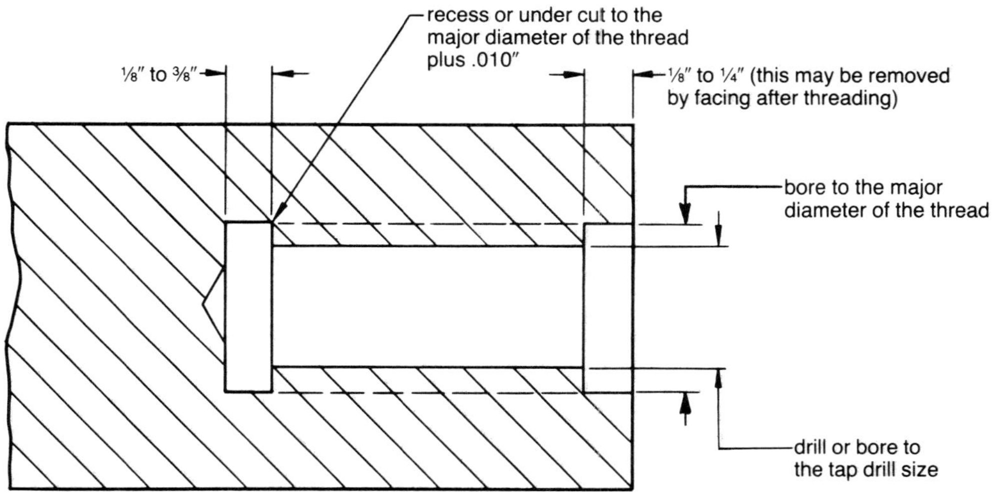

PREPARING THE WORK FOR CUTTING A THREAD WITH A BORING TOOL

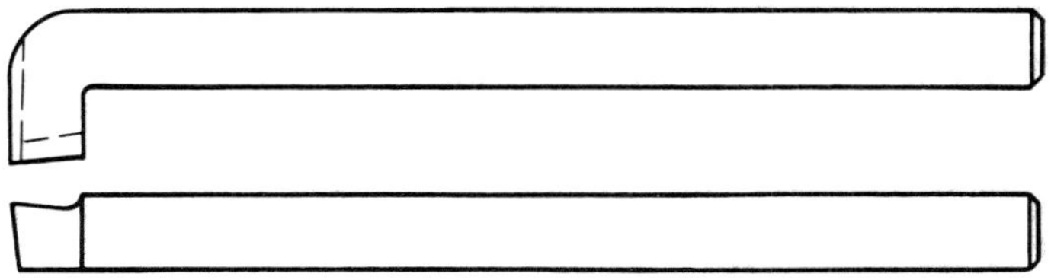

STANDARD BORING BAR

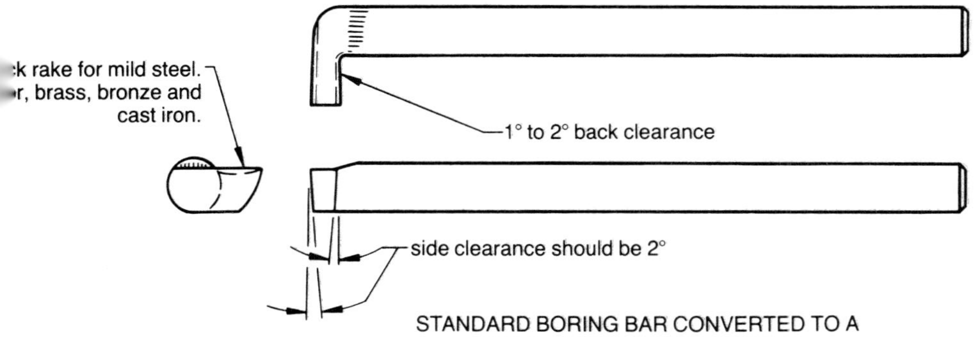

STANDARD BORING BAR CONVERTED TO A RECESSING TOOL BY GRINDING

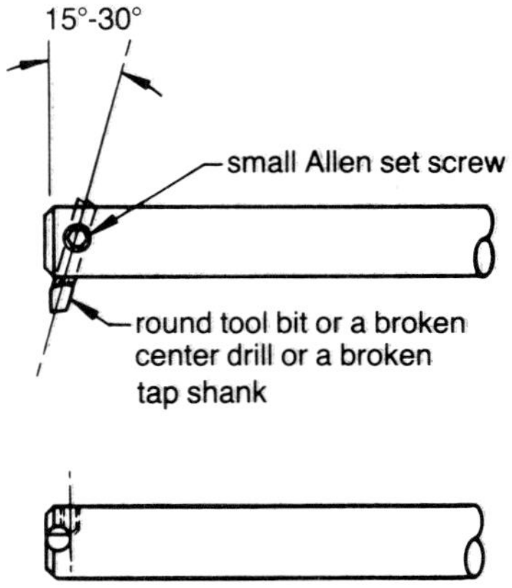

HOME-MADE BORING TOOL

for forming the recess. A smooth surface is required at 90° to the lathe axis so the drills will start on center and drill true (concentric) holes. Use several drills, each one about 1/8" larger than the previous drill. Drilling large diameter holes is liable to cause the drill chuck shank or the Morse taper drill shanks to twist in the tailstock spindle socket and score the Morse taper. Remember, drilling a 1/2" diameter hole from the solid without a pilot hole is similar to taking a 1/4" depth of cut with a turning tool on the lathe. The drill takes a balanced cut, i.e., cutting on both sides of the lathe axis; however, the cutting action exerts considerable torque on the drill shank. If the drilled hole starts to run off center (eccentric) or a special size hole is required, then the hole will have to be finished by boring with a boring bar.

To find the correct size drill (or hole) to use before tapping a National thread, subtract one thread from the major diameter (outside) of the thread.

Example: 3/8" – 16NC
3/8 – 1/16 = 6/16 – 1/16 = 5/16" drill size.

For many threads, especially the special threads, calculations will be easier if the fractions are converted into decimals.

Example: 1/4" – 28 NF

$1/4" = 4\overline{)1.000}^{\,.250"}$ and $1/28" = 28\overline{)1.000"}^{\,.0357"}$

therefore .250"
−.035"
.215" Choose the nearest drill size – no. 3.

Example: 3/8" – 40 NS

3/8" = .375" and 1/40" = .025"

therefore .375"
−.025"
.350" Choose the nearest drill size – S.

To convert fractions into decimals, think back to your primary school days, and divide the numerator of the fraction by the denominator.

The next operation will be to bore a short part of the outer end of the hole (1/8" to 1/4") to the major diameter of the thread. There are two reasons for doing this: first, to set the recessing tool so the recess will be the correct diameter; second, the tip of the threading tool will just touch this major diameter when the thread has been cut to the correct size.

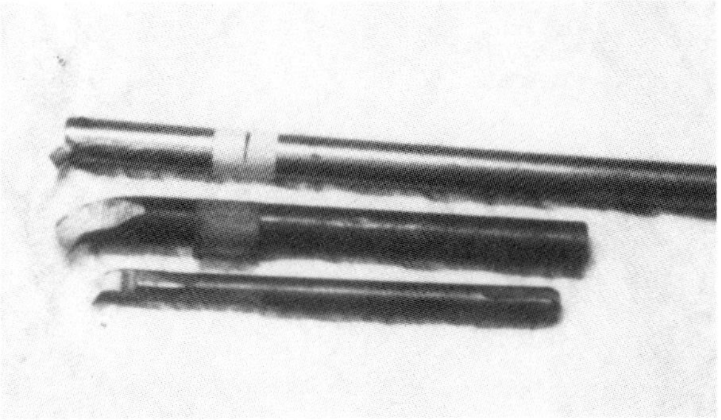

1 ***Three recessing tools.***

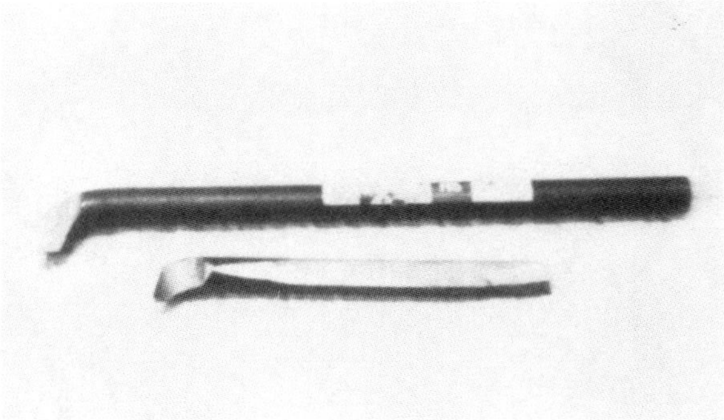

2 ***Two boring bars.***

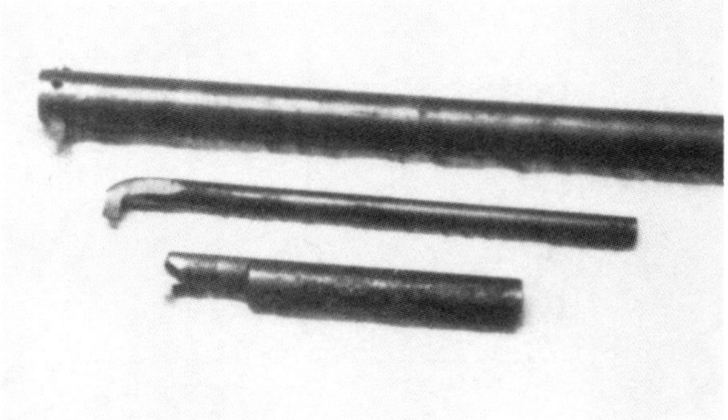

3 ***Three 60° vee threading bars.***

4 ***Wrap masking tape on the boring bar and make a line showing the depth. Use a ball point pen or a pencil.***

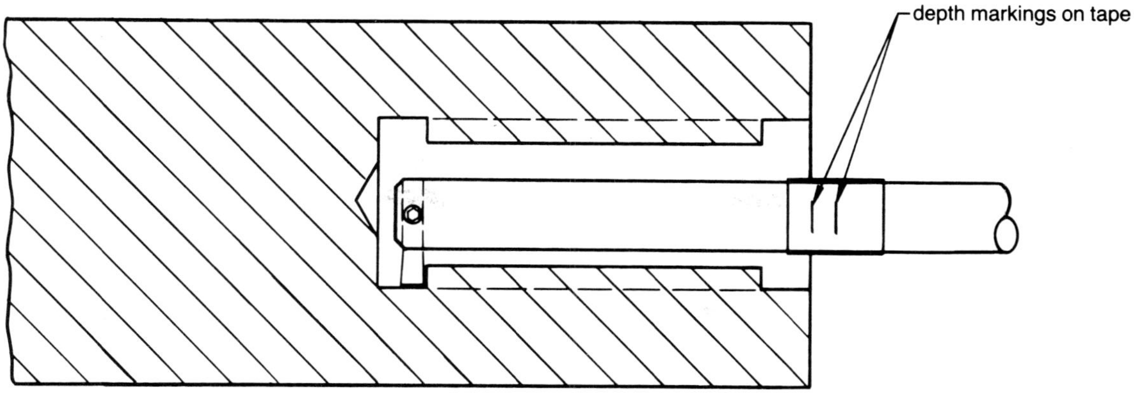

FORMING A RECESS OR UNDERCUT

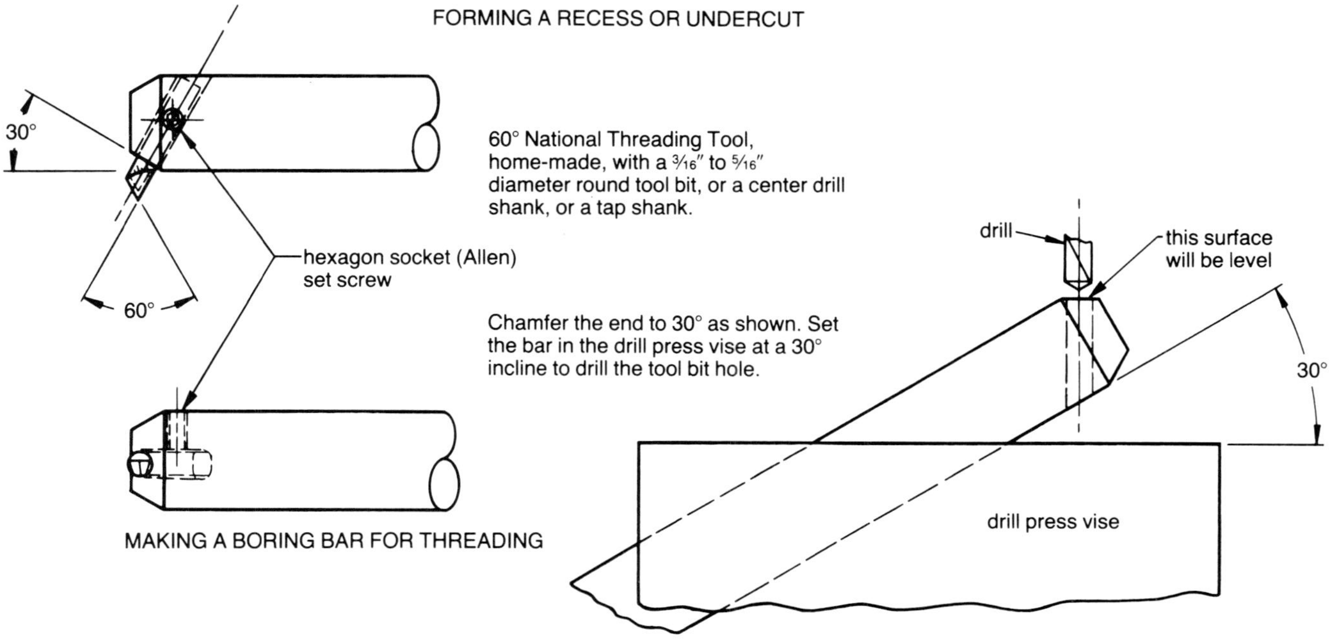

MAKING A BORING BAR FOR THREADING

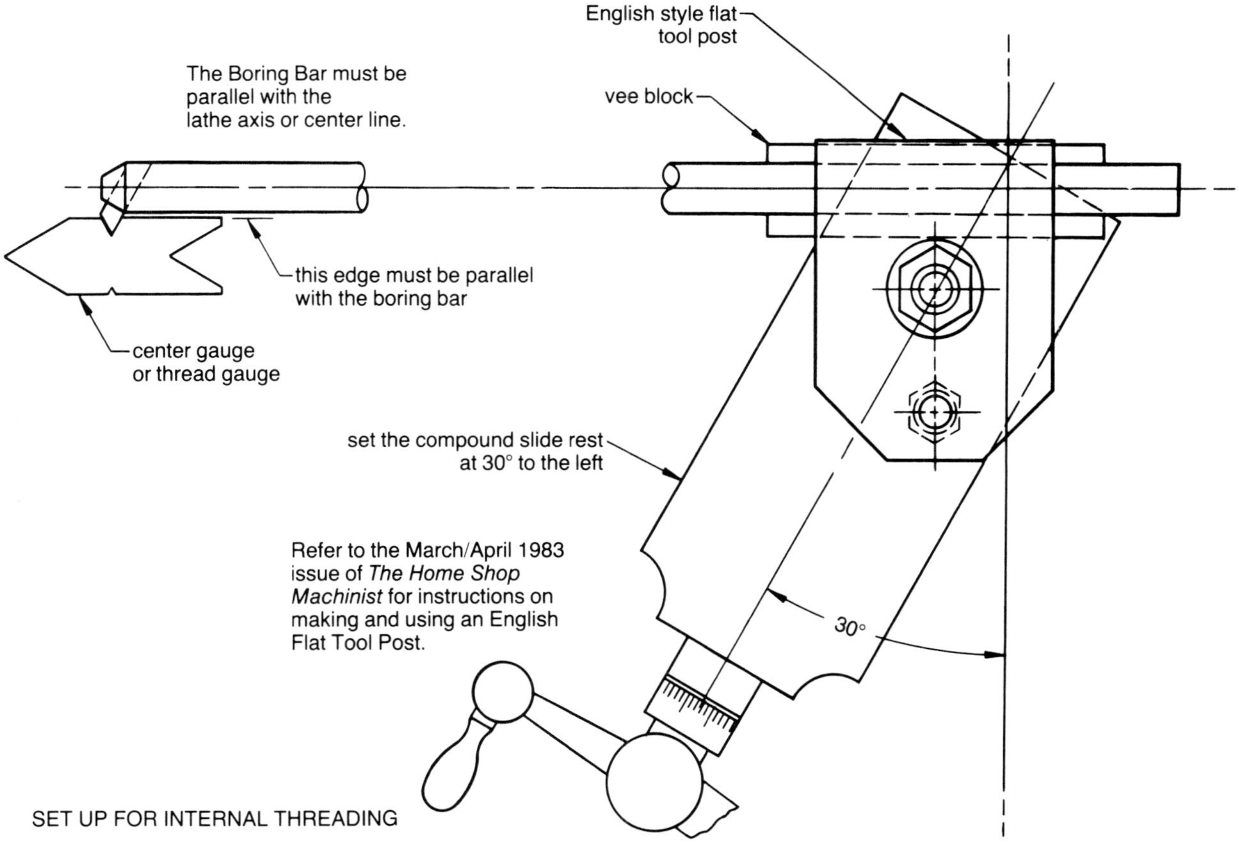

SET UP FOR INTERNAL THREADING

A recessing tool will have to be ground from a standard boring bar or made up from a bar of cold drawn steel and drilled or broached to suit a tool bit.

Set the recessing tool up on the lathe with the bar parallel with the lathe axis and the cutting edge on the center height. Measure the depth of the recess from the tip of the cutting edge and make a mark on the boring bar with a grease pencil or, better still, wrap a piece of masking tape around the bar and then use a pencil or a ball point pen to indicate the depth (see drawing). Bring the cutting edge of the recessing tool to touch the major diameter of the thread and then set the cross slide dial to zero.

To form the recess, set the recessing tool to the correct depth in the hole and then clamp the carriage to the bed. Use a moderate speed and feed the tool slowly into the work. Make the recess diameter of the work. This is a difficult operation so some chattering is almost inevitable. Squirt some cutting oil into the hole with an oil can. If you don't have any cutting oil, use old motor oil. This is usually available for the asking from your local automobile service station.

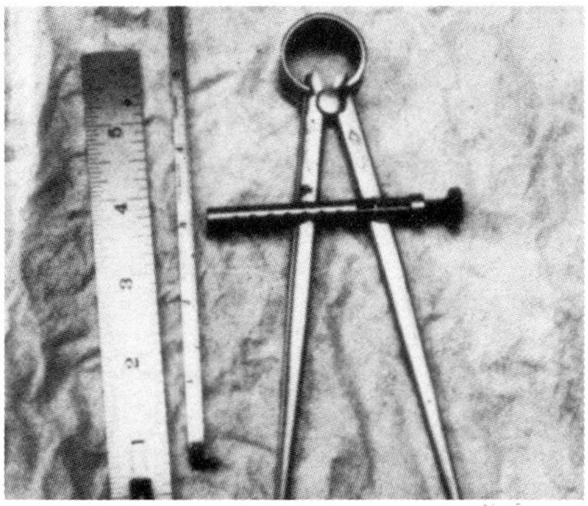

5 ***Two hook rules and a transfer inside caliper.***

The depth of the recess can be checked with a hook rule. To check the diameter of the recess, a special type of inside caliper called a "transfer" caliper is required. The caliper is adjusted so the tips just fit the diameter of the recess. Then the legs are squeezed together to withdraw the caliper from the bore. When the legs are released they will spring back to their former position and the measurement is taken with a micrometer.

Take care when setting up for threading. The compound slide rest is set 30° to the left. The boring bar must be parallel to the lathe axis. The tool bit must be ground so the gage is parallel to the bar. Be sure there is sufficient front clearance on the tool bit or the tool will rub on the flank of the thread producing a rough, ragged thread. The top of the threading tool should be located at the lathe center height. Carefully try the boring bar in the hole and locate the threading tool in the recess by rocking the carriage gently back and forth. Set the carriage stop (clamp) or the micrometer carriage stop or make a mark on a piece of masking tape wrapped around the bar. Some old-time machinists used to twist a piece of wire around the bar with a pair of pliers.

To start the threading operation, bring the threading tool to touch the minor diameter of the bore and set the cross slide and compound slide dials to zero. The tool will be advanced into the cut with the compound slide as for external threading, except the setting will be out towards the operator.

6 ***In setting a boring bar at the lathe center height, measure from the lathe bed to the tailstock center. Set the boring tool to the same reading on the scale.***

7 ***Parallels are shown resting on the lathe bed flatways to support the surface plate. My lathe bed has vee ways of different heights so the surface plate must be raised above the vee ways.***

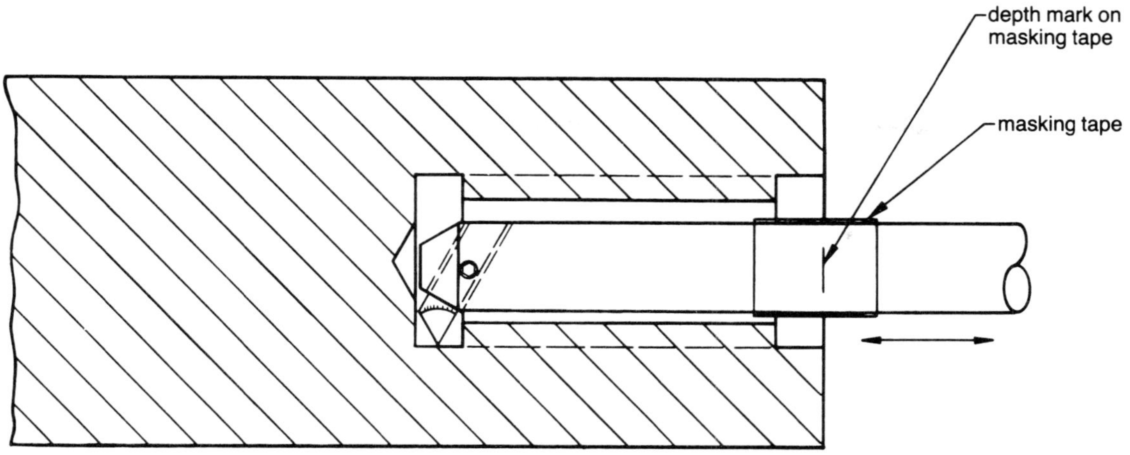

LOCATING THE THREADING TOOL IN THE RECESS

The bar is not as rigid as the tool holder for external threading so light cuts are necessary. Start with a cut of .005" and reduce the cuts to .002" and then .001" as required. Disengage the half nuts as soon as the threading tool is in the recess. Withdraw the tool bit about 1/2 a turn of the cross slide dial for returning the carriage to the starting position. When the tip of the threading tool scrapes the major diameter of the thread, the thread should be the correct size. Make several passes at the same setting to permit the spring in the boring bar to make light scraping cuts.

Test the thread for size with the mating part or a plug gage. The plug gage can be used as a chucking piece to hold the work for any secondary operations on the lathe, drill or milling machine.

If the short length of the major diameter is not required, this should be faced off before the work is removed from the lathe chuck.

9 ***Locate the threading tool at the center height using a surface gage.***

8 ***Measure the center height from a surface plate resting on parallels, on the flat ways of the lathe bed.***

10 ***A micrometer carriage stop is used to stop the carriage at the correct depth in the bore.***

External Threads

Cutting threads on a lathe with a single point threading tool seems to be an operation that most novice machinists try to avoid. There is no need to face a threading operation with dread and trepidation, for the cutting of a thread is a simple exercise when the operation is fully understood. You must be able to start a threading operation with confidence, for thread-cutting is frequently the last operation on a part, and you will have to perform the operation without making a mistake.

Turn the major (outside) diameter of the thread .002" to .005" undersize so that any cold flow or burr from the threading tool will not make the major diameter oversize and then give a false reading when testing the thread with a nut (.002" for machine screws and .005" for bolt sizes). One caveat here: Be certain that the lathe is turning parallel; you don't want a tapered thread (Figure 1).

Turn a short length, say 1/4", to the minor diameter of the thread. This can be removed by facing after the thread is finished by holding the thread in a collet or in the three-jaw chuck with a piece of cardboard wrapped around the thread (Figure 4). If possible, make a recess (undercut) at the end of the thread for the threading tool to turn into before withdrawing the tool bit for the next cut. The short portion at the end of the work is turned to the minor diameter first. Bring the end of a parting tool to touch the minor diameter, and then set the cross slide dial to zero. Withdraw the parting tool and make the undercut with the parting tool. Advance the parting tool until the cross slide dial reads zero. The recess should be the minor diameter of the thread (Figure 10).

The threading tool must be ground to the correct angle (60°). If the tool bit is to be held in a toolholder with 15°-20° back rake, the top of the tool bit must be ground level (horizontal) before grinding the 60° angle, or that angle will be incorrect when the tool bit is inclined in the toolholder. Use a center or thread gage to check the angle of the tool bit by holding the gage and tool bit up to a window or a light. You should not be able to see any light on the flanks of the threading tool. If there is any error, make sure the angle is less than 60° and not more. Although this is not shown in most books, a

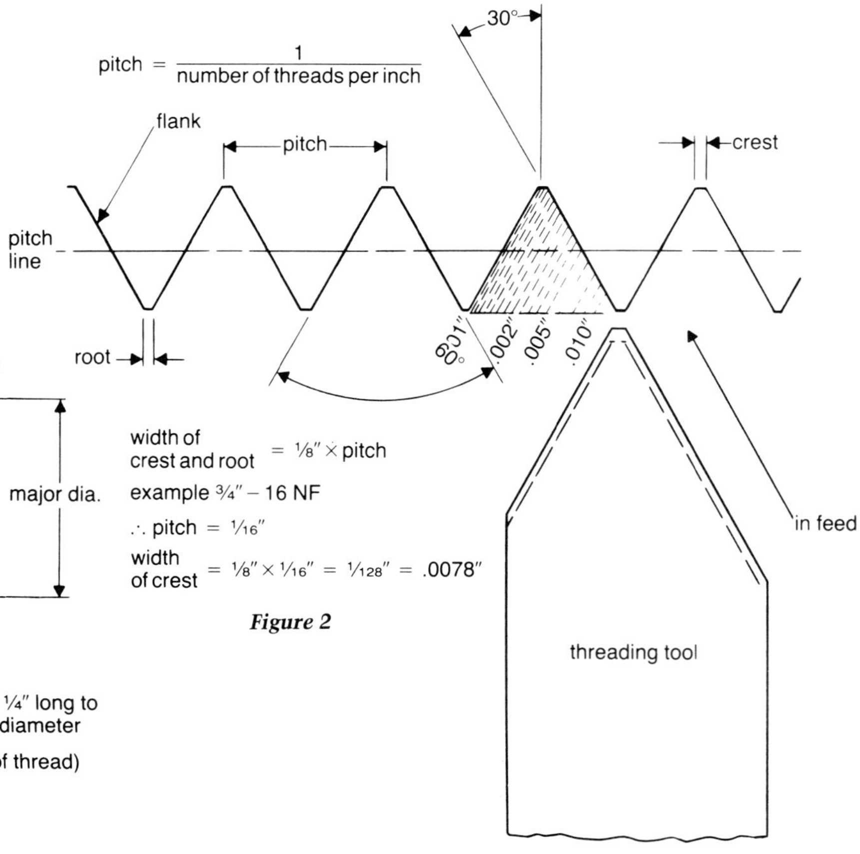

Figure 2

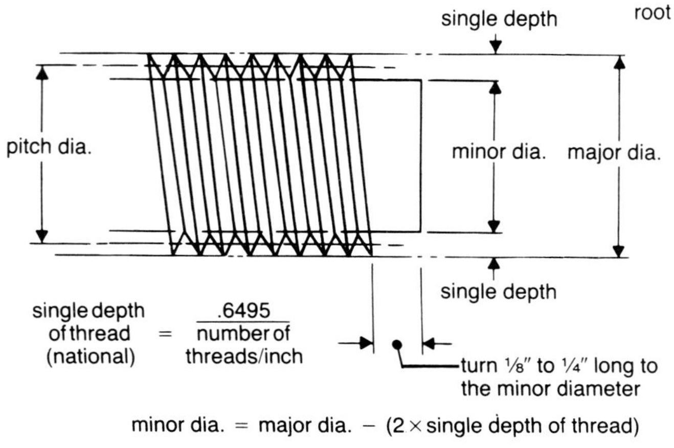

Figure 1

side rake from the leading flank may be ground for easier threading (Figures 4, 5, and 6).

The size of the flat (width of the root) can be calculated for large threads. This flat is made with an oilstone and measured with a micrometer or vernier caliper and a magnifying glass to 10 to 15 power. For fine threads, one or two strokes of a fine oilstone on the end of the tool bit will suffice.

Try to turn the major diameter of the thread, and then thread the work without removing it from the chuck. It will be very difficult to place the work back in the average universal chuck and have it run true. If the thread is very short, the work will not have to be supported with the tailstock center. All longer threads must be supported on the tailstock center. Failure to do this will permit the work to spring away from the threading tool. This means the thread will be tapered with the larger diameter at the tailstock end of the work.

The lathes with a quick-change gearbox can be easily set to the required number of threads per inch for all the common threads.

Lathes that use change gears will have to have them set up according to the chart supplied with the lathe. If the chart is not available, or some of the gears are missing, or a special

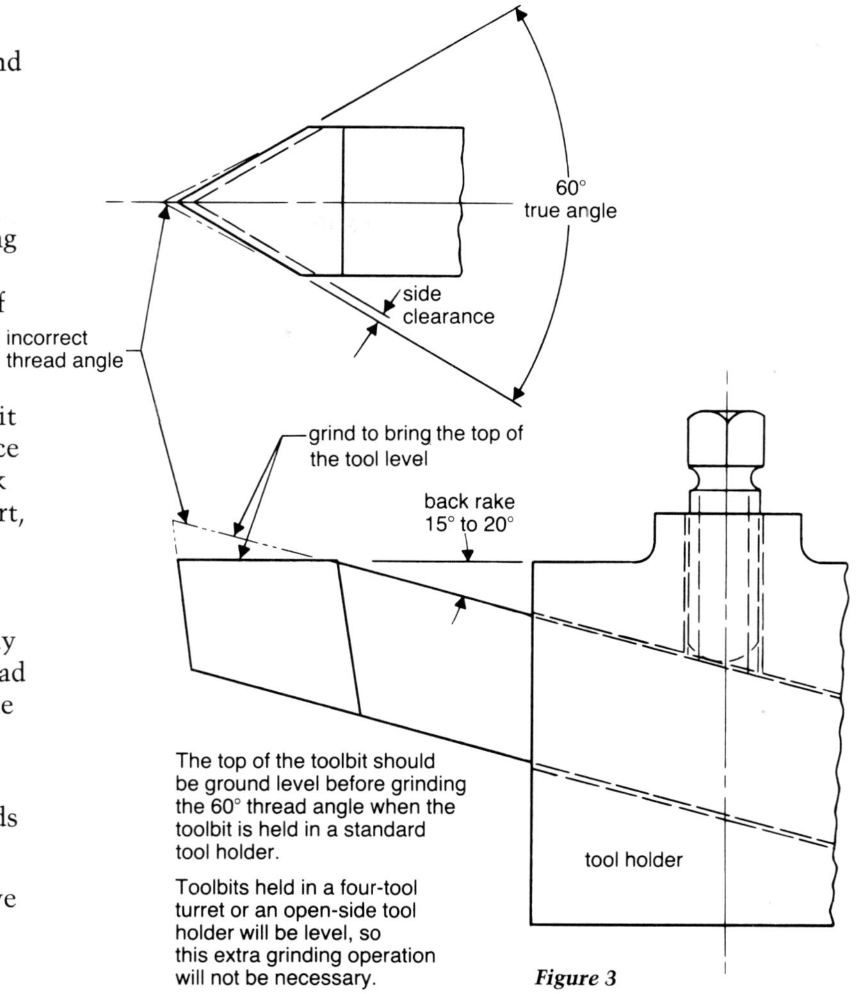

Figure 3

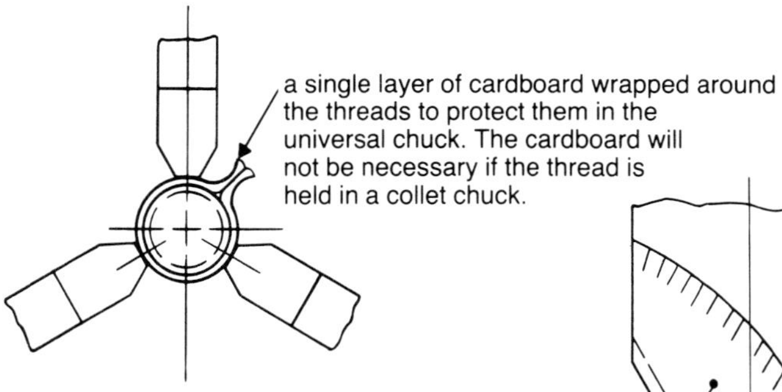

Figure 4

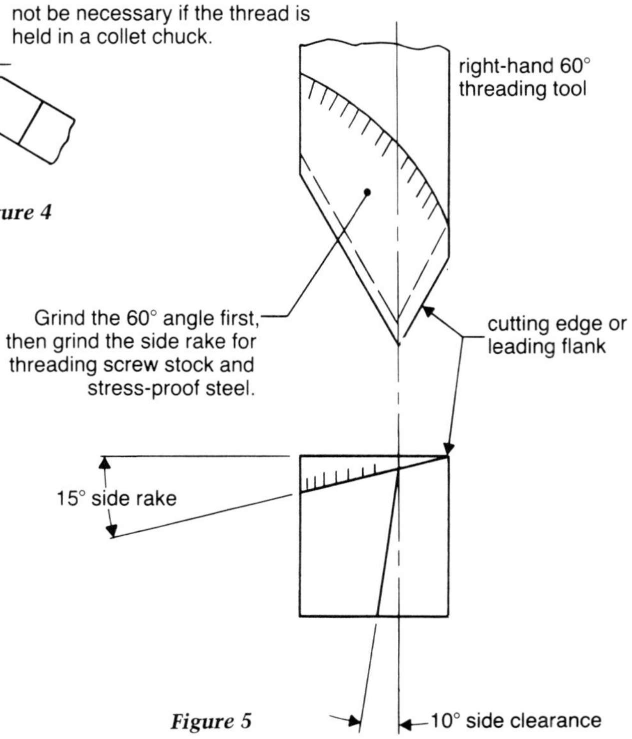

Figure 5

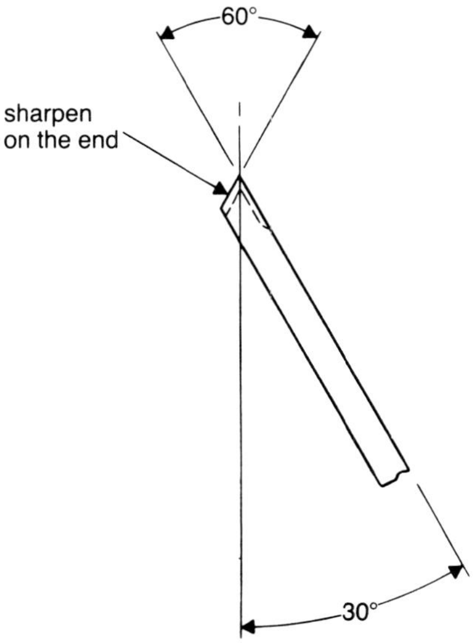

Another method of grinding and setting a threading tool for maximum tool life. This works best with a large tool held directly in an American toolpost or with an open-side tool holder.

Figure 6

thread is to be cut, then you will have to make some simple calculations. Some of the older lathes had a leadscrew with six or some other odd number of threads per inch. Most small lathes now have leadscrews with eight threads per inch, and larger lathes have four threads per inch. Most sets of change gears have an increment of 4 and a pair of identical gears.

A typical set of change gears will consist of the following gears: 20, 24, 28, 32, 36, 39 (for 1/2-13 NC), 40, 44, 48, 52, 56, 60, 64, 68, 72, 76, 80, etc. If some of the gears are missing, or some special gears are required, they are available from the Boston Gear Division (North American Rockwell), 14 Hayward Street, Quincy, MA 02171, or from one of their many distributors.

These change gears are a uniform thickness and have standard size bores and keyways. There is a gear made on the end of the spindle. These used to be separate, but are now part of

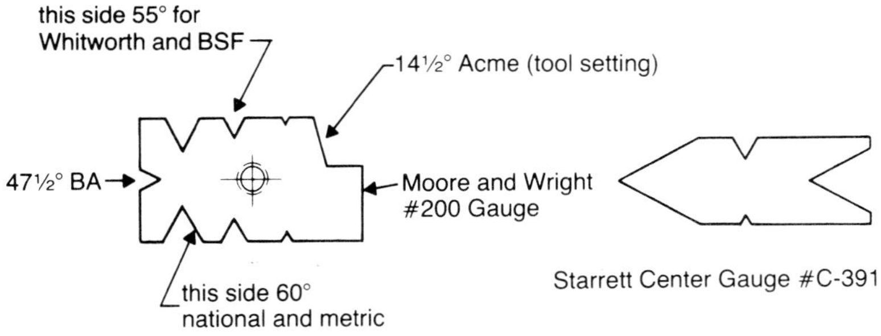

Two types of gauges used to check the shape of the threading tool during the grinding operation and to align the tool correctly during the set up.

Figure 7

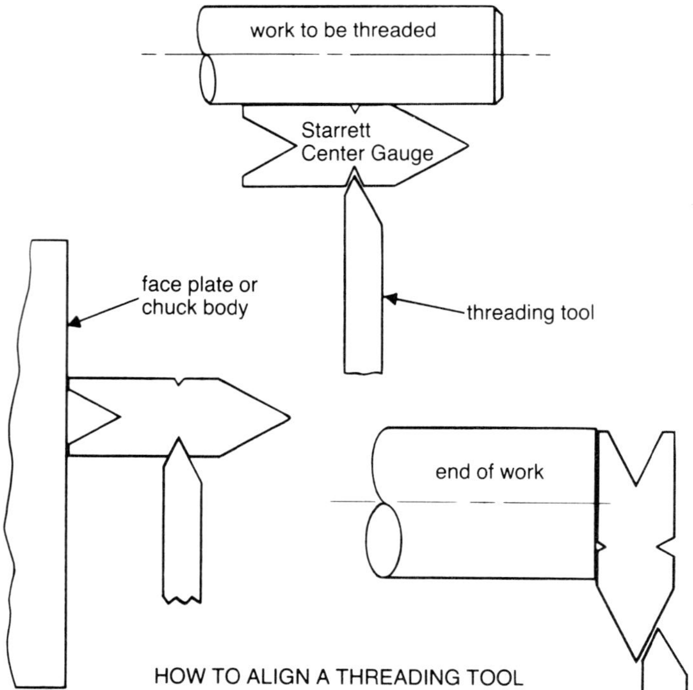

HOW TO ALIGN A THREADING TOOL

Figure 8

the spindle so a larger bore can be made in the headstock spindle.

This gear drives the tumbler reverse gears. These change the direction of rotation but not the gear ratio. The tumbler gears drive a gear mounted on a stud. This gear usually has the same number of teeth as the gear on the headstock spindle. The outer end of the stud has a place for a change gear.

This gear placed on the outer position of the stud is the main driver gear and will drive the remainder of the gear train.

If your lathe is an old model, count the number of teeth on the stud gear and the number of teeth on the headstock spindle gear. If the number of teeth on each is not the same, you will have to determine the "lathe constant" for the change gear calculations.

To find the lathe constant, place the same size gears on the stud and leadscrew, and connect them with any single gear as an idler. Take a light trial cut on a piece of scrap bar with the leadscrew nuts engaged, and measure the number of threads per inch the lathe has cut. This number of threads will be the lathe constant and will be used in place of the number of threads on the leadscrew for change gear calculations.

$$\frac{\text{Driver Gear (on stud)}}{\text{Driven Gear (on leadscrew)}} = \frac{\text{TPI on the leadscrew}}{\text{TPI to be cut}}$$

Example: ½-13 NC Choose a multiplier to match the change gears available by trial and error.

$$\frac{\text{Driver}}{\text{Driven}} = \frac{8}{13} = \frac{8 \times 3}{13 \times 3} = \frac{24}{39}$$

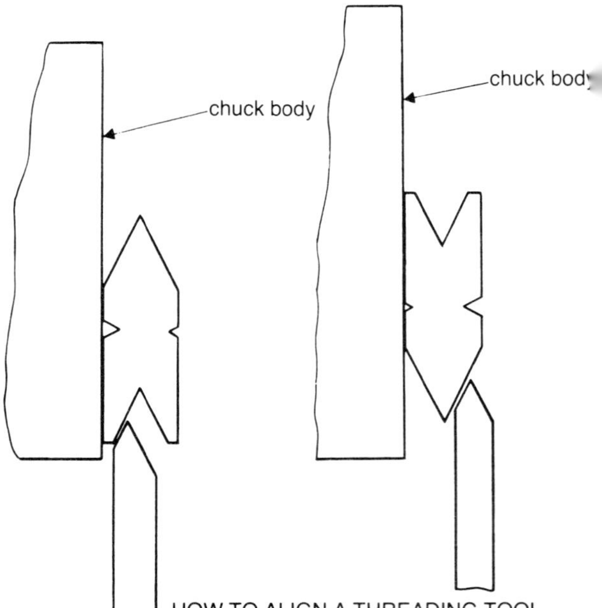

HOW TO ALIGN A THREADING TOOL

Figure 9

Use any single suitable gear as an idler to connect the driver and driven gears. (Refer to Figure 13.) These calculations are only simple fractions.

To set up the change gears, set the 24-tooth gear on the inside position of the stud and the 39-tooth gear on the inside position of the leadscrew. Add the spacer washer and nuts or lock the washer with setscrews according to the system provided. Now place a suitable idler gear on the "banjo" to mesh with the gear on the leadscrew. Place a thin piece of paper between the teeth of these two gears to have a slight clearance between the gears. Lock the idler gear in position on the banjo with the nut at the back of the spindle (or stud). Now rotate the banjo until the idler gear meshes correctly with the 24-tooth gear (use the paper again). Try turning the lathe spindle by pulling the belt by hand to be sure the gear train is turning freely.

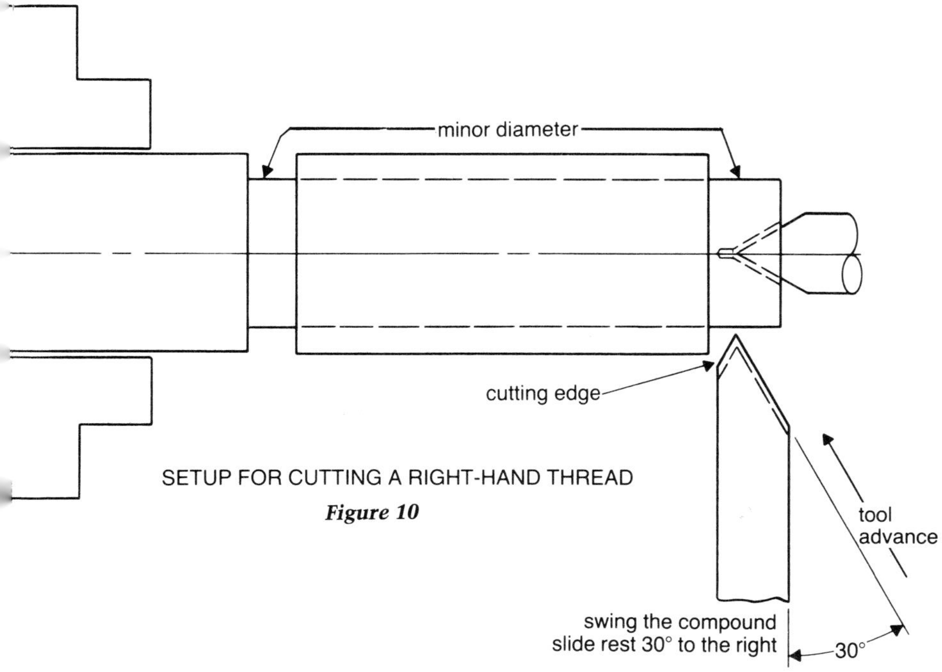

SETUP FOR CUTTING A RIGHT-HAND THREAD

Figure 10

TOOL LAYOUT FOR A FOUR-TOOL TURRET

1. Right-hand tool for turning and facing.
2. Round nose tool for forming a fillet at a shoulder or use a forming tool for a particular operation.
3. Threading tool. Set the toolbit in the turret so the toolbit is aligned with the gauge when the side of the turret is parallel with the lathe axis.
4. Parting tool.

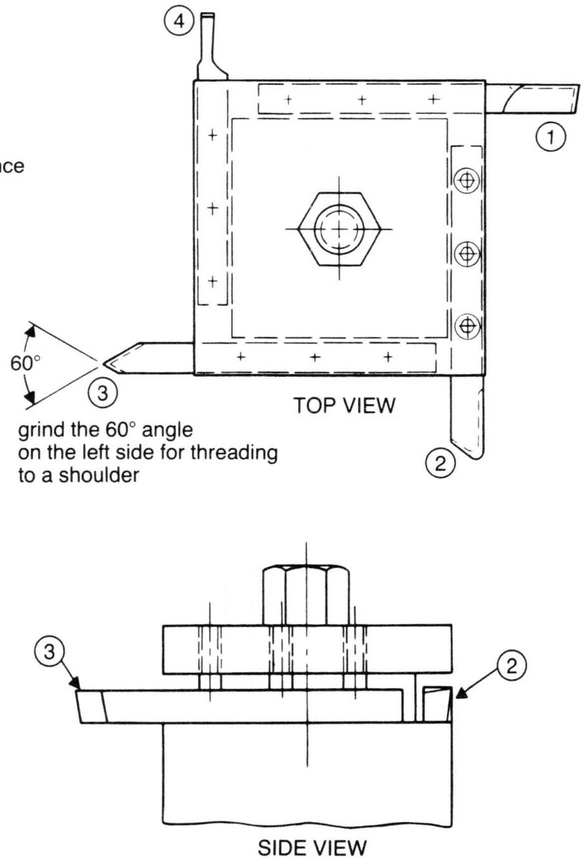

Figure 12

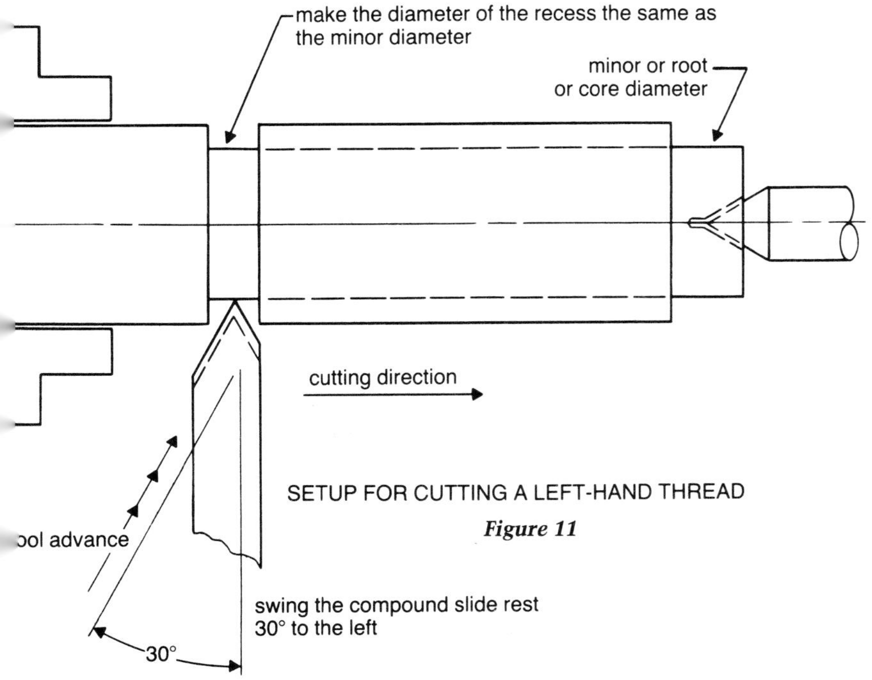

SETUP FOR CUTTING A LEFT-HAND THREAD

Figure 11

Other examples of simple gearing are:

(a) 16 TPI

$$\frac{\text{Drivers}}{\text{Driven}} = \frac{8}{16} = \frac{1}{2}$$ choose any suitable gears for a ½ ratio

$$\frac{1 \times 10}{2 \times 10} = \frac{20}{40} = \frac{8}{16} = \frac{20}{40} \text{ or } \frac{32}{64} \text{ or } \frac{36}{72} \text{ etc.}$$

(b) 12 threads per inch

$$\frac{\text{Driver}}{\text{Driven}} = \frac{8 \times 3}{12 \times 3} = \frac{24}{36} \text{ or } \frac{8 \times 6}{12 \times 6} = \frac{48}{72}$$

(c) 11 threads per inch

$$\frac{\text{Driver}}{\text{Driven}} = \frac{8}{11}$$ choose 4 as a multiplier so that the denominator will match the 44 gear

$$\frac{\text{Driver}}{\text{Driven}} = \frac{8 \times 4}{11 \times 4} = \frac{32}{44}$$

When the gear ratio becomes larger than 1/4 (or 1:4), then compound gearing will have to be used.

Example: 40 threads per inch

$$\frac{\text{Drivers}}{\text{Driven}} = \frac{8}{40} \text{ factor these numbers to } \frac{4 \times 2}{8 \times 5}$$

$$= \frac{4 \times 10}{8 \times 10} \times \frac{2 \times 12}{5 \times 12} = \frac{40 \times 24}{80 \times 60}$$

The numbers 40 and 24 are driving gears and are interchangeable in the stud or idler positions. The numbers 80 and 60 are driven gears and are interchangeable in the leadscrew and idler positions (see Figure 14). You may have to repeat these calculations several times to find the right multipliers to match the change gears.

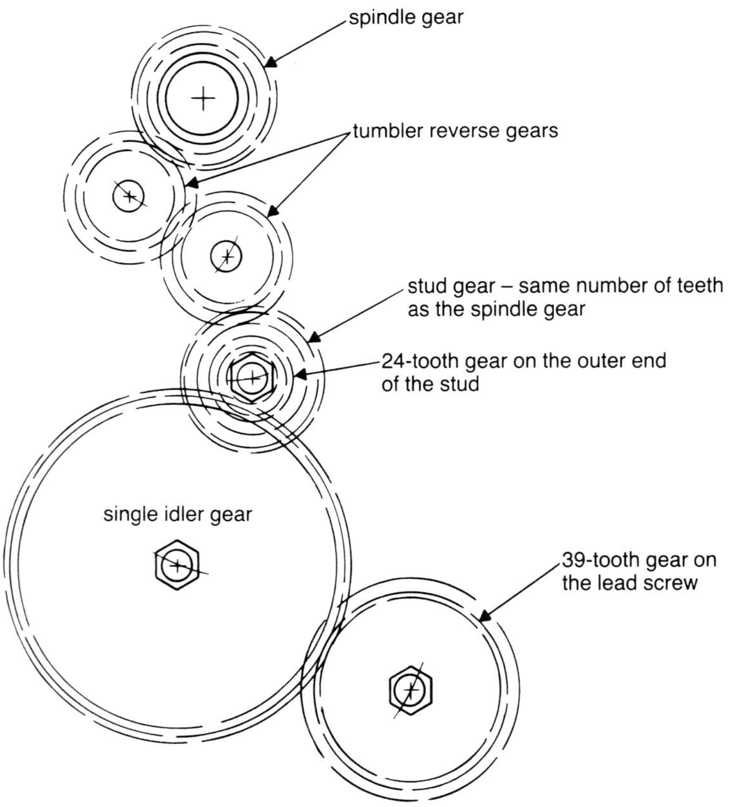

SIMPLE GEAR COMBINATION TO CUT A ½″-13 NC thread

Figure 13

This exercise will take you back to the multiplication tables you memorized in primary school. If you "jog" your memory a bit, they will come back to you. Remember, the old-time machinists usually had only a primary school education; yet when you look back upon what they accomplished, this lack of higher education never hindered them.

Here are some more examples of the calculations for compound gearing:

(a) 64 threads per inch

$$\frac{\text{Driver}}{\text{Driven}} = \frac{8}{64} = \frac{2 \times 4}{8 \times 8} = \frac{2 \times 10}{8 \times 10} \times \frac{4 \times 6}{8 \times 6}$$

$$= \frac{20 \times 24}{80 \times 48}$$

(b) 56 threads per inch

$$\frac{\text{Drivers}}{\text{Driven}} = \frac{8}{56} = \frac{2 \times 4}{8 \times 7} = \frac{2 \times 10}{8 \times 10} \times \frac{4 \times 8}{7 \times 8}$$

$$= \frac{20 \times 32}{80 \times 56}$$

Now we shall set up the lathe for threading. The major diameter is slightly undersize, and a short portion at the tailstock end is turned to the minor diameter. When the tip of the threading tool touches the minor diameter, the thread should be the correct size or very close, depending upon the size of the flat on the end of the threading tool.

Set the compound slide rest to 30° to the right. Some books recommend setting the compound to 29°. This means the back side of the threading tool will take a slight scraping cut if the tool angle is exactly 60°.

I prefer 30°; then if the angle of the threading tool is slightly less than 60°, it won't matter as long as the leading edge of the tool bit is aligned correctly with the threading gage.

Locate the tool post on the left-hand side of the compound slide rest. This will prevent "chewing up" the side of the compound slide rest with the chuck jaws when threading close to the chuck. Set the threading tool to the lathe center height and align the leading edge of the tool bit with a thread or center gage (see Figures 7, 8 and 9). A piece of white paper placed on the cross slide under the tool will help you to see if the tool is aligned correctly.

Run the lathe carriage up and down the bed

by hand to make sure the tool holder does not touch the lathe dog or the chuck jaws and that the tailstock spindle is extended enough for the tool bit to clear the end of the work. When using a four-tool turret, be sure to check to see that the other tools in the turret do not touch the tailstock spindle or the tailstock casting.

For lathes with a power cross feed, now is the time to set the apron controls for threading.

The threading dial is no mystery. Each numbered line corresponds to one inch of the leadscrew, so there are eight teeth to match the eight threads on the leadscrew. Most dials are numbered 1 to 4; there are 32 teeth on the gear. For even-numbered threads the half nuts can be engaged at any numbered line or half-line. For odd number threads such as 13 and 11, the numbered lines must be used. Threads such as a 4-3/4 will use only one line, usually number 1. Alternately, the half nuts may be kept engaged and the motor reversed at the end of each cut (after the toolbit has been withdrawn from the cut). This is the usual method when cutting metric threads.

Now for the actual threading operation. Bring the tip of the threading tool to just touch the work. The compound slide rest may have to be adjusted so that the knob on the cross slide handle is at the ten o'clock position and the tool bit may be quickly withdrawn at the end of each cut. Set the cross slide and the compound slide dials to zero. Set the spindle speed for about 100 rpm. This speed will be suitable until you have more experience. Start the motor, engage the half nuts at a numbered line and take a light scraping cut along the work. Withdraw the tool bit at the end of the cut and disengage the half nuts, in that order. Return the carriage to the starting position and reset the cross slide to zero. Now check the number of threads per inch on the work with a thread pitch gage.

Now to start the actual threading. Advance the compound slide rest .010". Engage the half nuts and take a cut. Withdraw the cross slide and disengage the half nuts when the tool bit reaches the undercut. Return the carriage to the starting position and take a cut of .005". Take the next cuts of .002", and then .001" when the chip becomes large or discolored. When the tip of the tool bit is near the minor diameter at the end of the work, take several passes at the same setting to bring the thread to size with a good finish. Use a sulfurized or clear cutting oil for a good finish and a longer tool life. To test the size of the thread with a nut or the mating part, *do not remove the bar from the chuck,* but back the tailstock from the work for enough room to try the nut on the thread. When the thread is the correct size, chamfer the end of the work with a tool bit set to an angle about 45° to the lathe axis*. This chamfer will prevent the end of the thread from being damaged. If a recess is not made, the tool bit will have to be withdrawn quickly at the end of each cut. This is easy to do with a little practice. Withdraw the tool just as the end of the thread comes into view at the tip of the work. The old machinists were experts at this. The thread always came up to the major diameter in a nice, even slope. Use a slower spindle speed (say 50 rpm) until you become used to this method.

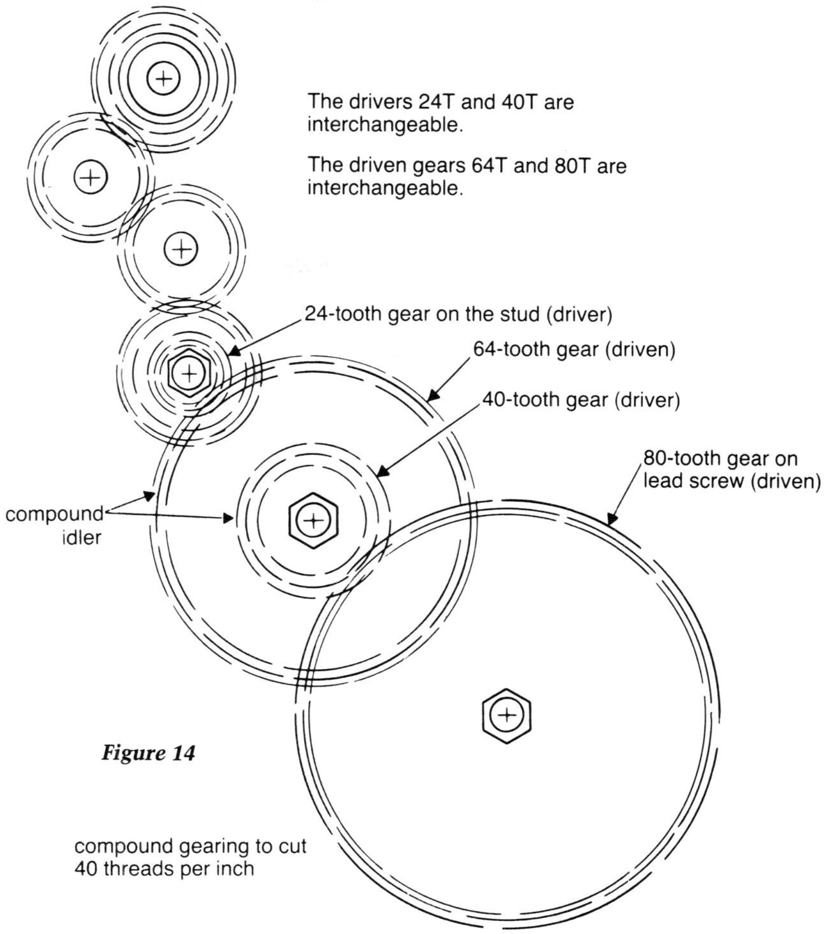

Figure 14

To cut a left-hand thread, swing the compound slide rest 30° to the left. The threading tool will be cutting on the right-hand flank now, so be sure to grind enough side clearance. Make a recess or undercut in the work near the chuck. The bottom of the recess should be the minor diameter of the thread, as for a right-hand thread. Place the threading tool in the recess to start each cut. A left-hand thread is easier to cut than a right thread, because the tool bit runs out at the end of the work (tailstock end) rather than up to a shoulder.

*(*The editor's preference is to chamfer before threading to minimize burrs and the necessity of retouching thread after chamfering. Ed.)*

Simple Pulley Calculations

Have you ever wondered what the speeds (RPM) are for the various steps on the pulleys of your drill press?

The circumference of a pulley is directly proportional to its diameter. For ease of calculation, we will use the diameters of the pulleys instead of the circumferences.

The revolutions per minute (RPM) of one pulley multiplied by its diameter must be equal to the revolutions per minute of the matching pulley multiplied by its diameter.

$$\frac{\text{RPM} \times \text{Dia.}}{\text{(of one pulley)}} = \frac{\text{RPM} \times \text{Dia.}}{\text{(of the other pulley)}}$$

EXAMPLE: We have an 1800 RPM motor with a 2" diameter motor pulley driving a 6" diameter pulley on the spindle. We want to know the spindle speed (RPM). (For easier calculation, place the unknown on the left-hand side.)

$$\text{RPM} \times \text{Dia.} = \text{RPM} \times \text{Dia.}$$

$$\text{Spindle RPM} \times 6\text{" Pulley} = 1800 \text{ RPM} \times 2\text{" Pulley}$$

$$\text{Spindle RPM} = \frac{1800 \times 2}{6} = 600 \text{ RPM}$$

For fraction-size pulleys, change the dimensions to decimals to simplify calculation.

EXAMPLE: A motor of 1800 RPM with a 2-1/4" diameter pulley is driving a 4-1/2" diameter pulley on the spindle and we want to know the spindle speed (RPM).

$$\text{RPM} \times \text{Dia.} = \text{RPM} \times \text{Dia.}$$

$$\text{Spindle RPM} \times 4.5\text{"} = 1800 \text{ RPM} \times 2.25\text{"}$$

$$\text{Spindle RPM} = \frac{1800 \times 2.25}{4.5} = 900 \text{ RPM}$$

These pieces of information, along with the choice of correct speed for various materials covered in the previous *Beginner's Column*, will let you proceed "like a pro" with your next drilling operation!!

Pulley Ratio and Cutting Speed Problems

Suppose you are at the store, about to purchase a bench grinder with two 6" diameter grinding wheels and a 2" diameter spindle pulley mounted between them.

Look at the cardboard washer on one of the grinding wheels and see if a speed is recommended. Sometimes the washer has the maximum safe speed (RPM) or a maximum safe cutting speed in feet per minute.

Let us assume a cutting speed of 5400 rpm is recommended. Use the cutting speed formula to find the spindle speed:

$$\text{RPM} = \frac{\text{CS} \times 4}{\text{Dia.}} = \frac{5400 \times 4}{6} = \frac{21600}{6} = 3600$$

You have a suitable motor (1/3 or 1/2 HP) at home but you must buy a suitable pulley to give the correct spindle speed:

$$\text{Motor RPM} \times \text{Pulley Dia.} = \text{Spindle RPM} \times \text{Pulley Dia.}$$

$$1800 \times ? = 3600 \times 2\text{"}$$

$$\text{Motor Pulley Dia.} = \frac{3600 \times 2}{1800} = 4\text{" Dia.}$$

Once you have memorized these two simple formulae, it will be easy to do these calculations at the store on a pocket note book or, more usually, on the back of an envelope.

Compound Pulley Ratios

The formulas used for simple pulley ratio problems can be used for compound ratio problems by considering one ratio at a time.

The compound reduction occurs when we use a countershaft between the motor and the machine being driven, as in Figure 1.

Remember that a countershaft has both pulleys fixed to the same shaft and, therefore, they turn at the same speed.

Solving for no. 1, we find:

$$1800 \times 2'' = 8'' \times \text{RPM}$$

$$\text{RPM} = \frac{1800 \times 2}{8} = 450$$

Taking this a step further, we can quite easily move on to countershaft no. 2:

$$450 \text{ RPM} \times 2'' = 12'' \times \text{RPM}$$

$$\text{RPM} = \frac{450 \times 2}{12} = 75$$

And the final step brings us to no. 3:

$$75 \text{ RPM} \times 3'' = 12'' \times \text{RPM}$$

$$\text{RPM} = \frac{75 \times 3}{12} = 18.75$$

We can also use the same formula to find the diameter of a pulley, as shown in Figure 2.

To figure the diameter of pulley **B**, the formula will be:

$$\text{RPM} \times \text{dia.} = \text{RPM} \times \text{dia.}$$
$$1800 \times 2'' = 360 \times \mathbf{B}$$

$$\mathbf{B} = \frac{1800 \times 2}{360} = 10'' \text{ dia.}$$

We then can figure the diameter required for pulley **C**, in order to arrive at our final speed of 90 RPM on pulley **D**.

$$\mathbf{C}\ 360 \text{ RPM} = 12'' \text{ dia.} \times 90 \text{ RPM}$$

$$\mathbf{C} = \frac{12 \times 90}{360} = 3'' \text{ dia.}$$

By using this formula with common sense, the pulley diameter or pulley speed can be found for any drive system. When using V-belts, always measure the diameter of the pulleys at the top of the V-belt or the maximum diameter of the pulley.

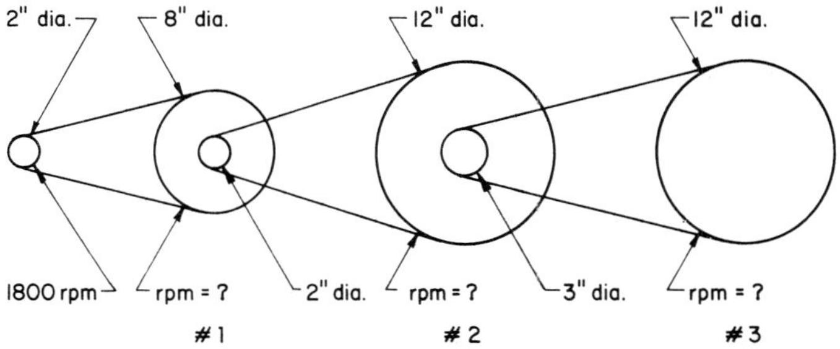

Figure 1

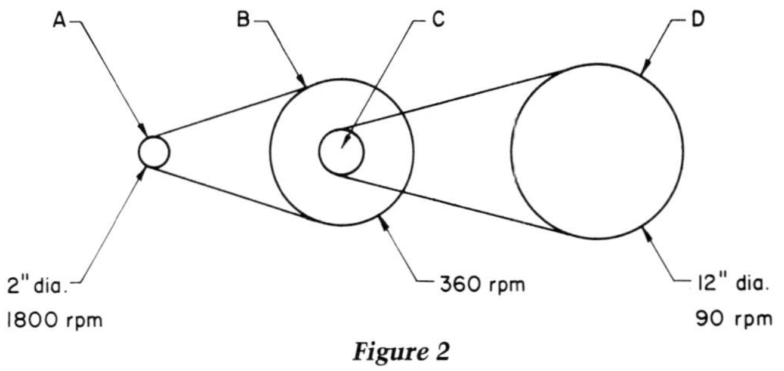

Figure 2

Choosing the Correct Speed

Do you have difficulty choosing the correct speed for setting your lathe or drill press to machine mild steel, tool steel, brass, bronze or cast iron? Here is a simple method of calculating the correct speed to use.

CUTTING SPEED

An automobile traveling 60 miles per hour covers a mile (5,280 feet) each minute. The circumference of the wheels travel 5,280 feet in one minute. If a trailer is being towed, the trailer wheels must also travel 5,280 feet, but will turn more revolutions per minute if they are smaller than the automobile wheels.

The cutting speed is the distance in feet that the circumference of the work or the cutting tool travels in one minute.

COMMON CUTTING SPEEDS

Mild Steel	80 feet per minute
Tool Steel	40 feet per minute
Cast Iron	60 feet per minute
Bronze	100 feet per minute
Brass	300 feet per minute
Aluminum	200 feet per minute

Note: These cutting speeds are less than those used in industry where the machinery is much more massive than the machinery in home work shops. Use this chart as a guide. The values given are a starting point. Choose a cutting speed 1/3 less for milling at first until you have determined if higher speeds can be used.

CALCULATION

Suppose you want to set the lathe to turn a cast iron driving wheel 6" in diameter. Use the following formula:

$$\text{RPM} = \frac{\text{CS} \times 4}{\text{Dia.}} = \frac{60 \times 4}{6} = 40 \text{ RPM}$$

RPM = revolutions per minute to set the machine.

CS = cutting speed in feet per minute for the material being machined.

Dia. = diameter in inches of the material being turned or the diameter of a drill or a milling cutter. Choose the nearest speed available on the machine. Now let's compare the speed required to machine the same diameter in aluminum:

$$\text{RPM} = \frac{\text{CS} \times 4}{\text{Dia.}} = \frac{200 \times 4}{6} = 133 \text{ RPM}$$

You can use the same formula for the drill press or lathe for drilling a 1/2" hole in mild steel.

$$\text{RPM} = \frac{\text{CS} \times 4}{\text{Dia.}} = \frac{80 \times 4}{1/2} = \frac{80 \times 4 \times 2}{1} = 640 \text{ RPM}$$

When fraction size drills are used, they must turn faster (remember the trailer wheels) so the fraction must be inverted.

Use the same procedure to calculate the speed to set the lathe for end milling with the work set on the milling attachment or on a milling machine.

Let's try another example:

You are to mill a slot with 3/8" diameter end mill in bronze. (Subtract 1/3 of the cutting speed for milling.)

$$\text{RPM} = \frac{\text{CS} \times 4}{\text{Dia.}} = \frac{60 \times 4}{3/8} = \frac{60 \times 4 \times 8}{3} = 640 \text{ RPM}$$

These simple calculations will provide a safe and efficient speed for operating your shop equipment.

Simple Gear Ratios

When gear ratios and speeds are to be calculated, we use the number of teeth on the gears instead of the diameters, as was used for the pulley calculations.

Let's assume we have a small gear with 20 teeth meshed with a larger gear with 60 teeth, as shown in Figure 1. The small gear is 1/3 the size of the large gear and must then turn 3 times as fast. If the small gear is mounted on an 1800 RPM motor, how fast will the 60-tooth gear turn?

$$\text{Motor RPM} \times \text{No. of Teeth} = \text{RPM} \times \text{No. of Teeth}$$
$$1800 \text{ RPM} \times 20 \text{ Teeth} = \text{RPM} \times 60 \text{ Teeth}$$
$$\text{RPM} = \frac{1800 \times 2}{60} = 600$$

If we add an intermediate gear between the two gears, as in Figure 2, then we will change the direction of rotation of the large (driven) gear. A simple way to determine the direction of rotation of a train of gears is to follow the direction changes with a pencil. See Figure 3.

When an intermediate, or idler, gear is added to a train of gears, the direction of rotation is changed, but *not the ratio* between the driving (first) and driven (last) gear.

When the first gear makes one complete turn, as in Figure 4, the intermediate gear is turned the same number of teeth (20 in this case) as the driving gear. The size of this idler gear thus has no influence on the simple ratio of driver to driven gear.

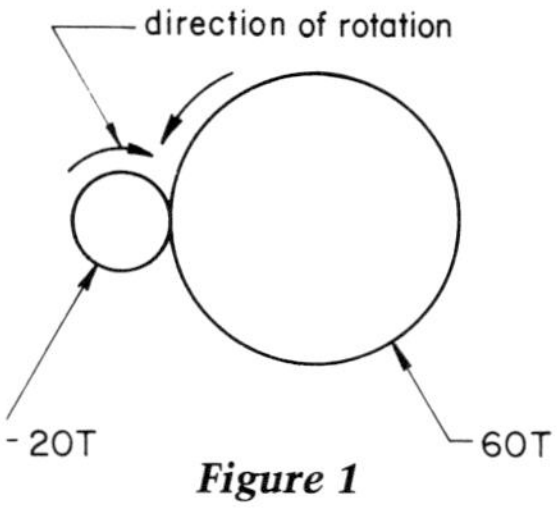

Figure 1

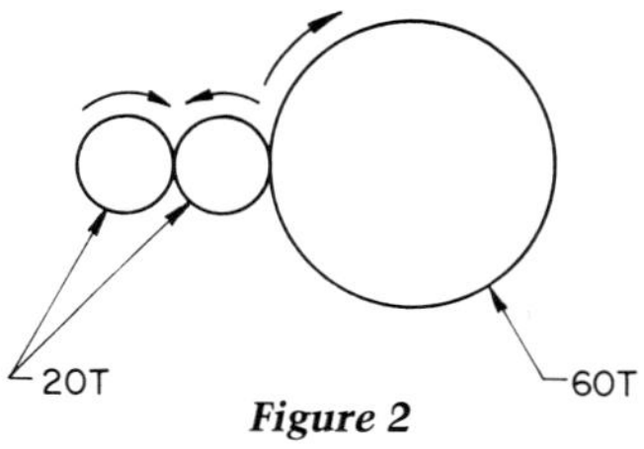

Figure 2

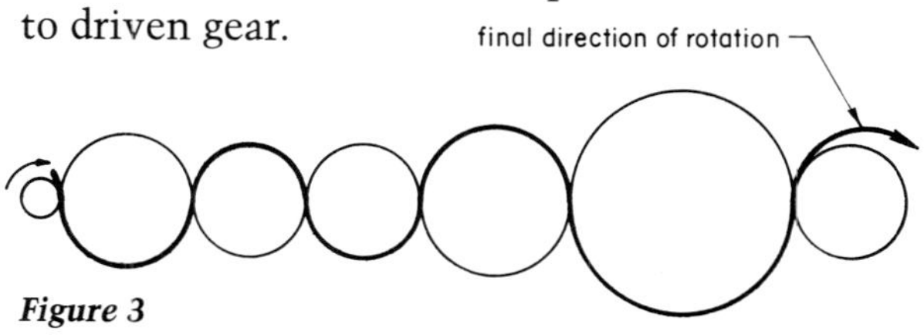

Figure 3

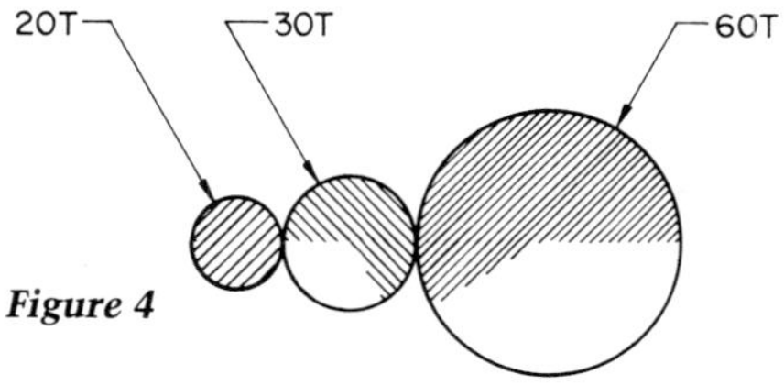

Figure 4

Compound Gear Ratios

Whenever a gear reduction greater than 1:4 is needed, compound gearing is used. The formula is the same as for a simple gear problem but, like the compound pulley problems, the formula is individually applied to each separate step in the reduction.

Remember that a simple idler will not change the ratio of a gear train, regardless of size. A compound idler (two gears of different sizes fixed on the same shaft and, of course, turning at the same speed) *will* change the gear ratio.

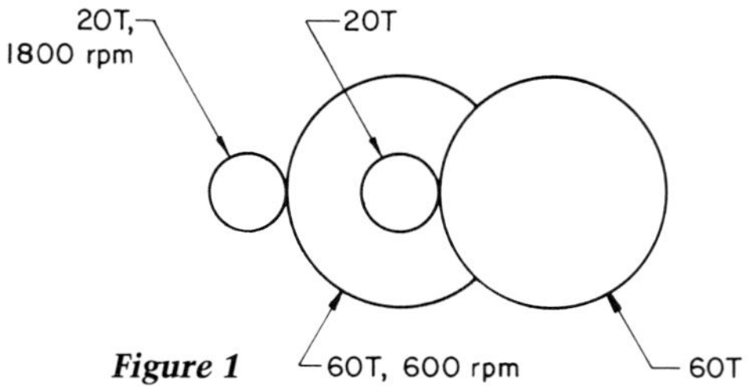

Figure 1

Let's figure the speed of the driven gear in Figure 1. Remember, the basic formula is:

$$\text{RPM} \times \text{No. of Teeth} = \text{RPM} \times \text{No. of Teeth}$$
$$1800 \text{ RPM} \times 20 \text{ Teeth} = \text{RPM} \times 60 \text{ Teeth}$$
$$\text{RPM} = \frac{1800 \times 2}{60} = 600$$
$$600 \text{ RPM} \times 20 \text{ Teeth} = \text{RPM} \times 60 \text{ Teeth}$$
$$\text{RPM} = \frac{20 \times 600}{60} = 200$$

The ratio between each set of gears is 20:60 or 1:3. Gear ratios are multiplied, not added, to check your calculations:

$$1/3 \times 1/3 = 1/9 \text{ or } \frac{200 \text{ RPM}}{1800 \text{ RPM}} = 1/9$$

This procedure can be used for each set of matching (meshing) gears. I once used this method to work out the speed and the power for the winch of a special tow truck with five gear reductions. The truck builder was an excellent mechanic, but short on schooling and wanted to tell his customer what the towing speeds and estimated power would be.

Tapping Large Pulleys

Tapping a setscrew hole in a large pulley for the first time will present a problem for the novice. The rim of the pulley will interfere with the drill chuck. This problem is easily overcome by using one of two methods.

The quickest method is to incline the pulley about 10° to 15° to clear the drill chuck. File a small flat on the hub so the drill will start on a surface perpendicular to the drill axis or use a two-lip end mill in the

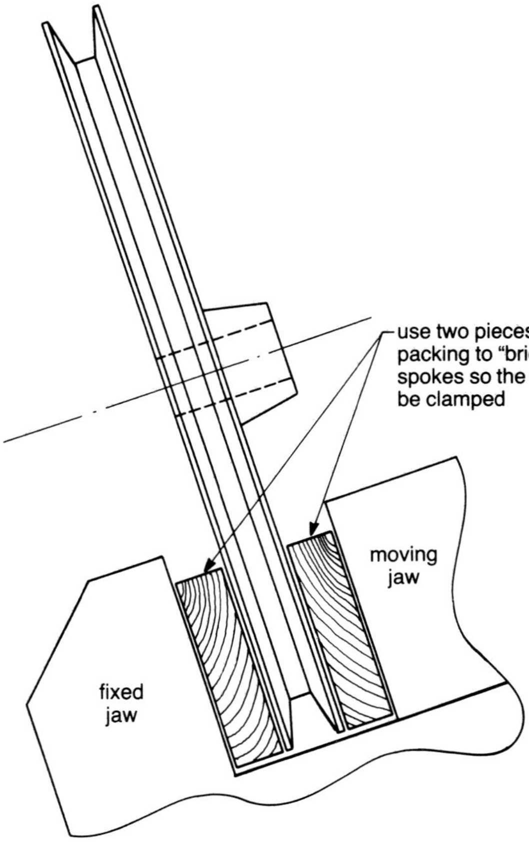

HOW TO HOLD A LARGE PULLEY IN A TILTING VISE

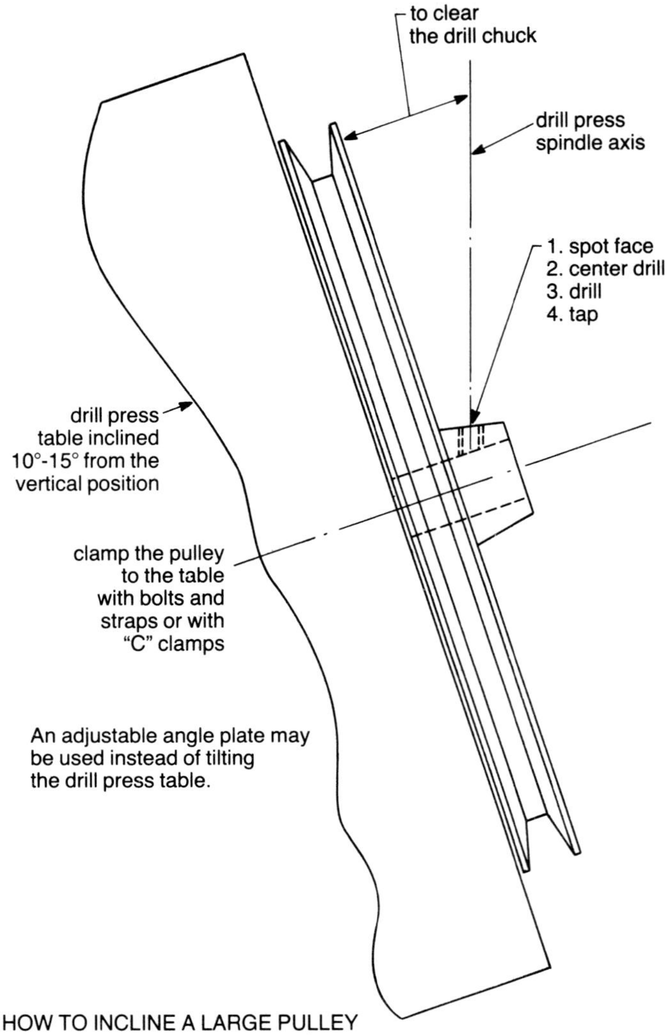

HOW TO INCLINE A LARGE PULLEY TO DRILL AND TAP A SET SCREW HOLE

drill chuck to spot face a flat surface. Drill and tap as usual. The tap wrench should be the tee-type with a sliding handle.

The other method is to use extra long drills and a pulley tap or one extension tap wrench.

Instead of buying these special tools, an ordinary drill can be used in a homemade extension drill holder and an ordinary tap wrench can be modified as shown on the drawings.

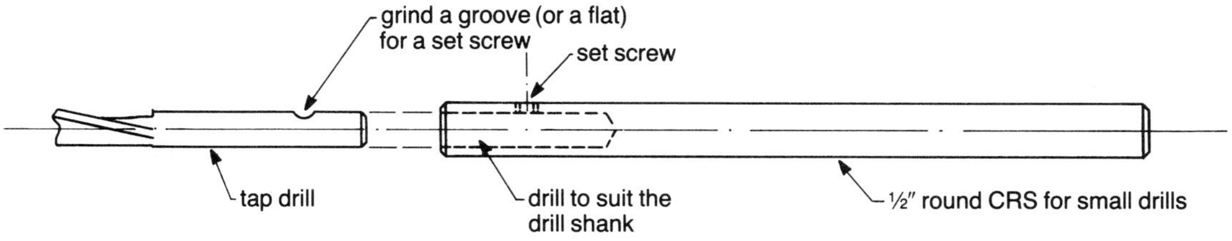

HOMEMADE EXTENSION DRILL

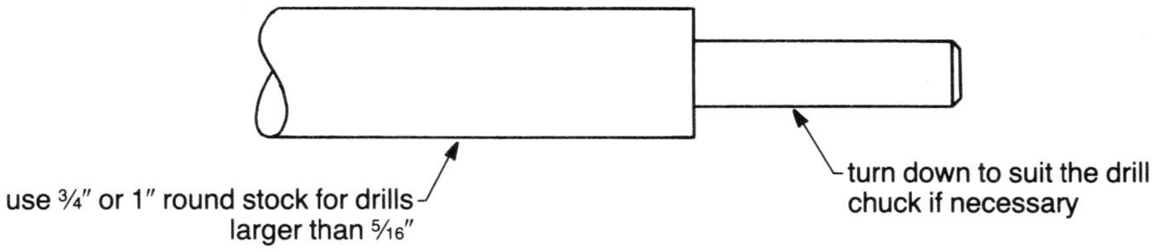

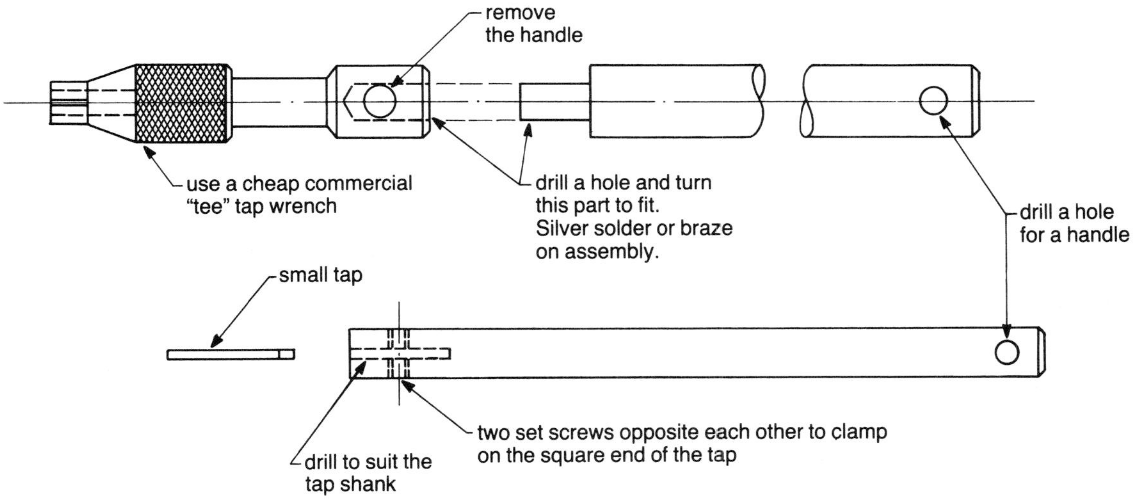

HOMEMADE EXTENSION TAP WRENCHES

Replacing Broken Gear Teeth

When a second-hand lathe is purchased, the proud owner often finds that the back gears or some of the change gears have some teeth missing. Spare parts may be unobtainable or too expensive, so the broken teeth will have to be replaced.

Prepare the gear by removing all the oil and grease with a solvent, such as *Varsol,* and a wire brush. File the areas where the teeth are broken to remove the rough surface and to expose clean metal. Now file some slots or drill shallow holes to anchor the new teeth to the gear. Gear teeth of No. 10 diametral pitch and larger may have some threaded rods screwed into the gear to provide extra support (see drawing).

The teeth will be replaced by bronze welding. Set the gear vertically between two fire bricks or in a vise. Preheat the gear all over and then concentrate on the area where the teeth are broken. This area will have to be a dull red color when you start to add the bronze welding rod. For small teeth, fill in the area between the teeth adjacent to the broken teeth. Large gear teeth can be built up to approximately the correct shape. For those people without an oxy-acetylene or propane gas outfit, the local garage should be able to do the welding successfully. Allow the gear to cool slowly by covering the gear with some insulating material such as granulated asbestos or other insulation.

Hold the gear on an arbor or in the chuck to turn off the excess bronze.

The new teeth may be cut with a proper gear cutter on a milling machine. However, most people will have to use other methods. I shall describe the method to use on the lathe or vertical milling machine. Planing on the lathe is the most common method used by most people with limited shop equipment and this is the only method to recut teeth in a cluster gear (several gears joined together). The new teeth can be easily filed to shape if a template is made to check the shape of the teeth (see drawing).

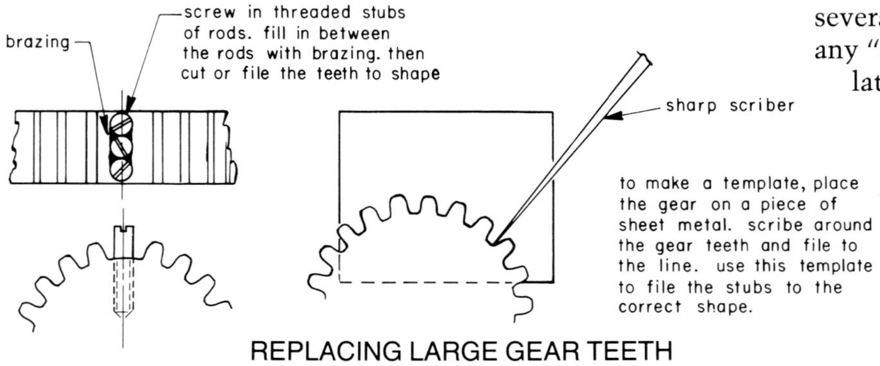

REPLACING LARGE GEAR TEETH

PLANING ON THE LATHE

Grind a tool bit using the corners of the grinding wheel until the shape fits exactly in one of the gear teeth. Don't forget to have about 5° to 10° side clearance. Final fitting may be done with small oil stones. Take care and this tool can be kept for future use. Place the tool bit in a toolholder. Set the tool bit at 90° to the lathe axis and exactly at the center height. Set the tool so that the cutting action will take place towards the lathe spindle.

Now place the gear in the lathe on an arbor mounted in the universal chuck or between centers. Locate the tool bit in a good tooth of the gear at the full depth. Reset the cross slide dial to zero. Now set up a piece of hacksaw blade at the rear of the gear to index the teeth. This hacksaw blade is fastened to a block which is clamped to the lathe bed. This will have to be made to fit each lathe. Wrap a cord around the chuck or faceplate and hang a heavy weight from the cord to keep the gear teeth on the hacksaw blade. Instead of using a weight, the back gears could be engaged and the spindle rotated by pulling on the belt to index the teeth (see drawing).

The teeth are cut by traversing the lathe carriage back and forth by the carriage handwheel. Take light cuts, .0005" to .001" each pass. When the full depth has been reached, make several passes at the same setting to allow for any "spring" in the arbor or looseness in the lathe. Actually, the setup takes far longer than cutting the teeth.

VERTICAL MILLING MACHINE

The fortunate owners of a vertical milling machine can grind up a tool bit as for planing in the lathe. Use a fly-cutter arbor. A special gear arbor will be required, but the same indexing (dividing) method as for the lathe can be used if a dividing head is not available (see drawing). Remove the burrs with a small file after the teeth are cut. Try the new teeth by rotating the gears together. Some corrective filing may be necessary.

The old-time millwrights made all the gears in the windmills and water mills. Some were made of cast iron with the teeth cast to shape from a wooden pattern. Most of the gears were wooden. The teeth were laid out, sawn out, and finished by chisel and files (wood rasps). Some of the large gears had wooden teeth with a dovetail base to fit into dovetail slots in the rim of the gear, much like the front sight of a rifle. The millwright would leave spare teeth with the miller so they could be easily replaced when one broke in service. The millwright would visit the mill once a year to take care of any maintenance work required and would make up any spare teeth needed. These wooden gears were lubricated with tallow.

In the old shop where I started, some broken teeth in cast iron gears were mended by drilling and tapping the gear and screwing in threaded rods. The rods (stubs) were then filed to shape. All this was done without removing the gear from the machine.

So don't be intimidated by a broken gear tooth. Go ahead and repair it. You will be successful, and will gain confidence in your ability to cope with anything that comes along in your shop.

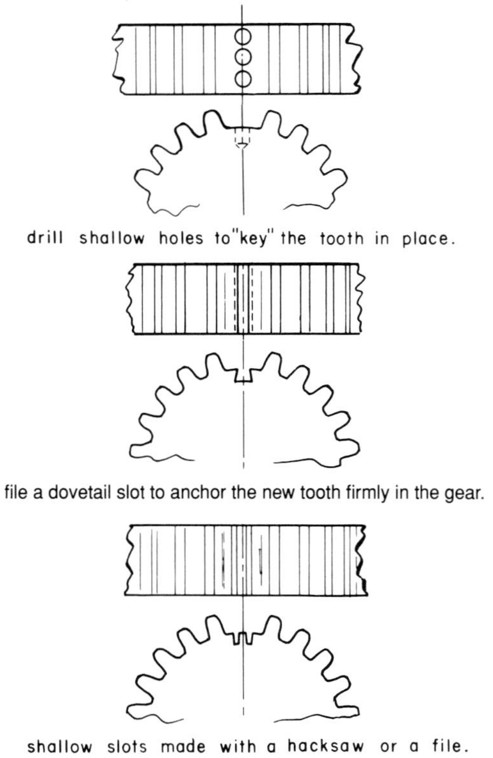

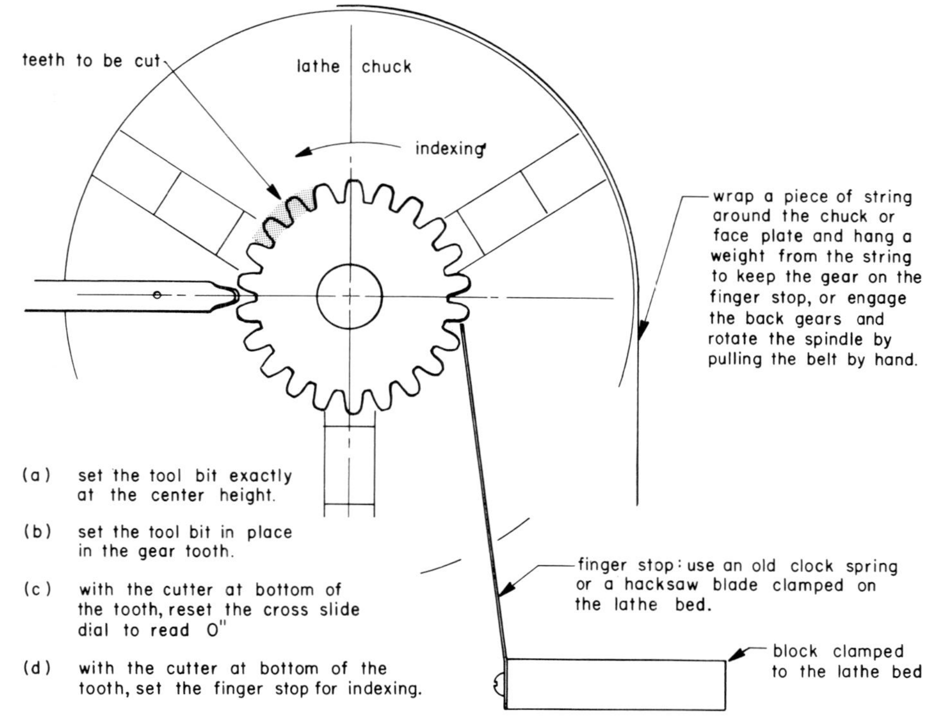

METHODS OF ANCHORING THE BRONZE TEETH IN THE GEAR

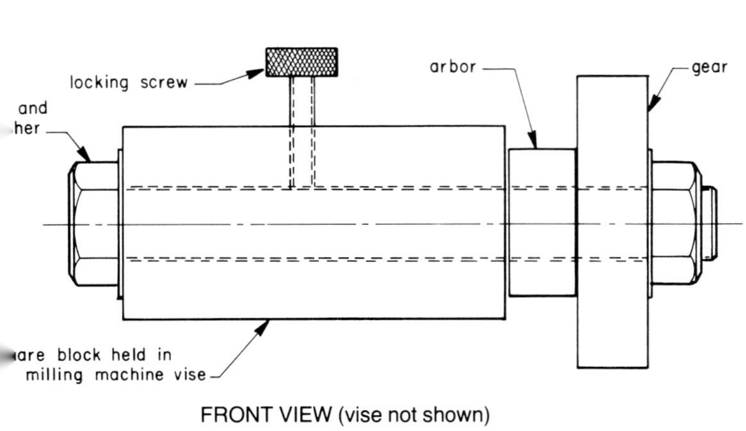

FRONT VIEW (vise not shown)

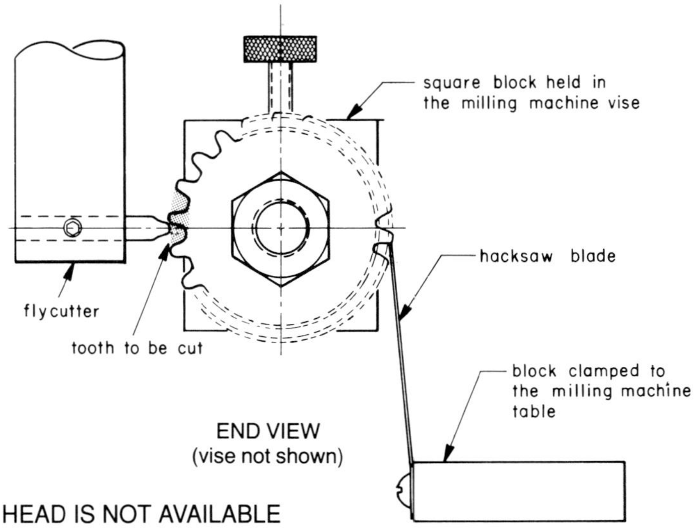

END VIEW
(vise not shown)

SET UP FOR A VERTICAL MILLING MACHINE WHEN A DIVIDING HEAD IS NOT AVAILABLE

A Few Tips on Knurling

Good diamond knurling should be well formed into points, like the checkering on the stock of a fine shot gun or rifle. Unfortunately, the knurling done by amateurs often has small flats instead of points. This detracts from the appearance and does not provide a slip proof surface.

Knurling is a "swaging" or pressing operation. If you press the tip of your fingers into a piece of soft butter, the butter displaced by your finger will rise above the surface around the crater.

The same action takes place during the knurling operation and the diameter of the work increases. The use of straight knurling to increase the size of a shaft so that a press fit is obtained for a ball bearing, wheel, or gear is often seen on cheap toys and similar classes of work.

Set the knurling tool on center so the top and bottom knurls will exert the same force on the work.

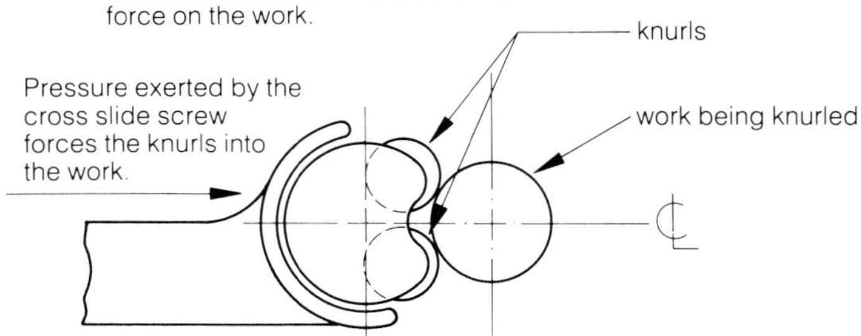

The knurls can swivel slightly to follow if the work is eccentric or deflects from the knurling pressure.

TAIL STOCK END VIEW

Figure 1

Setting the knurling tool is critical for good results. The knurling tool should be placed so that the knurl holder can rotate up or down slightly to exert equal pressure by each knurl (Figure 1). The knurling tool should also be placed at a slight angle with the work so that the full width of the knurls are used to form the knurling (Figure 2). The left-hand edges of the knurls will just start to make an impression and the right-hand sides of the knurls should be down to the full depth of the knurling.

Small diameter work will deflect away from the knurls, so several passes will be required. Simply back out the cross slide at the end of the pass and return the carriage to the starting position. With the lathe running, feed in the knurling tool with the cross slide and the knurls will "pick up" like gear teeth on the work. Apply more pressure with the cross slide screw and continue until the knurling is properly formed.

Use oil to lubricate the knurls and the work. Cutting oil or machine oil will do. Be *very careful* not to allow the oil brush, oil can spout or *your fingers* to be pinched between the knurls and the work.

Commercial knurling tools are available that have the knurls diametrically opposite, and the knurls are forced into the work by a nut that is tightened as the work rotates between the knurls (Figure 3). This type of tool is better for light lathes where the normal knurling tool puts a severe load on the cross slide, screw and nut.

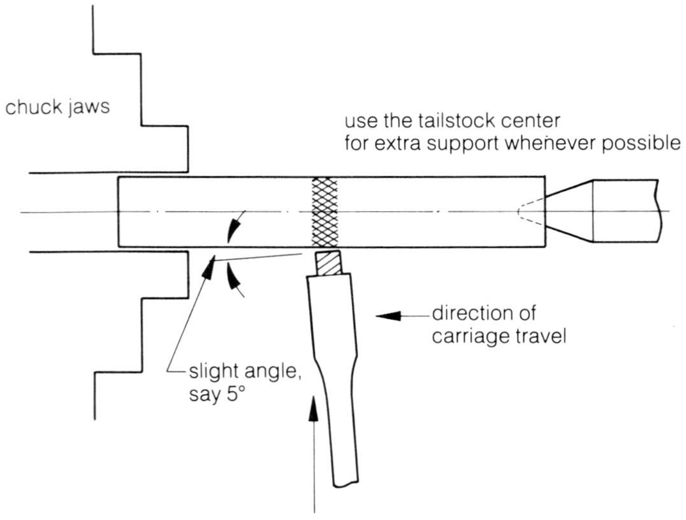

Figure 2

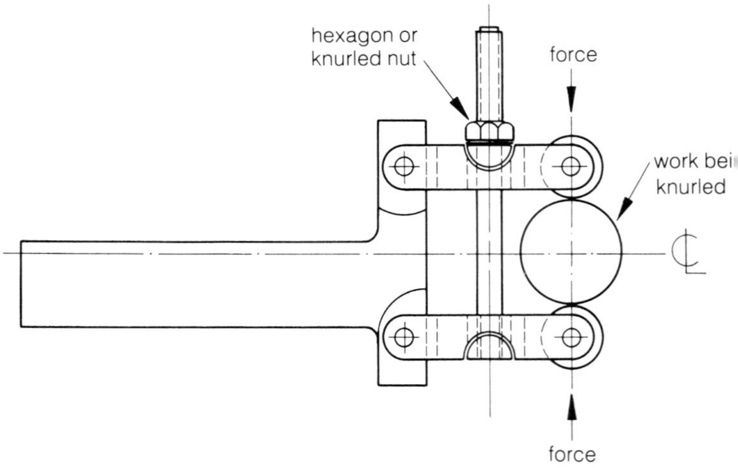

KNURLING TOOL SUITABLE FOR LIGHT LATHES AND LONG SMALL DIAMETER WORK

Figure 3

1 ***An adapter plate attached to the bottom of a Myford milling attachment. The standard plate suitable for a Myford cross slide is shown on the left-hand side of the picture. Note the new adapter plate has been made identical with the underside of the lathe compound slide rest.***

2 ***Another view of the adapter plate and the washer. The washer and the adapter plate are held with two bolts or studs.***

Milling on the Lathe

The beginner who has just bought a lathe will benefit from these photographs showing how to set the milling attachment quickly. I prefer the Myford type of milling attachment for its greater versatility with a tee-slotted table and ability to be set for compound angles. However, the South Bend and Atlas types with the vise as an integral part are very rigid, and are in common use. I have used both types many times.

The beginner should try to buy a milling machine as soon as possible. Some people think a Bridgeport or other knee type is the only kind to buy, but such a mill is too expensive for many folks, and cannot be brought down to a basement workshop easily.

This is the advantage of the mill/drill type of milling machine. It is not so expensive, can be dismantled easily for moving, and can be used as a drill press or for milling operations. I would like a Bridgeport, or its clone, but I'm thankful to have my small mill/drill, since almost anything is better than milling on the lathe.

The lack of rigidity, combined with anxiety that the work or the milling attachment will shift during the cut, takes all the pleasure out of the operation. You may also feel a little

3 ***A quick method of aligning the milling attachment with a faceplate***

4 ***Another method of quickly aligning the milling attachment***

5 *A quick method of aligning the milling attachment when the vise is in place. This method is very suitable for the South Bend and Atlas milling attachments.*

6, 7 *The usual method of aligning a vise or angle plate – slow, but accurate*

awkward standing there with your head cocked to one side like a bird listening for worms as you are trying to see the cutter.

Many fine models – especially those made in England – have been made by milling on the lathe. However, they have been made by very competent machinists who are able to cope with the inherent problems.

The beginner who had to "lick the spoon" to buy the lathe will have to cope with a milling attachment for a while, and this is a worthwhile accessory that will be used from time to time even after a milling machine has been purchased.

Myford lathe cross slides have tee slots for bolting the milling attachment. American type lathes will require an adapter plate to clamp the milling attachment to the cross slide in place of the compound slide rest. Photos 1 and 2 show such an adapter plate. The clamping is done with two bolts or studs.

Myford lathes have a micrometer handwheel on the end of the lead screw for advancing the carriage in thousandths of an inch. I suggest using a micrometer carriage stop, or use a rule to measure the depth of cut. Remember when milling on the lathe: take light cuts, use slow feeds and always feed against the direction of the cutter rotation.

8 *A quick and accurate method of clamping a round bar for cutting a keyseat with a slot drill or a Woodruff cutter.*

9 *Some homemade straps and tee nuts for 1/4-20 bolts for use on the milling attachment.*

13 *A collet for end mills for the lathe spindle. The back end has been tapped 3/8-16 NC for a draw rod to make sure the collet doesn't come loose. The sleeve is an adapter to hold a smaller end mill.*

14 *A 3/4" diameter end mill turning at 500 rpm. The depth of cut is .025" set with a micrometer carriage stop. A slow feed is required. This is about the maximum cut for such a setup.*

10

11

12

10, 11, 12 *Three views of a rectangular plate with many holes tapped 1/4-20 NC, and an angle plate made from a piece of 1/4 x 2 x 2" structural steel.*

A New Vertical Milling Machine

In 1983 I located my shop in a small (10 × 12), well-lighted and convenient room in my basement. In my previous shop I had an Arboga (Swedish) geared head vertical mill. It was a fine, accurate machine that could be used for some jig boring work; however, it was too heavy to safely move into the basement, even if it were partly disassembled. There was a two-speed motor built into the mill that would have been very costly to rewind or replace, and there just wasn't enough space in the new shop for the machine.

I bought one of the smaller milling and drilling machines that are becoming so popular in home shops. Mine is called a "First" and is model LC-15 made by the Long Chang Machinery Company Limited in Taiwan, Republic of China. This model is typical of the several kinds available from the machinery dealers who advertise in *HSM*.

Most of these machines are made in very modern factories and are good values. The best assurance of good quality and after-sales service is to make sure of the supplier's integrity. Also, consult others who have bought similar machines before you order one. Either your supplier or his wholesale importer should have inspected the factory in Taiwan.

Most of these mills have plastic handwheels and cast tee slots in the table to reduce costs. I am in favor of this, because it means that many more people can afford to buy a milling machine instead of trying to mill on the lathe, an operation for which the lathe was never designed in the first place.

My mill came with a No. 3 Morse taper spindle socket. However, some of the larger models and the medium size knee-type vertical mills are supplied with the R-8 spindle taper. The Morse taper will be more useful if you plan to use the machine for drilling holes larger than 1/2" with taper shank drills. The R-8 taper is an excellent choice if you will only be milling. Taper shank adapters are available but not very convenient if you need to use several sizes of taper shank drills in succession.

I would have bought one of the knee-type vertical mills except for the weight and space restrictions. However, I expect to buy one later. The small mill-drill machines are much more common, so articles dealing with these machines will interest more readers, particularly those who are in the process of setting up a shop.

My machine came equipped with a 2-1/2" diameter tungsten carbide-tipped cutter, a 1/2" drill chuck, and a 3-1/4" tilting drill press vise. I altered the vise for use on the milling machine while I waited for a proper 4" milling machine vise, which finally arrived several months after I received the mill.

After I uncrated my mill, I removed the preservative coating with a solvent (Varsol). After removing the cap on top of the column, I removed the head casting from the column by lifting the casting straight up. This was to reduce the weight so the machine could be easily carried by two men. Also, it allowed for the removal of the rack for drilling and countersinking two holes in its ends, thus enabling the rack to be fastened to the column. I removed the head casting alone, but if your mill comes equipped with a motor, you should ask someone to help you.

Do not remove the column from the base unless you have to carry the machine alone. My neighbor and I were able to put the machine in place, eliminating the need to lift it from the floor later. I had previously drilled and countersunk two holes in the ends of the rack for 1/4"-20 NC flathead screws (Photo 1), so he helped me lift the head and slide it down the column a few inches while I replaced the rack. I then located the rack with a precision square (Photo 2).

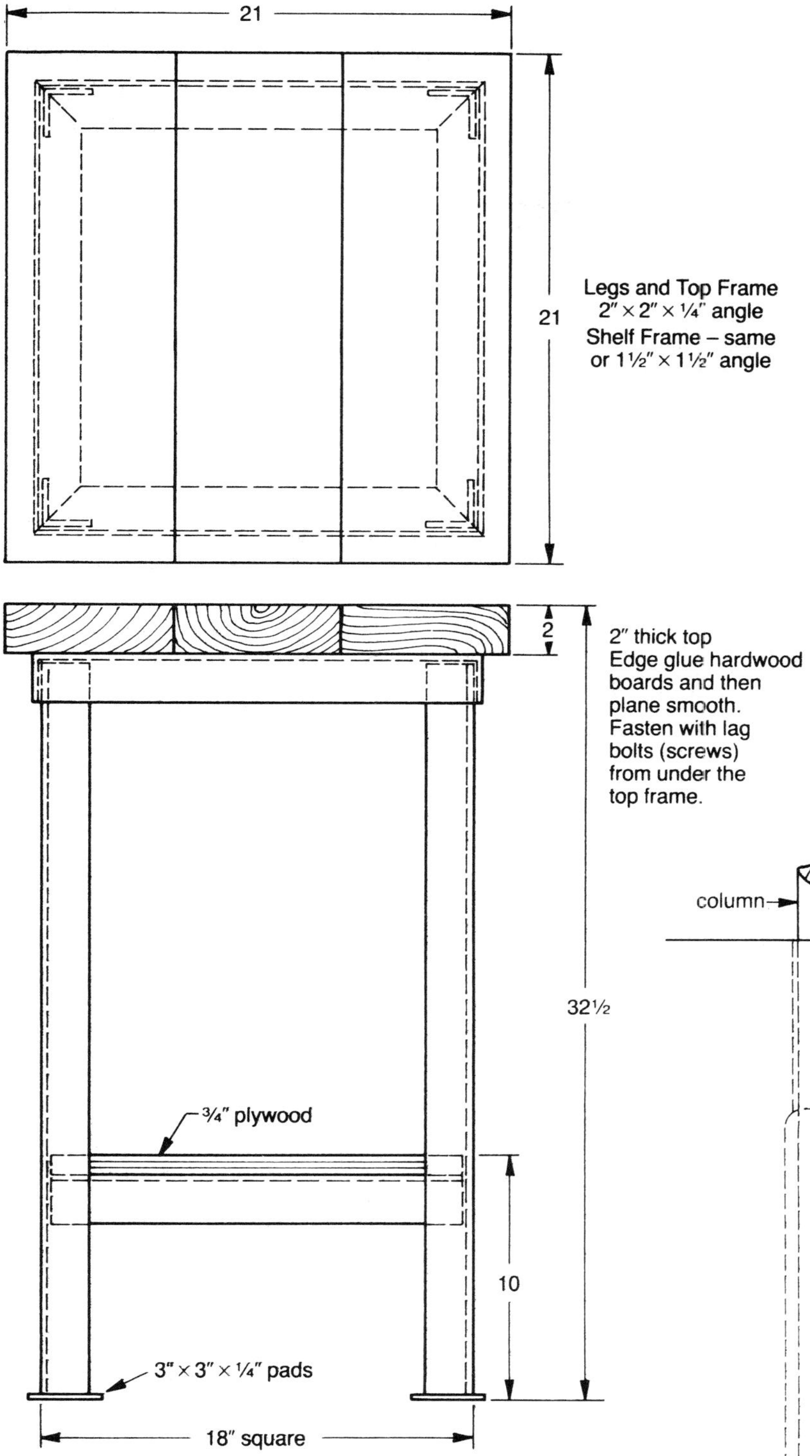

STEEL STAND FOR A MILLING-DRILLING MACHINE

1

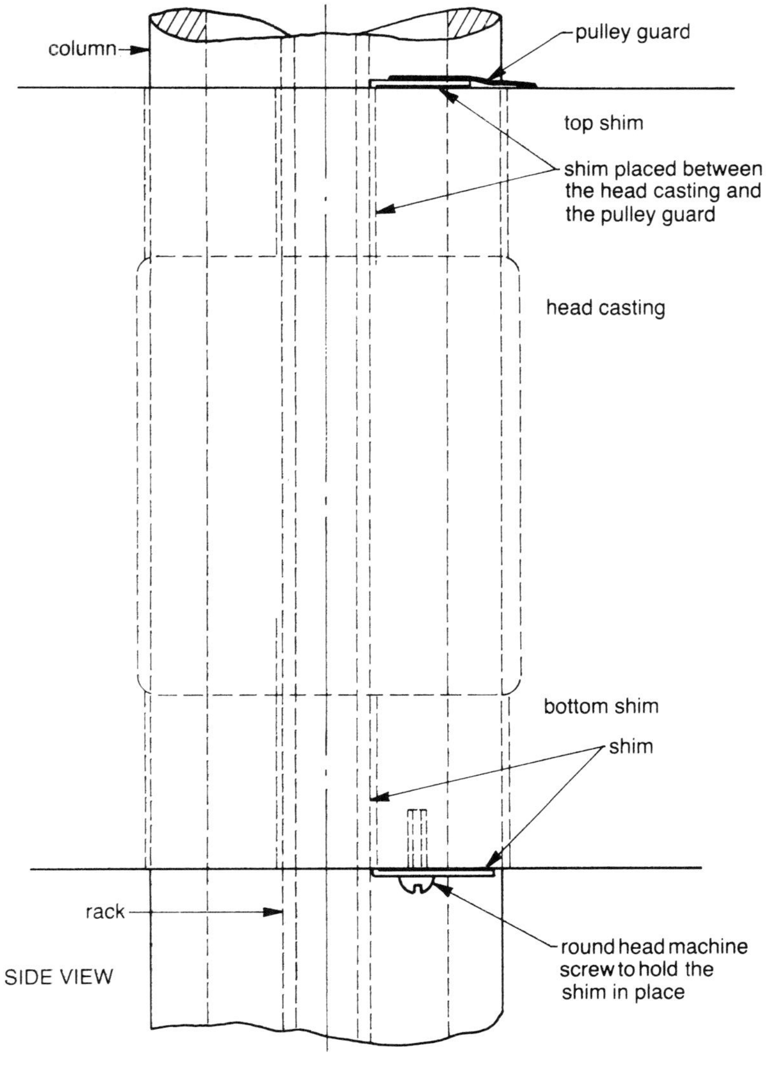

2

My mill was bolted to the wooden top of the stand and then leveled by use of cedar shingles under the feet. Use the best machinist's level you can buy or borrow, and persist until the table is level from side to side (X axis) and from front to back (Y axis) as illustrated in Photos 3 and 4. Then you can set up some work with a level when it may not be possible to use parallels.

I made a Turn Around Test (Photo 5) with a dial indicator. The machine gave the same reading from front to back but was up .001" on one side and down .001" on the other side. I didn't think there was enough error to warrant shimming, but if this is necessary, loosen the bolts holding the column to the base and use cigarette paper or brass shim stock.

My mill came without a motor, so I used a spare 1/2 hp Delco motor (with ball bearings). I had to rebend the hinge tabs on the motor mounting plate to enable the plate to move easily, and I ground off a few turns of the spring that controls the belt tension.

I had to drill and ream the motor pulley supplied with the machine to fit the motor shaft. Note in Photographs 6 and 7 how to use a lathe dog sliding on a plate resting on top of the compound slide rest. This prevents the drill from twisting in the tailstock spindle socket. It's an old machinist's trick. The plate on top of the compound slide rest is to prevent the lathe dog from scratching the paint.

I placed the switch for the motor at the back on the left-hand side of the head casting so the table would be free on both sides of the machine base to hold wrenches, measuring instruments and similar items. I usually sit on the right-hand side in front of the machine, and I have no trouble reaching the switch.

I fastened the rack to the column to prevent the head from moving from side to side while being raised or lowered. My Arboga used this system with an adjustable gib against the rack.

My new mill has a space of .037" between the rack and the edge of the groove. At present, I pull the head toward me to take up the slack when I tighten the clamping bolts, after moving the head. Later, I expect to use shims, as shown in the drawings.

I used hexagon socket flathead screws (Allen-type) because I happened to have them, but the ordinary slotted flathead screws could be used (1/4"-20 NC). The rack was aligned with a 12" precision square placed on a 1 × 2" ground parallel resting on top of the table (see Photo 5).

3

I suggest that you try to borrow the parallel and the square, because you may not need them again. My square is an unhardened Moore & Wright and is much cheaper than the hardened type. The parallel was ground from a piece of 1 × 2" mild steel. Use pieces of cardboard between the clamps and the column to avoid scratching the column. Use a 1/4" drill to make a "dimple" in the column.

This will guide the No. 7 drill; then tap 1/4-20 NC. Do not remove the rack after drilling, but use the hole in the rack to guide the tap straight. You may want to use a split pin in the middle of the rack if you think that the rack may bend sideways. Mine seems to be very rigid, so I omit the split pin to allow easy removal of the two flathead screws in the event I want to swing the head around the column.

I have drilled and tapped holes in the cap on top of the column to hold a flexible lamp. This light enables me to easily see the layout lines on the work.

My mill was not supplied with a chart showing the spindle speeds for the various belt positions. To calculate the spindle revolutions per minute (spindle speeds) for each belt position can be done quite easily in one evening.

Prepare a drawing like Chart 1 showing the diameters of the pulleys on the motor, idler and spindle. Draw the pulleys in line to avoid errors. Now use the formula to calculate the revolutions per minute of the idler pulley for each of the belt positions from the motor pulley to the idler pulley.

Formula:

$$\text{Idler Pulley Dia.} \times \left(\begin{array}{c}\text{Idler Pulley}\\ \text{RPM}\end{array}\right)$$

$$= \left(\begin{array}{c}\text{Idler Pulley}\\ \text{RPM}\end{array}\right) \times \text{Motor RPM}$$

4 5

6 7

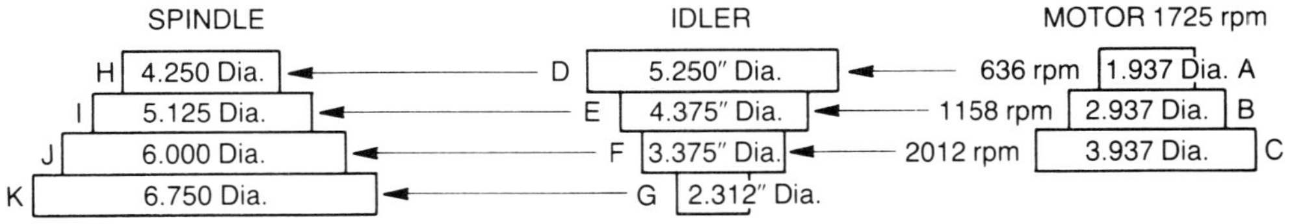

Formula: Dia. × RPM = dia. × rpm

Chart 1

A to D =
5.250″ × RPM = 1.937″ × 1725 rpm

$$\text{RPM} = \frac{1.937 \times 1725}{5.250} = \frac{3341.325}{5.250} = 636 \text{ rpm}$$

B to E =
4.375″ × RPM = 2.937″ × 1725 rpm

$$\text{RPM} = \frac{2.937 \times 1725}{4.375} = \frac{5066.325}{4.375} = 1158 \text{ rpm}$$

C to F =
3.375″ × RPM = 3.937″ × 1725 rpm

$$\text{RPM} = \frac{3.937 \times 1725}{3.375} = \frac{6791.325}{3.375} = 2012 \text{ rpm}$$

Formula for all Pulley Calculations: Idler Pulley Diameter × RPM = Motor Pulley Diameter × RPM of the Motor

Example:
5.250″ × RPM = 1.937″ × 1725 rpm

$$\text{RPM} = \frac{1.937 \times 1725}{5.250} = 636 \text{ rpm}$$

Chart 2

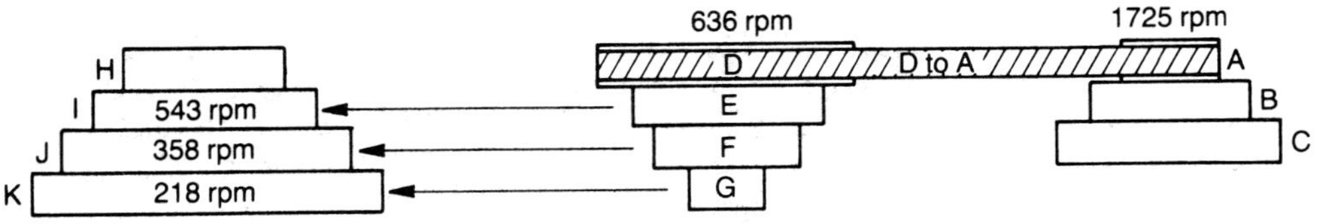

I to E =

$5.125'' \times \text{RPM} = 4.375'' \times 636 \text{ rpm}$

$$\text{RPM} = \frac{4.375 \times 636}{5.125} = \frac{2782.5}{5.125} = 542.926 \text{ or } 543 \text{ rpm}$$

J to F =

$6.000'' \times \text{RPM} = 3.375'' \times 636 \text{ rpm}$

$$\text{RPM} = \frac{3.375 \times 636}{6} = \frac{2146.5}{6} = 357.75 \text{ or } 358 \text{ rpm}$$

K to G =

$6.750'' \times \text{RPM} = 2.312'' \times 636 \text{ rpm}$

$$\text{RPM} = \frac{2.312 \times 636}{6.750} = \frac{1470.432}{6.750} = 217.84 \text{ or } 218 \text{ rpm}$$

Chart 3

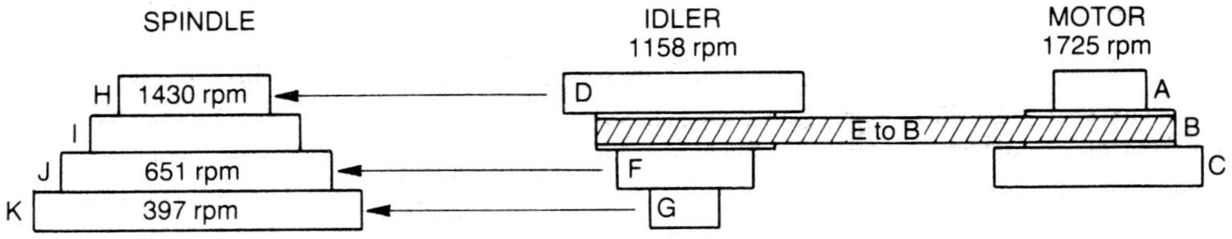

H to D =

$4.250'' \times \text{RPM} = 5.250'' \times 1158 \text{ rpm}$

$$\text{RPM} = \frac{5.250 \times 1158}{4.250} = \frac{6079.5}{4.250} = 1430.47 \text{ or } 1430 \text{ rpm}$$

J to F =

$6.000'' \times \text{RPM} = 3.375'' \times 1158 \text{ rpm}$

$$\text{RPM} = \frac{3.375 \times 1158}{6} = \frac{3908.25}{6} = 651.37 \text{ or } 651 \text{ rpm}$$

K to G =

$6.750'' \times \text{RPM} = 2.312'' \times 1158 \text{ rpm}$

$$\text{RPM} = \frac{2.312 \times 1158}{6.750} = \frac{2677.29}{6.750} = 396.636 \text{ or } 397 \text{ rpm}$$

Chart 4

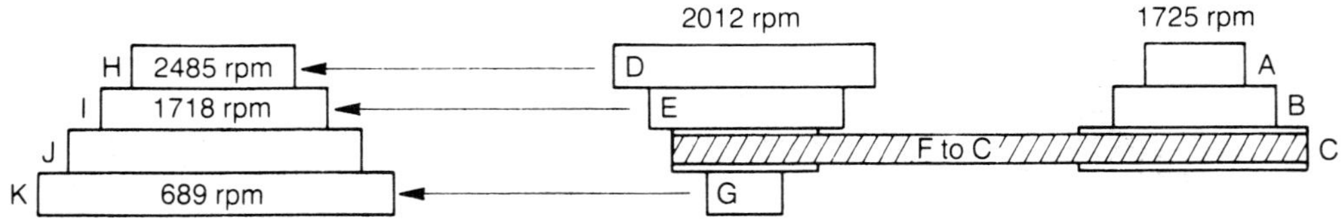

H to D =

$4.250'' \times \text{RPM} = 5.250'' \times 2012 \text{ rpm}$

$$\text{RPM} = \frac{5.250 \times 2012}{4.250} = \frac{1056.3}{4.250} = 2485.41 \text{ or } 2485 \text{ rpm}$$

I to E =

$5.125'' \times \text{RPM} = 4.375'' \times 2012 \text{ rpm}$

$$\text{RPM} = \frac{4.375 \times 2012}{5.125} = \frac{8802.5}{5.125} = 1717.561 \text{ or } 1718 \text{ rpm}$$

K to G =

$6.750'' \times \text{RPM} = 2.312'' \times 2012 \text{ rpm}$

$$\text{RPM} = \frac{2.312 \times 2012}{6.750} = \frac{4651.744}{6.750} = 689.147 \text{ or } 689 \text{ rpm}$$

Chart 5

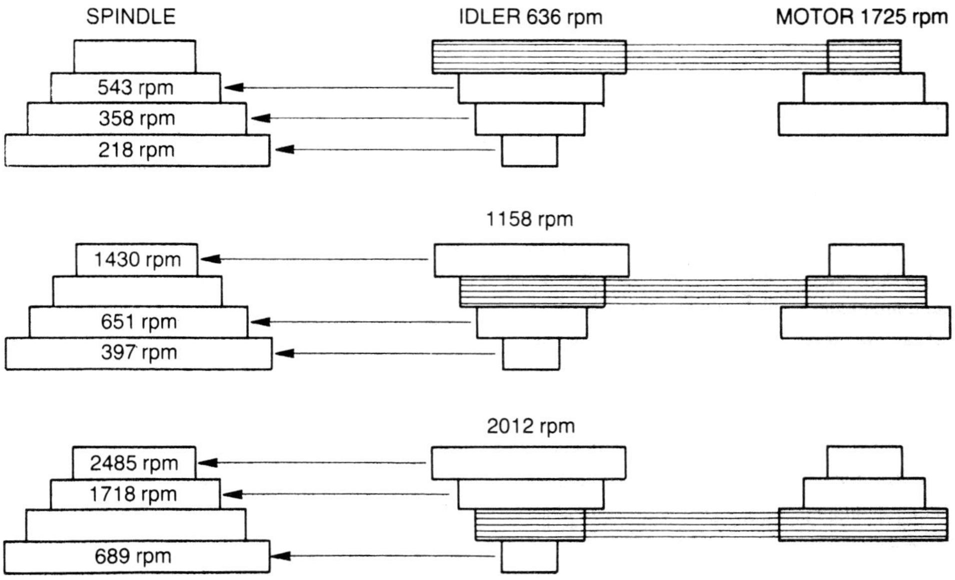

Chart 6

Example: 5.250" × RPM = 1.937" × 125 rpm

$$RPM = \frac{1.937 \times 1725}{5.250} = 636 \text{ rpm}$$

Mark down the RPM for each belt position between the two pulleys as shown on Chart 1. Now that you have calculated the RPM for each belt position between the motor and idler pulleys, you can calculate the spindle revolutions per minute for each belt position between the idler and spindle pulleys. Use the same formula each time and mark down the spindle RPM as I have shown on Charts 3, 4 and 5.

Now prepare a permanent chart on a piece of cardboard or plywood incorporating all the belt positions and spindle speeds as shown on Chart 6.

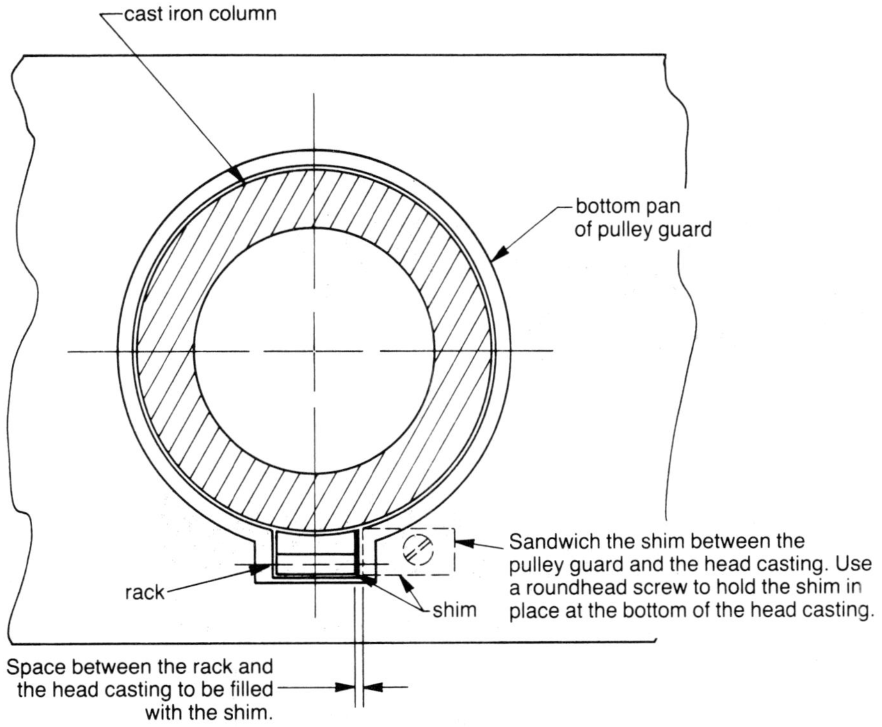

TOP VIEW OF THE
COLUMN AND THE HEAD CASTING
SHOWING HOW TO PLACE THE SHIM ON TOP

Improving Your Vertical Mill

1

While setting up and inspecting my own brand new mill (Photo 1), I discovered that the tee slots were out of alignment with the table ways by .013" from one end to the other. These tee slots are cast and not machined, and are entirely suitable as is for clamping a vise on the table and for clamping an angle plate, and so forth.

However, I use the tee slots to align a dividing head (Photo 2), so they should be in correct alignment with the table ways and parallel with the other tee slots. Figure 1 shows how to test to be sure that the front of the milling machine table is parallel with the table vee ways.

These slots can be milled out on the machine if a larger milling machine is not available. Mill the slots out on the edges with a suitable end mill to the full length of the table travel. Remove one end plate, the dial and hand wheel and align the milling cutter in the tee slot, and then mill to the end of the table.

2

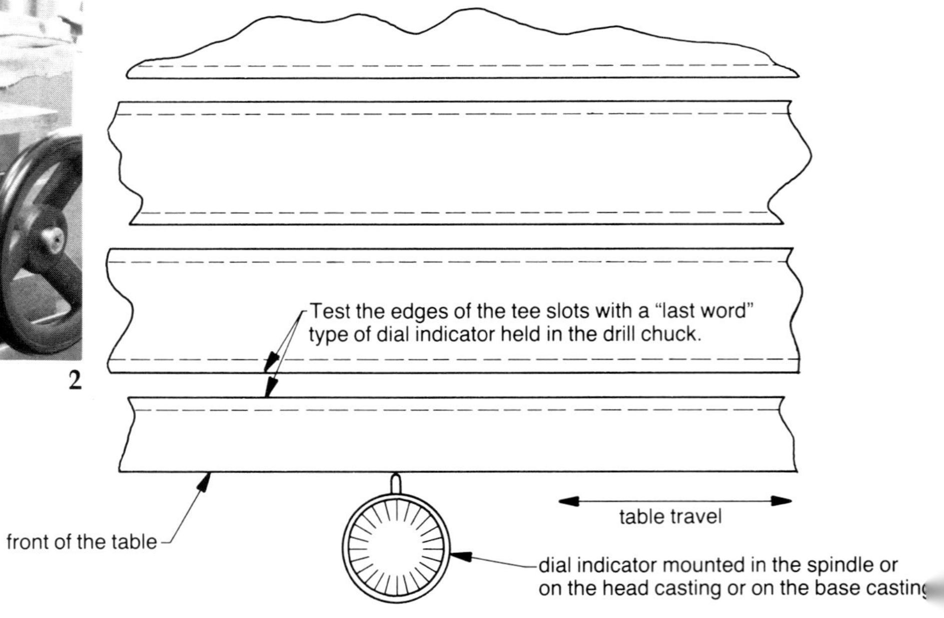

Testing to be sure the front of the milling machine table is parallel with the table vee ways.

Figure 1

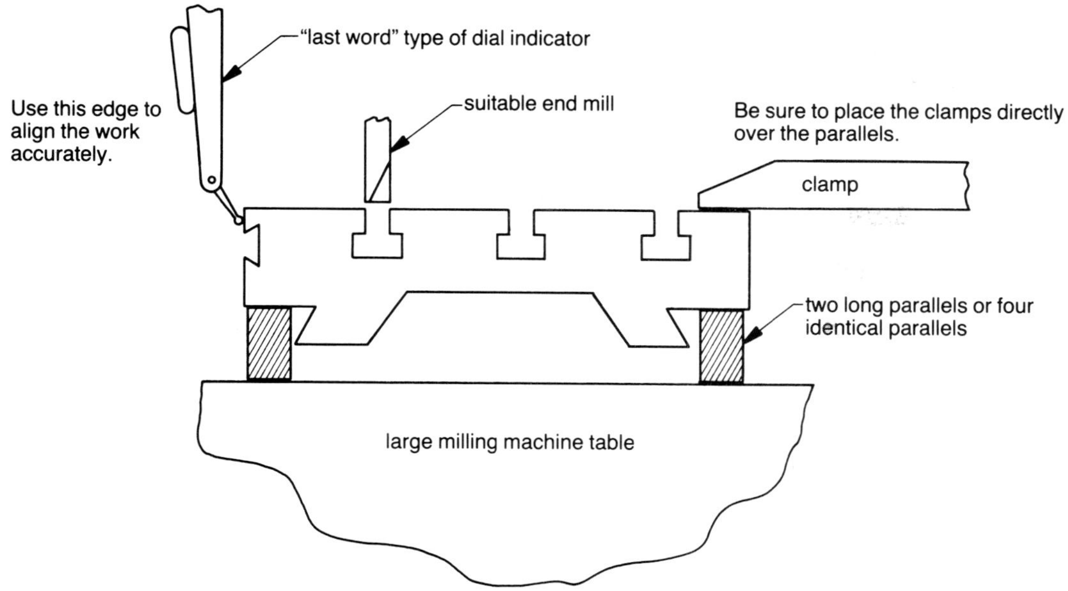

How to set up the milling machine on a larger milling machine to "true" the table tee slots.

Figure 2

Many readers will have the use of a large vertical or horizontal milling machine at work or at an evening shop class. If suitable parallels are not available, buy an 18″ length of ground flat stock and cut it into four pieces. Then support the table on the parallels as shown in Figure 2.

If you have the use of a surface grinder, buy several sizes of cold rolled steel and grind "pairs" of parallels in various sizes, such as ½ × 1″, ⅜ × ⅝″, ¾ × 1½″, all in lengths to suit your milling machine vise.

Milling out the tee slots will probably bring the slots to an odd width. Mine cleaned up at .512″. You can either make wider keys for your machine vise or push the vise hard against one side as you tighten the clamping bolts.

3

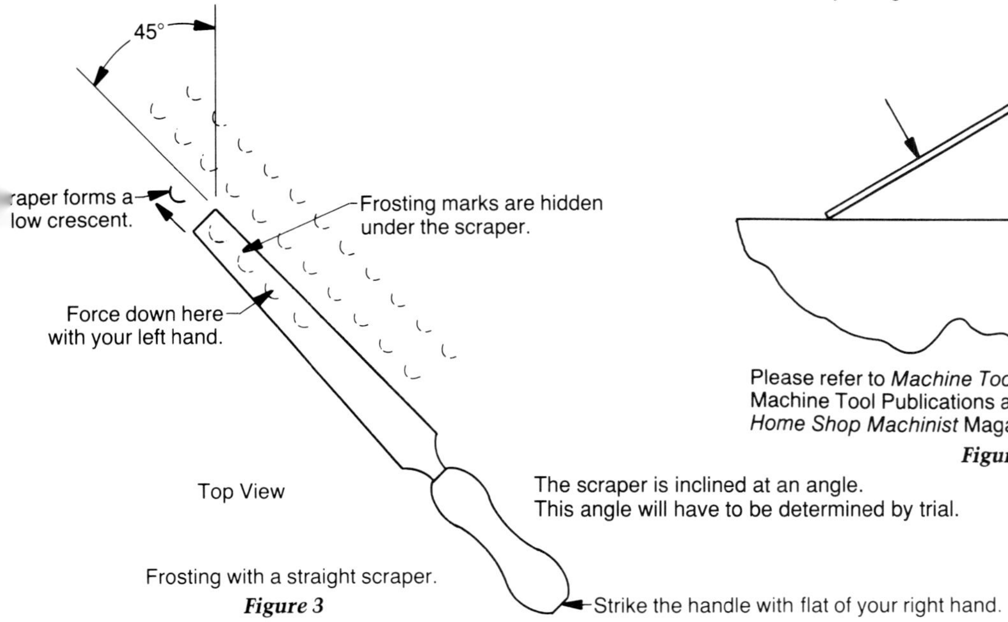

Frosting with a straight scraper.

Figure 3

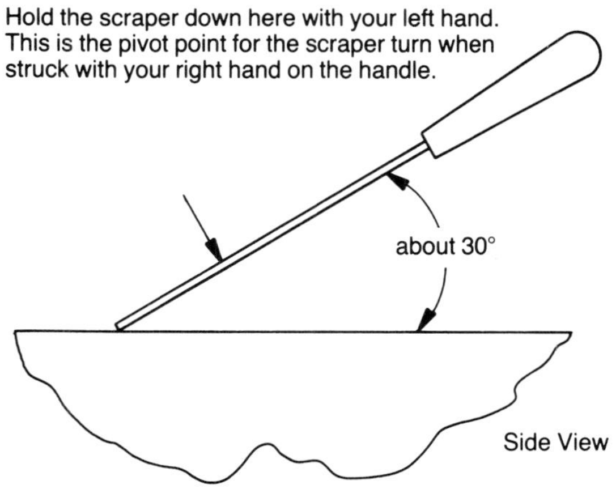

Please refer to *Machine Tool Reconditioning* from Machine Tool Publications and advertised in *The Home Shop Machinist* Magazine.

Figure 4

While the table was off the machine, I decided to "frost" the flat and vee ways of the table to provide better oil distribution and to prevent a condition known as "stick slip." This occurs as the sliding surfaces become slightly worn and tend to wring together like Johannson gage blocks, then suddenly release, making accurate positioning of the table difficult. Note that the gibs on a milling machine should be adjusted so there is a slight resistance or drag when moving the table, because the milling cutter gives an intermittent cut and chattering can easily occur. Photo 3 shows the results of frosting on my slide ways.

There are two common methods of frosting. One requires a straight scraper and the other uses a bent, or hook scraper. Both types of scrapers can be made from an old worn out flat (8″ or 10″) file. Grind all the teeth off, being careful not to overheat the file.

The hook scraper is then heated, and ½″ to ¾″ is bent over at 90° to the flat surface. Harden and temper to a light straw color. Run the scraper back and forth on an oil stone with the flat of the file in a horizontal position. The scraper is used by drawing toward you while giving the tool a slight rocking motion.

The straight scraper does not need to be heat-treated, since it is already very hard. Grind the end of the scraper square with the sides and then finish honing on an oil stone by drawing it back and forth while holding it in a vertical position. Place your hand so your thumb is in a vertical position and your fingers grasp the scraper tightly to the palm of your hand.

The scraper is used as shown in Figures 3 and 4, but be sure to practice on a piece of scrap metal before scraping the milling machine ways.

Always align the vise fixed jaw with a dial indicator. Do not rely on the graduations if you expect to do accurate work. When the vise is placed on the milling machine table for the first time, do not take for granted the accuracy of the vise. Mine had to be reground on a surface grinder before I could use it. Follow the testing procedures shown in Figures 5 and 6.

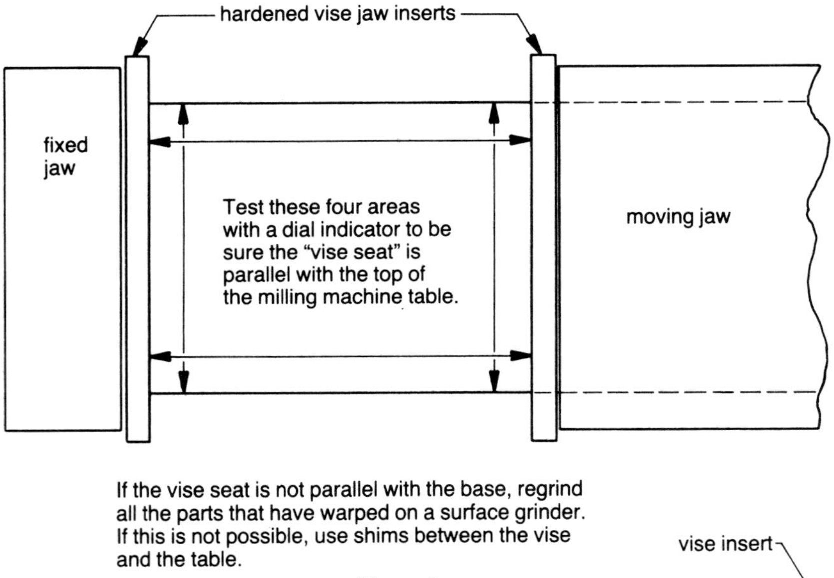

Figure 5

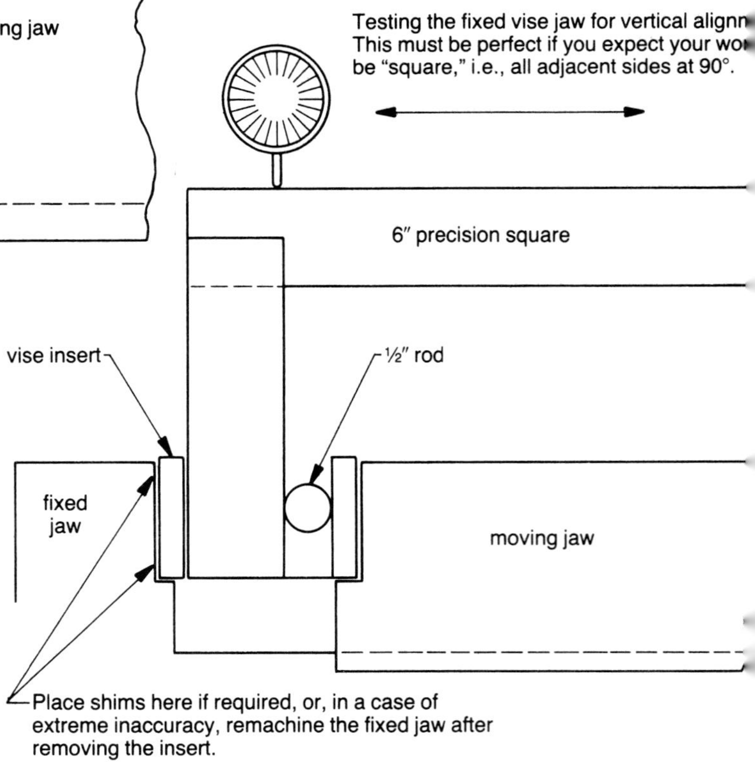

Figure 6

Tips for Silver Soldering

Most of us use too much silver solder. With the high prices today, this is downright foolhardy. To hold silver solder while we are working with it, we either hold the solder several inches back from the hot end to avoid burning our fingers or else we try to hold the solder in a pair of pliers. Both methods leave something to be desired, for they lead to excessive solder use and inaccurate placement, both of which are wasteful.

For a simple remedy, take a cheap ballpoint pen, remove the ball and drill a hole through the plastic handle. Use 1/32" diameter silver solder wire and feed it down through the handle. A two-foot length is about right. With 1" to 1-1/2" of the silver solder projecting, you can place the solder exactly where you want it without burning your fingers. Keep feeding the wire through the pen as needed. This method works very well for small boiler fittings and valve gear parts.

If you use 1/16" diameter silver solder as well as the 1/32" diameter, make a pen holder for each size and type of solder.

One important thing in silver soldering many assemblies, such as valve gear parts, where a rod or bushing passes through a plate or lever, is the need for strength. You don't want the solder all in a glob at one end of the joint. Instead, the solder should penetrate right through the joint to have the maximum possible strength with the neatest possible joint, using a minimum of solder. To achieve this, it is necessary to drill the hole for the rod or bushing large enough so there is a clearance of .002" to .003" on all sides.

This leads to a problem in that the shaft or tubing must be located in the center of the hole, both to have a uniform thickness of silver solder around the shaft and also to maintain the dimensional accuracy of the parts.

Make up a pyramid punch as shown in the sketch, from drill rod. Use the center left by facing the punch in the lathe as a guide to mill or file the four flats that make the pyramid point.

To harden the punch, heat with a torch until the punch becomes a dull red. At this point, heat gently and test frequently with a magnet. When the punch becomes non-magnetic, the critical temperature has been reached and the punch is ready for quenching.

If you are using water-hardening drill rod, plunge the punch straight down into a pot of clean water at room temperature and move it in a figure-8 pattern. This will contribute to uniform cooling and minimize warping.

Remove the punch from the water once it has cooled. Be careful not to drop it, for it is very brittle and could easily break or crack. Test the hardness with a file.

Polish the punch with emery cloth, then temper it. To temper, heat the shank gently with a propane torch. When the point reaches a brown color, quench again. The punch is now ready for use.

I use oil-hardening drill rod myself, because the quench is not so severe and there is less warping. For quenching oil, I use old motor oil. The same procedures apply as for water-hardening steel.

To use the pyramid punch, place the punch in a vertical position over the hole and strike with a hammer. Reverse the work and repeat on the other side.

Clean the shaft or tubing with emery cloth and make sure the hole is free of grease, oil or dirt. Press the shaft through the nicks in the hole and into correct position.

Apply flux and then silver solder using the pen holder. You will see just a thin line of silver solder around the joint. Pickle the part in a weak solution of sulfuric acid to remove the flux and scale.

You will be pleasantly surprised how little silver solder is needed with this method.

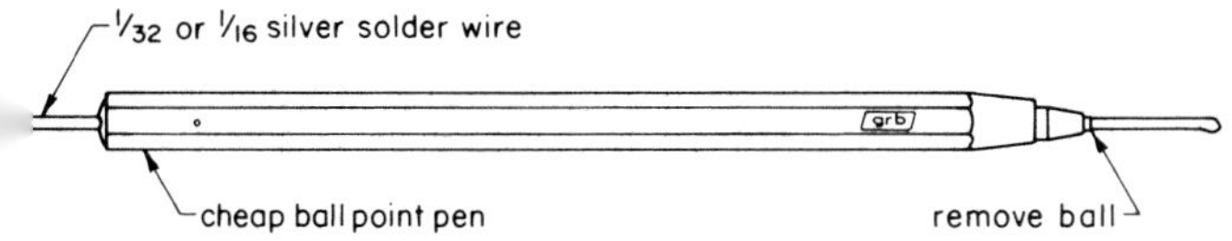

SILVER SOLDER HOLDER

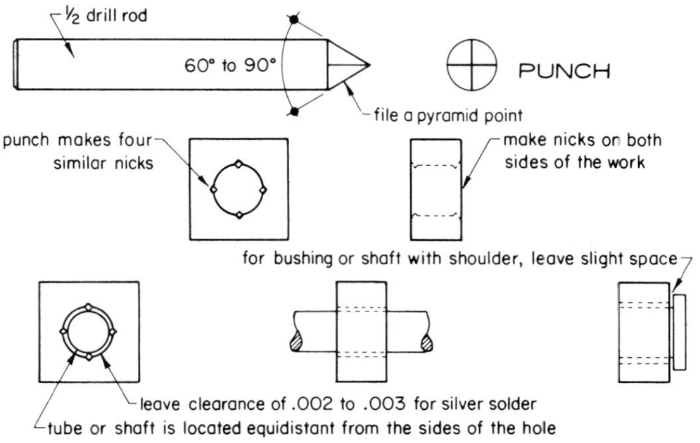

SHAFT CENTERING

Installing a Power Feed Unit on a Vertical Mill

A small mill/drill machine does not need a power feed to the table; the work is usually of no great size. There is a real advantage to operating the mill from both ends of the table or from behind the mill (if it is difficult to see the cutter) by hand. One doesn't want to sacrifice a handwheel and possibly a couple of inches of table travel; it's not a very good tradeoff.

However, a larger mill/drill machine with a table travel of 24" is quite another matter. The traditional milling machine has a table lead screw with a pitch of four threads per inch. The new mill/drill machines have table lead screws of 10 tpi, so the full table travel will require 240 turns of the table handwheel for the full length of table travel.

I usually operate the lathe or drill press when the milling machine is working since the stops on the table will turn off the table feed at the end of the cut. Then there is no danger of breaking a cutter or damaging the machine if I'm not there to stop the machine at the end of the cut.

The rapid traverse is very convenient to return the table to the starting position for the next cut. I use the rapid traverse when I'm aligning the milling machine vise with a dial indicator on the fixed jaw. This saves a lot of time and I'm in front of the machine where I can see the dial indicator easily rather than at the end of the table.

There were no instructions for installing the power feed unit on my mill, so I have prepared this set of instructions to help those of you who may not be sure of the proper procedure.

A. Unpack the power feed unit and check for damage. There should be a motor unit, two metal brackets, adapters to suit various size lead screws, and a limit switch. There may be other items, too. The unit should be plugged into the electric outlet and turned on to make sure it operates well.

B. Attach the "L" angle bracket to the front of the table with the bolts and dovetail nuts provided. Just tighten the screws sufficiently to hold the bracket in place. The bracket

General view of the power feed set upside down on the miling machine table.

General view of the power feed unit with the extra plastic coupling and the spring-type stops.

The "L" angle bracket with the dovetail nuts for the dovetail slot on the front of the mill table.

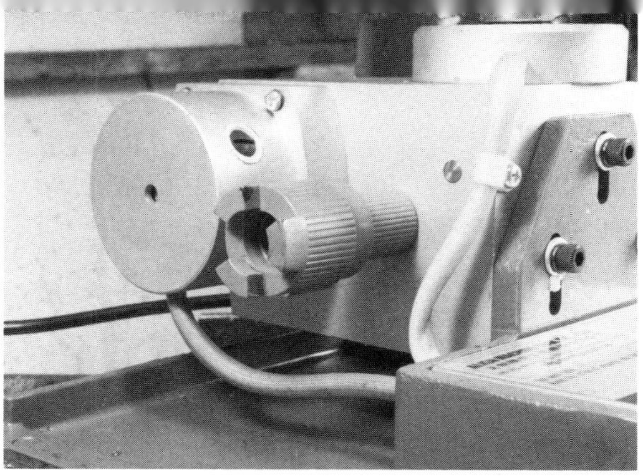

The plastic adapter coupling is in place on the drive shaft.

Use C-clamps to hold the power feed unit in place before using the transfer punch to locate the bolt holes.

Use a transfer punch to locate the center of the slots for drilling and tapping the bolt holes in the table.

Use a drill the same size as the holes in the casting to make a "dimple" or conical hole in the table to start the other drills exactly in the center.

The limit switch is fastened in place on the stop supplied with the machine.

The "L" angle bracket is shown bolted in place with two bolts screwed into the end of the table. Also shown is the homemade oil cup used to replace the ball oiler supplied on the machine.

The power feed is shown from underneath and the front in its mounted position.

should fit up against the end of the table and rest on top of the table. Make any alterations necessary to achieve this.

C. Now place the correct adapter on the motor shaft and tighten the setscrew just enough to hold the adapter in place.

D. Try the motor in place on the lead screw and clamp the motor to the "L" angle bracket with small C-clamps. Turn the lead screw by hand to be sure the motor turns easily.

E. Remove the motor unit and drill and tap through the two slots in the top of the "L" angle bracket into the milling machine table. I used 3/8-16NC bolts.

F. Place the 3/8-16NC bolts in position with a washer under the head of each. Now try the motor unit in position again. Use the bolts provided to fasten the motor unit to the "L" angle bracket on the table.

G. Check the alignment of the motor unit with the lead screw and then tighten the top bolts and the front bolts of the L-bracket to the table. Correct any misalignment and then tighten all the other bolts. Check to see that the motor turns freely by rotating the adapter with your fingers.

H. Remove the stop on the front of the saddle and bolt the limit switch in its place using one or two of the existing bolt holes. The round stops supplied with the milling machine were large enough to operate the limit switch so I didn't use the two stops provided with the power feed kit.

I. *Make sure not to turn on the power feed if the table has been locked so it won't move in the longitudinal "Y" direction!* The circuit breaker should prevent serious damage; however, the power feed is too expensive to take any risks of damage.

J. The lead screw bracket at the end of the table has an oil hole on top that is inaccessable when the power feed unit is in place. I made up the special oil cup shown in Figure 1. This oil cup just fits in the hole in the bracket after the original oil cup has been removed. The photographs show this in place before the power feed was fastened in place.

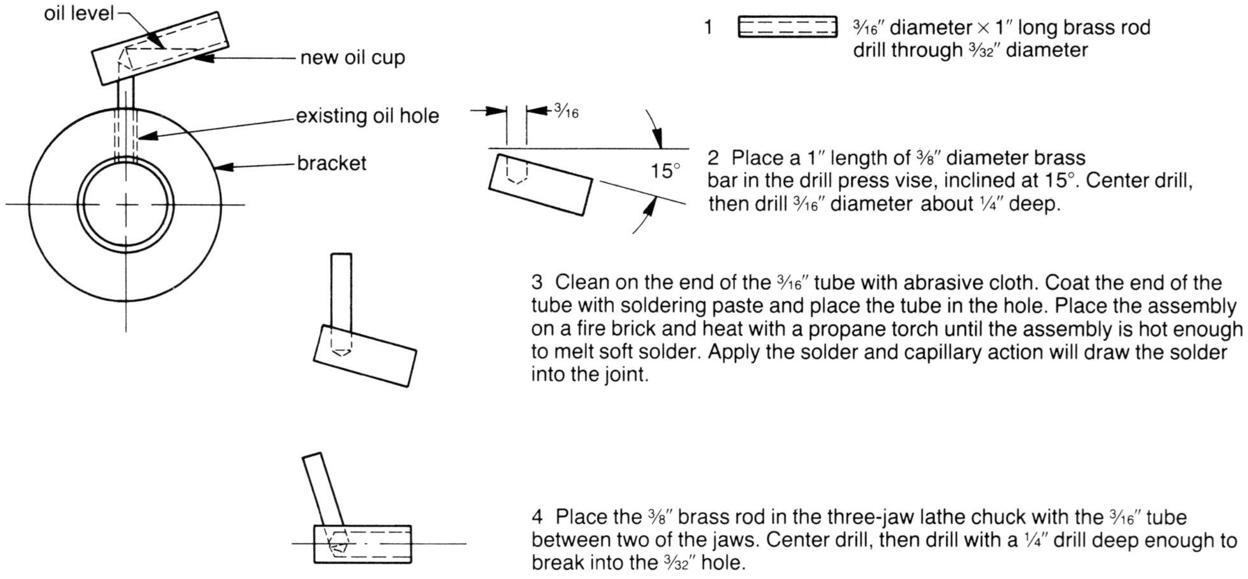

HOW TO MAKE THE OIL CUP
Figure 1

Milling Keyways

Sooner or later, most amateur machinists are going to need to make a keyway in a shaft. This is a common milling operation and may be done either on the lathe with a vertical milling slide or in a milling machine.

Let's assume we're going to do it on a lathe and look at the processes involved. First, we have to set the vertical slide 90° to the lathe axis. With the setscrews loose enough to permit it to turn, bring the slide up to touch the faceplate. If you don't have a faceplate or don't want to take the time to put one on, touch it to the chuck body between the jaws. This is much quicker than setting the slide with a dial indicator and nearly as accurate. Needless to say, tighten the setscrew while the slide is in contact with the chuck or faceplate.

Now, however, the vise will have to be aligned level with the lathe bed with a dial indicator if it has been moved since it was last aligned. Use a magnetic base or other holder and place the contact on the fixed jaw. Move the cross-slide back and forth. Adjust the vise until you have the same reading all along the jaw.

Once the vise is square and level, clamp the shaft in the vise. If you have a Myford-type milling attachment, large shafts may be clamped directly to the slide using a tee-slot as a vee-block.

Coat the top of the shaft with a china marking pencil, layout blue, felt-tip marker pen or a child's crayon. Start the lathe and raise the slide slowly until the cutter just scrapes away the wax and leaves a shiny mark showing through. Note the reading on the dial of the vertical slide or set the dial to zero. Now, raise the slide a distance equal to half the shaft diameter plus half the cutter diameter. This will bring the cutter in line with the axis of the shaft with an accuracy of .001" plus or minus, as shown in Figure 1.

To remove the arc of the shaft above the keyway, slowly feed the shaft into the cutter until the cutter just does cut its full width (Figure 2).

Also as shown in Figure 2, the depth of the keyway will be equal to half the key width plus .010" for clearance.

The Myford and similar types of lathe have a handwheel at the tailstock end of the leadscrew. Engage the half-nuts and bring the carriage along until the cutter touches the work. Use the graduations on the handwheel to set the depth of cut.

South Bend and similar types of lathe can use a micrometer carriage stop or a toolmaker's clamp and some packing equal to the depth of the cut as a gage. Bring the micrometer carriage stop to touch the lathe carriage and clamp it to the bed. Back the micrometer spindle the correct distance and clamp. Make shallow cuts of .020" to .030" until the depth is reached.

To use a toolmaker's clamp, place the correct thickness of packing between the carriage and the clamp, which is fastened to the lathe bed. Remove the packing and again take shallow cuts until the carriage touches the clamp.

If a very long keyway that exceeds the cross-slide travel must be cut, once the cut is to depth and as long as the machinery will allow, back the cross-slide up until the cutter is near the end of the slot. Loosen the vise and slide the shaft along, using the cutter in the slot as a guide. Reclamp the shaft and continue cutting the keyway as before.

The method of locating the cutter using the wax mark on the workpiece will be well-suited to those having vertical milling machines (Figure 3). The same method can be used in a horizontal milling machine with either an end mill in the horizontal spindle or with a standard Woodruff keyway cutter (Figure 4).

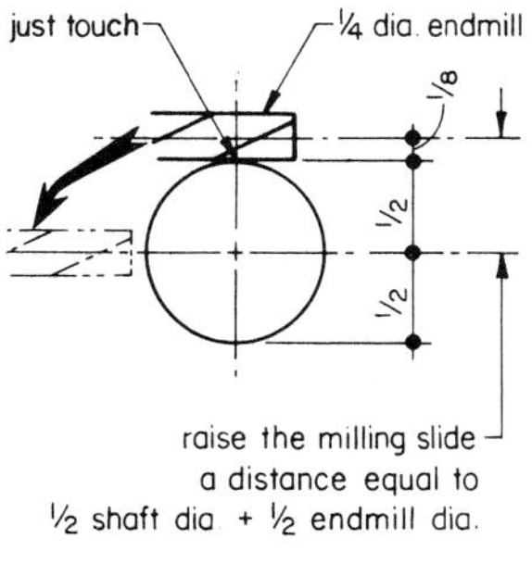

Figure 1

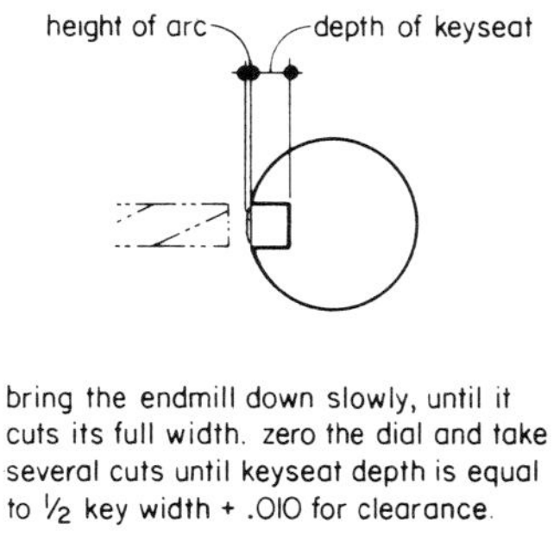

Figure 2

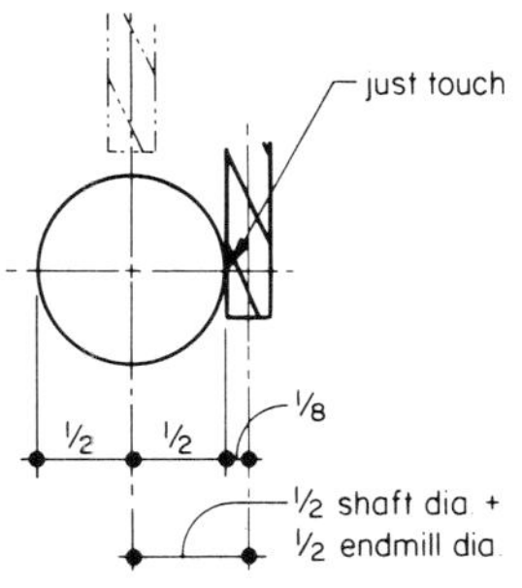

Figure 3

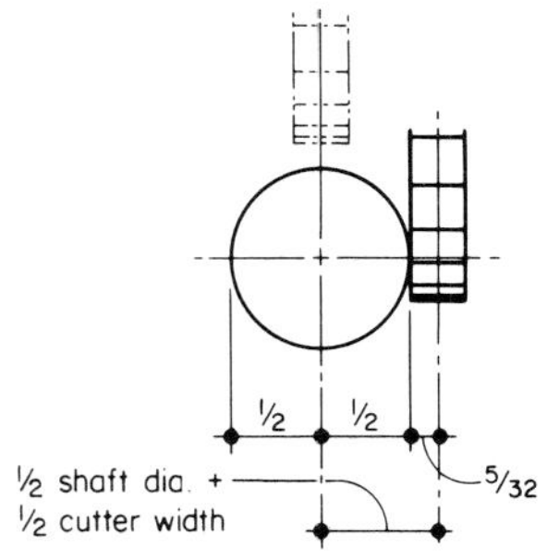

Figure 4

Cutting Left-hand Threads

There are several extra things to consider when cutting left-hand threads on a lathe, even though the threading operation is probably easier than right-hand threading, since you will not be threading up to a shoulder.

The compound slide rest will be set at 30° to the left, so the work will have to project farther from the chuck than normal to permit the compound slide rest to clear the chuck jaws. This means that the end of the work should be supported on the tailstock center.

The threading tool will have to start each cut from a recess. Follow the directions given in the drawings to make the recess.

The left-hand threading tool will be cutting on the right-hand side, so be sure there is sufficient clearance to clear the helix angle of the thread. Grind a special tool for left-hand threads, and identify it by grinding "LH" on the top of the tool shank.

The direction of the carriage travel will be toward the tailstock. There may be some play in the lead screw, so make one or two trial passes without the tool cutting.

I usually turn the work to the major diameter, then reduce the diameter by .002" to .005", depending upon the diameter of the thread so any burr produced by the threading tool will not interfere when testing the thread with a nut or the mating part.

I have bought several left-hand taps and dies from Kitts Industrial Tools, and have found them to be of good quality and very reasonable in price.

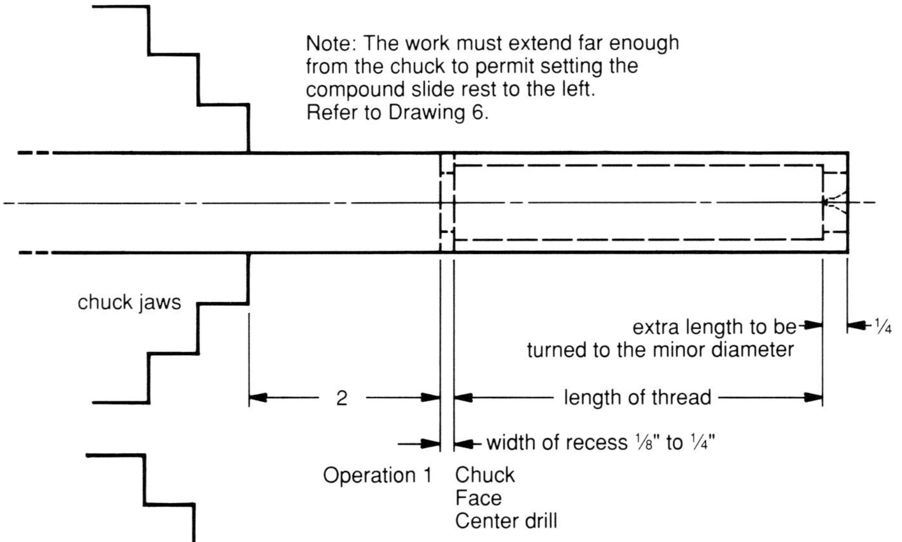

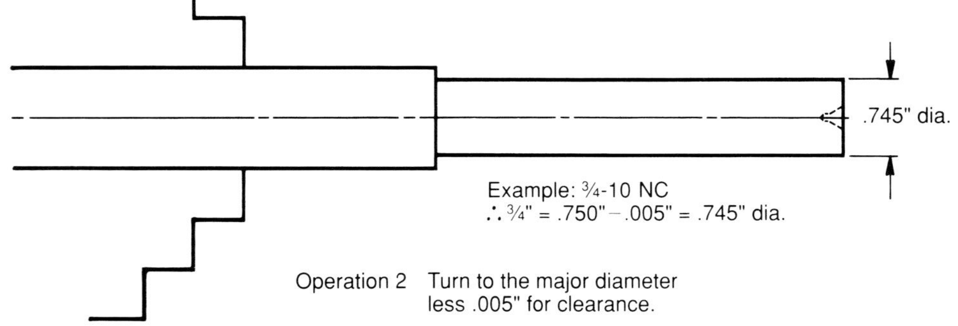

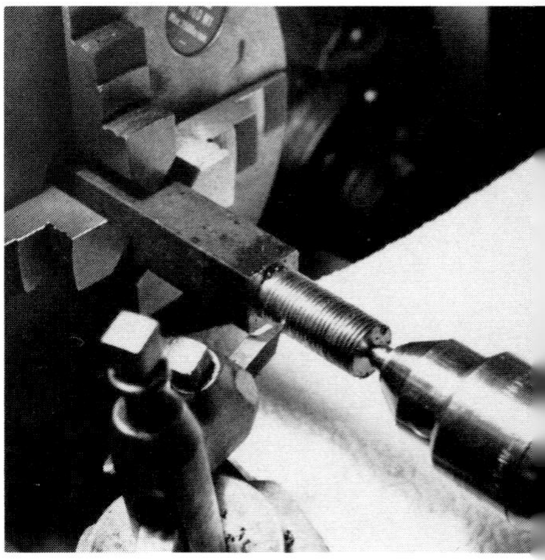

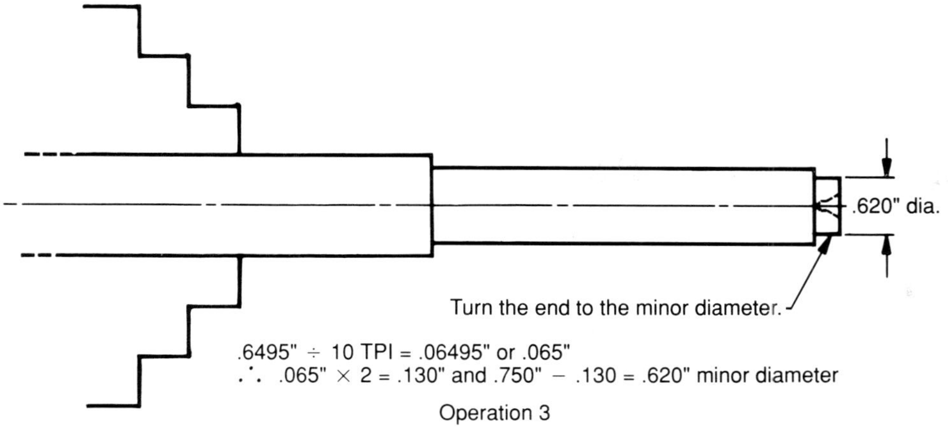

Operation 3

To find the single depth of a National thread, divide .6495 by the number of threads per inch.
To find the minor diameter, multiply the single depth by 2 and subtract the answer from the major diameter.

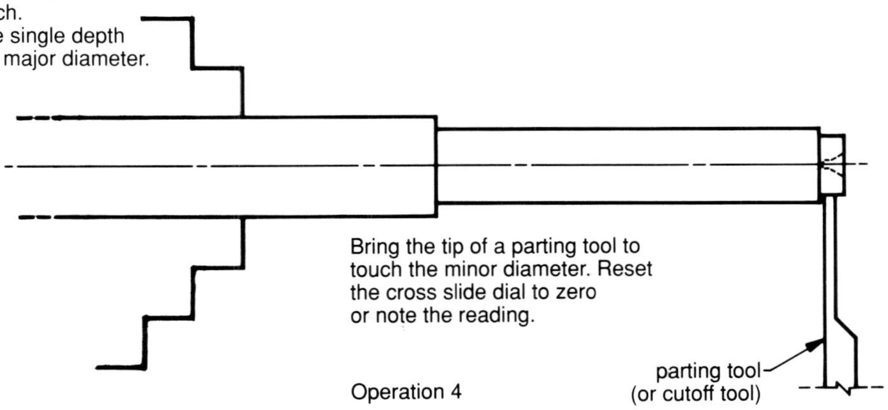

Operation 4

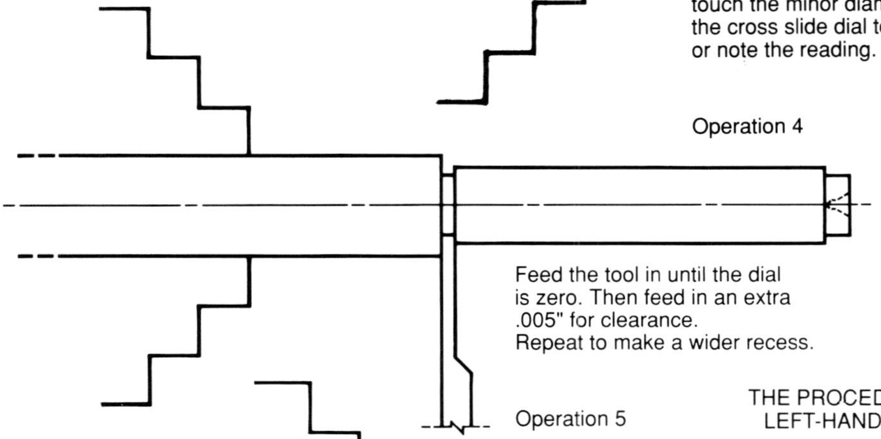

Operation 5

THE PROCEDURE TO FOLLOW TO CUT LEFT-HAND THREADS ON A LATHE

The work is now ready for threading.

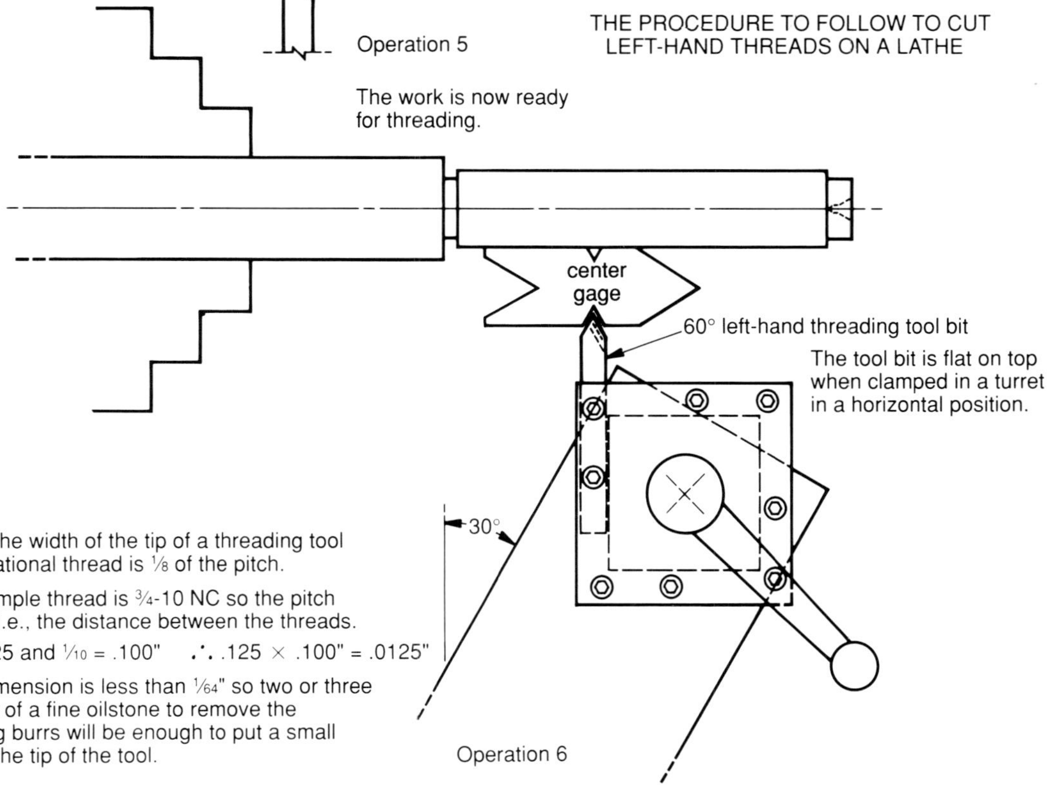

Note: The width of the tip of a threading tool for a National thread is ⅛ of the pitch.

The sample thread is ¾-10 NC so the pitch is 1⁄10", i.e., the distance between the threads.

⅛ = .125 and 1⁄10 = .100"   ∴ .125 × .100" = .0125"

This dimension is less than 1⁄64" so two or three strokes of a fine oilstone to remove the grinding burrs will be enough to put a small flat on the tip of the tool.

Operation 6

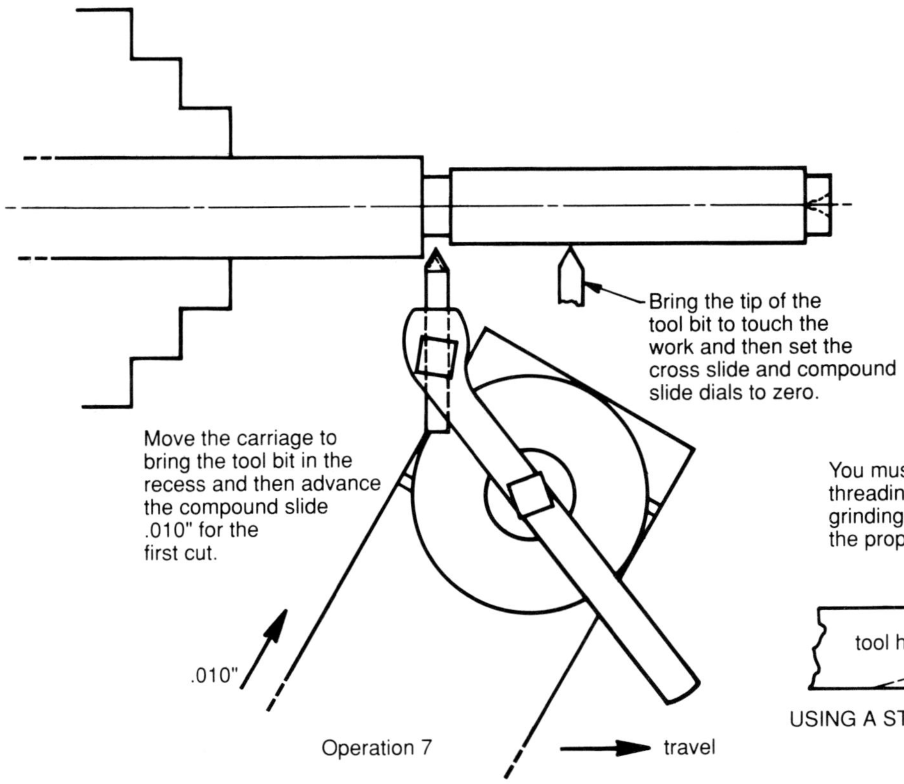

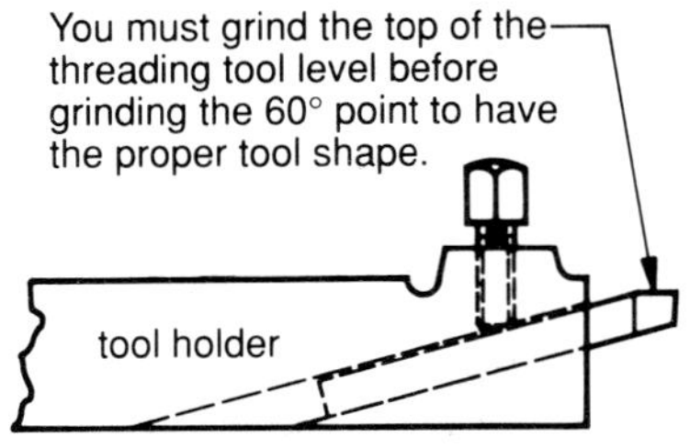

USING A STANDARD TOOL HOLDER

Protecting Your Mill from Shavings

The photographs show two covers I recently installed on my mill. The material is $1/16''$ red rubber gasket sheeting, but oil cloth or the rubber from a truck tire inner tube will do just as well – perhaps even better.

The bar screwed to the column casting in Photo 1 is $3/16 \times 3/4''$ by 12″ long. One the same size is fastened to the back of the saddle. 10-32 screws are used to fasten the bars to the machine, and 6-32 screws are used to hold the rubber to the bars in between the 10-32 screws.

The piece of rubber in front of the saddle in Photos 2 and 3 is held in place by the table stop and the nuts on the gib locating screws. I have to lift up the rubber to see the cross feed dial; however, this is a small inconvenience when I look under the covers and see nice clean ways covered with oil instead of steel shavings, or worse still, brass or cast iron ready to work under the saddle.

1

3

2

Machining Thin Plates on a Vertical Mill

Machining thin plates over their entire surface can be a problem for the work is usually too large to be held in the machine vise and the work will bend when the vise is tightened. Photo 1 shows a ¼″ thick steel plate on the 5 × 10″ magnetic chuck I sometimes use on my vertical milling machine. If light cuts of .005″ or less are taken with a fly cutter, the chuck will hold the work securely and there is no difficulty in making the two sides flat and parallel.

The other photographs show an aluminum plate measuring ⅜ × 4 × 16″ being machined flat with a fly cutter. The aluminum plate is held to the milling machine table by carpet tape. This is a tape with adhesive on both sides of the backing strip.

This method is used for non-ferrous materials (metals and plastics) or for ferrous plates that are slightly "cupped" or "twisted." The force of the magnetic chuck would draw them down to the surface of the chuck, and then, when released from the chuck, they would spring back again.

To use the tape method, care must be taken to remove all shavings and any traces of oil from the mill table or the tape will not stick properly. Place the tape on the table carefully to avoid any wrinkles and then peel off the separating strip as shown in Photo 3. If the plate is not located correctly it may be difficult to shift.

Next, tap the work lightly all over with a rawhide, plastic or rubber hammer to drive out the air and to seat the work down on the tape. If you feel that the tape might not hold the work securely enough to prevent the work from shifting during the machining operation, then clamp thin pieces of metal to the machine table with bolts in the tee slots. These pieces can be placed at the ends and sides of the plate.

The side of the work facing the tape (or magnetic chuck) must be reasonably flat or

1 ***A ¼″ steel plate is mounted on the magnetic chuck.***

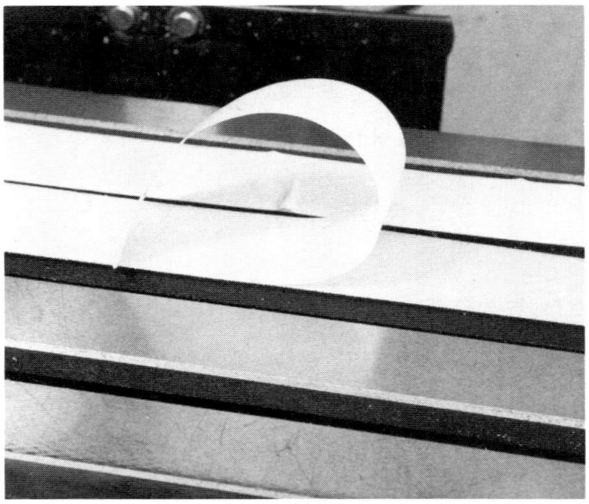

2 ***The protective strip is removed from the carpet tape.***

3 ***A fly cutter begins the cut on an aluminum plate.***

slightly concave but never convex or twisted. A little careful filing will soon produce two ¾" wide strips at the edges of the work and these will give an adequate holding surface for the tape or magnetic chuck.

To test the plate for flatness prior to machining, place the plate on a good surface plate and test all around the edges with feeler gages.

Small plates can be held in the milling machine vise by using a pair of Starrett hold-downs (see page 241 in the current Starrett catalog – the number of the hold-downs is No. 54). Figure 6 illustrates their use.

4 ***The depth of cut is .005" with a slow feed. The aluminum has a coating of soluble cutting oil to prevent its sticking to the tool bit.***

5 ***A fine feed, a light cut (.002"), a little cutting oil, and the result is a lovely finish.***

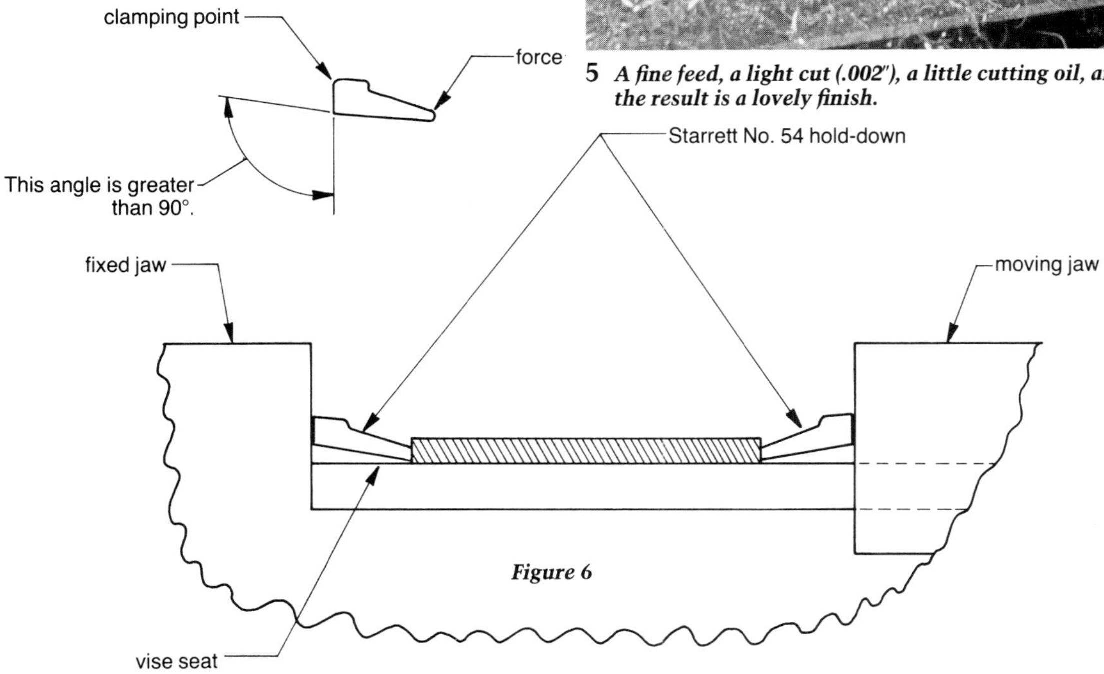

Figure 6

Finishing Plate Ends

EASILY AND ACCURATELY

A frequent task in the workshop is squaring the end of a piece of plate – for instance, the ends of locomotive frames or stretchers – prior to laying out holes or other work on it. Here are some methods which are suitable for various equipment found in the home workshops.

VERTICAL MILLING MACHINE

For those who own a vertical milling machine, it is only a matter of two or three minutes to finish the end of a plate square and smooth. Keep the vise accurately aligned as shown in the sketch. If you keep a 3/4" or 1/2" end mill in the spindle, the machine is always ready for use.

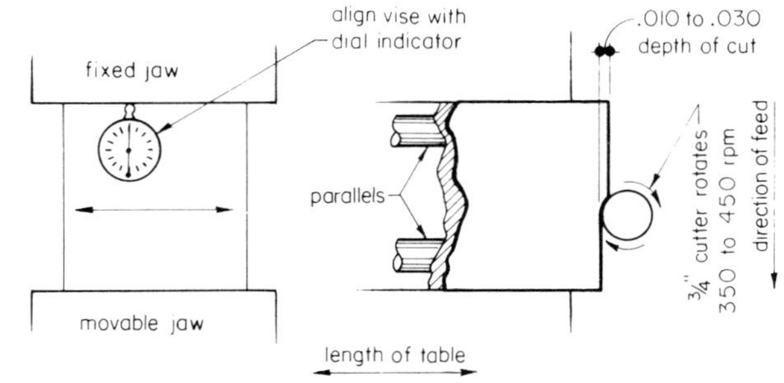

VERTICAL MILL

HORIZONTAL MILLING MACHINE

Align the vise with a square set against the column or with a dial indicator, as shown in the sketch. Set the plate on parallels and machine the work with a shell end mill or a plain end mill set in a collet. The work also can be machined with a side-and-face cutter set on an arbor in the usual way.

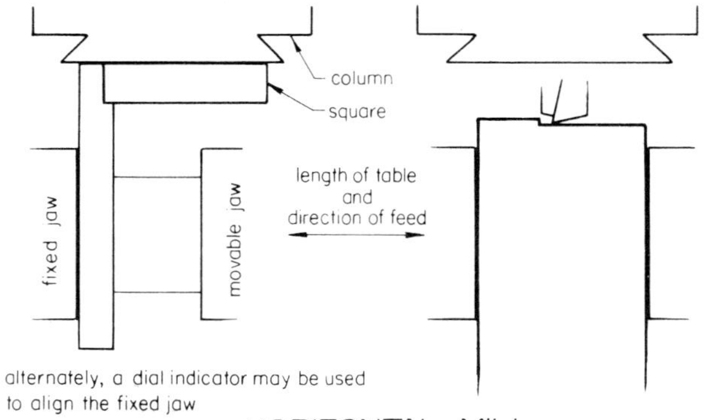

HORIZONTAL MILL

SHAPER

Set the vise in the same manner as for a vertical milling machine. Set the tool slide vertical by the graduations or with a square from the table. Keep the blade of the square against the side of the tool slide. Offset the clapper *away* from the cut and machine the end of the plate by feeding down with the tool slide. (Incidentally, feed on the return stroke.)

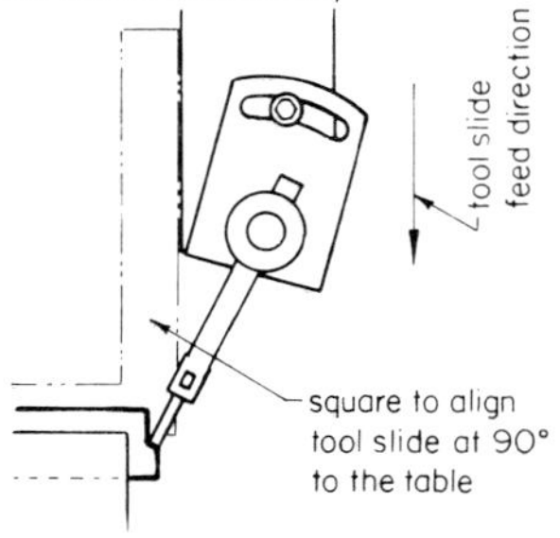

SHAPER

LATHE

Set the vertical slide on the lathe parallel with the lathe axis. Align the vertical slide with a bar in the 3-jaw chuck. Slide the tailstock all the way to the end of the bed or remove it, if necessary.

Hold the plate in the vise or clamped to an angle plate. Hold a 3/4" end mill in the 3-jaw chuck. Set the lathe for about 350 RPM and feed the plate against the cutter, as shown in the sketch. Take very light cuts, about .015" or 1/64".

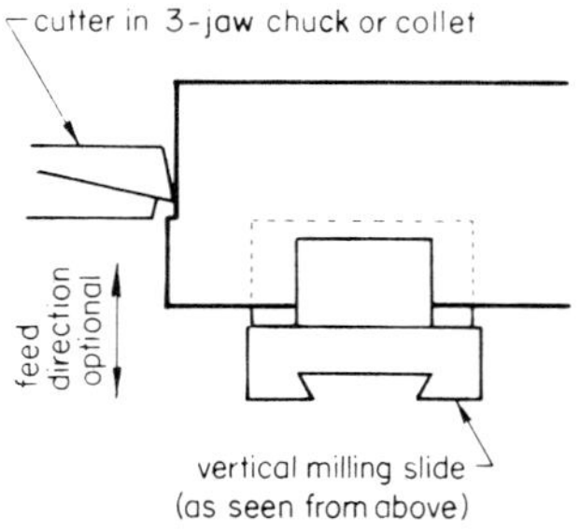

LATHE

FILING

To use the Armstrong milling machine, set the plate vertically in the bench vise. File in alternate directions with a double-cut file. Check frequently with a square. After rough filing with the double-cut file, finish by draw-filing with a single-cut (mill) file, as shown. It should not take more than ten minutes to finish each end of a plate fairly square and smooth.

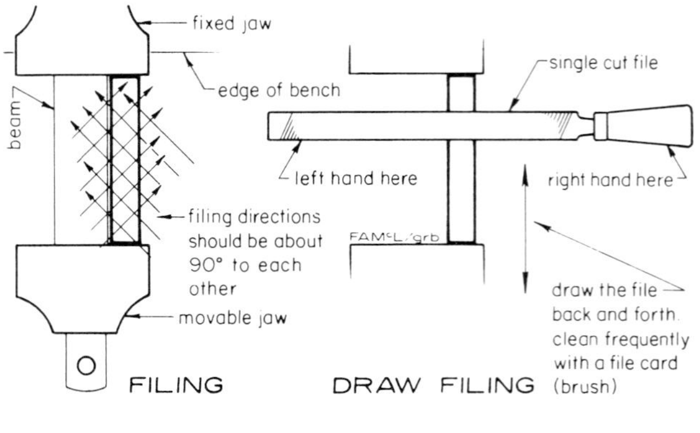

REMOVING AND INSTALLING

Bronze Bushings and Oilite Bearings

There are two basic methods of removing and installing bronze bushings. The first method employs a special drift mode to fit the bushing. This drift is used to force the old bushing out and to install the new bushing. Whenever possible, use an arbor press. If the work cannot be brought to the arbor press, then a heavy hammer may be used if the work is rigid enough to withstand the hammering.

If the bushing is seated in a blind hole, turn up the end of a 6" length of bar stock to a snug fit in the old bushing and at least twice as long as the length of the bushing. Put some grease or heavy oil in the bushing and then insert the drift. Strike the drift a smart blow with a heavy hammer and the bushing should pop out of the hole.

If this fails to remove the bushing, then you will have to drill three or four small holes down through the walls of the bushing and then break out the sections with a chisel.

The second method is the one to use to replace bushings in valve gears and other places where you don't want to dismantle the mechanism.

You will need a long Allen cap screw or a screw made to suit.

To remove the old bushing, place the screw in the bushing and assemble the pipe section, washer and nut. Tighten the nut and the bushing will be easily removed without straining the mechanism at all. To install a new bushing, place the screw through the bushing and the work. Put the washer and nut in place. Tighten the nut to draw the bushing into the work. Just make sure that the bushing is square with the work when starting.

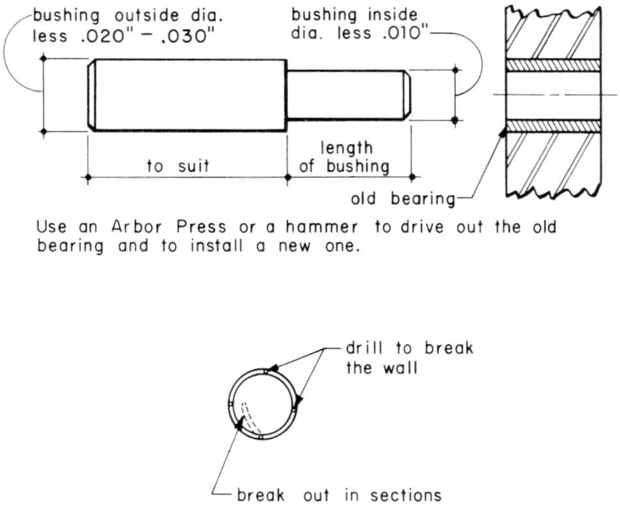

REMOVING AND REPLACING
BRONZE BEARINGS

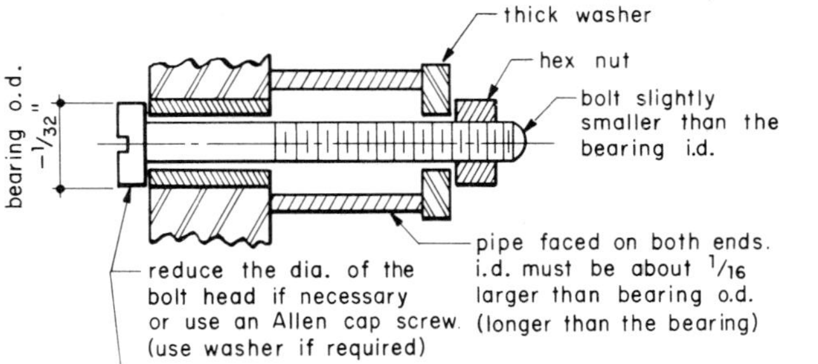

REMOVING BEARING

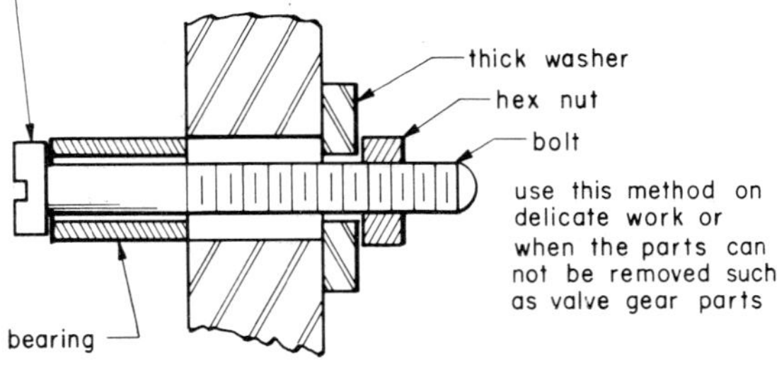

INSTALLING A NEW BEARING

How to Make a Layout with a Vernier Height Gage

Some people feel that an accurate layout is not necessary when the work is being located on the vertical milling machine or on the vertical milling attachment by the graduated dials. However, a major error, such as one too many turns of the leadscrew handwheel, can be avoided with a proper layout.

Also, the work can be repositioned on the machine later, if necessary.

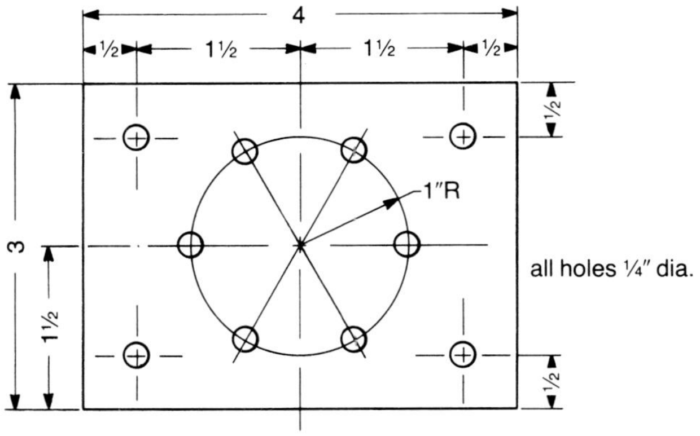

6¼″ holes equally spaced on a 2″ dia. circle

CONVENTIONAL DIMENSIONING

If the upper left-hand corner of the work is used as the reference or datum point, both the lengthwise (or X axis) and crosswise (Y axis) leadscrews will turn to the right (or positive direction).

This means that the dimensions will be added as the holes are "jig-bored," and any errors from backlash in the leadscrews will be avoided. This method is called "co-ordinate dimensioning or layout," and is similar to making a graph.

Clamp the work to an angle plate, if possible. Bring the scriber to just touch the top of the work and note the reading of the vernier. Now subtract each dimension from this reading to find the settings and scribe all the layout lines parallel with the top of the work.

Now turn the angle plate on its end and bring the scriber to just touch the end of the work and note the reading. Proceed as before to subtract each dimension and scribe the layout lines.

Convert the fractions to decimals and write them on the center lines of the holes.

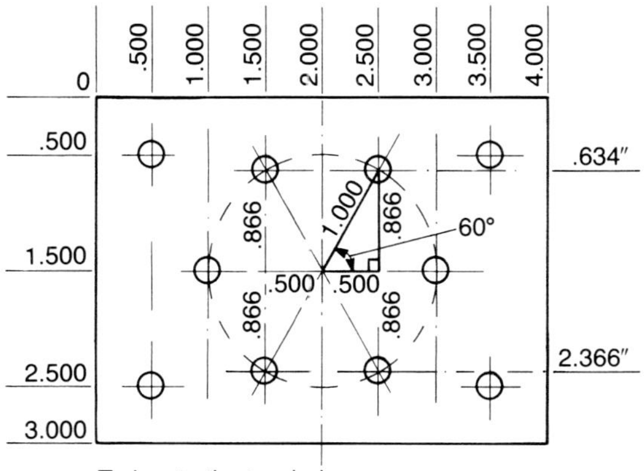

To locate the two holes on the 1.500 center line:

$$\begin{array}{r} 2.000'' \\ +1.000'' \\ \hline 3.000'' \end{array} \quad \text{and} \quad \begin{array}{r} 2.000'' \\ -1.000'' \\ \hline 1.000'' \end{array}$$

To locate 6 holes on a 2″ dia. circle we shall use the sine and cosine of a 60° right angle triangle.

$$\text{sine } 60° = \frac{\text{side opposite}}{\text{hypotenuse}}$$

$$.86603 = \frac{\text{side opposite}}{1.000}$$

$$\therefore \text{side opposite} = .86603 \times 1.000 = .866''$$

$$\therefore \begin{array}{r} 1.500'' \\ +.866'' \\ \hline 2.366'' \end{array} \qquad \begin{array}{r} 1.500'' \\ -.866'' \\ \hline .634'' \end{array}$$

$$\text{cosine } 60° = \frac{\text{side adjacent}}{\text{hypotenuse}}$$

$$.500 = \frac{\text{side adjacent}}{1.000}$$

$$\text{side adjacent} = .500 \times 1.000'' = .5000''$$

$$\therefore \begin{array}{r} 2.000'' \\ +.500'' \\ \hline 2.500'' \end{array} \qquad \begin{array}{r} 2.000'' \\ -.500'' \\ \hline 1.500'' \end{array}$$

A small vernier height gage will be very useful in the home workshop. The article "An Easy-to-make Vernier Height Gage" describes how to convert a vernier caliper into a height gage without altering the caliper. The vernier caliper can be used for measuring as usual, then the base and scriber can be clamped in place and the caliper becomes a vernier height gage. For occasional use, this will be an excellent substitute for an expensive vernier height gage.

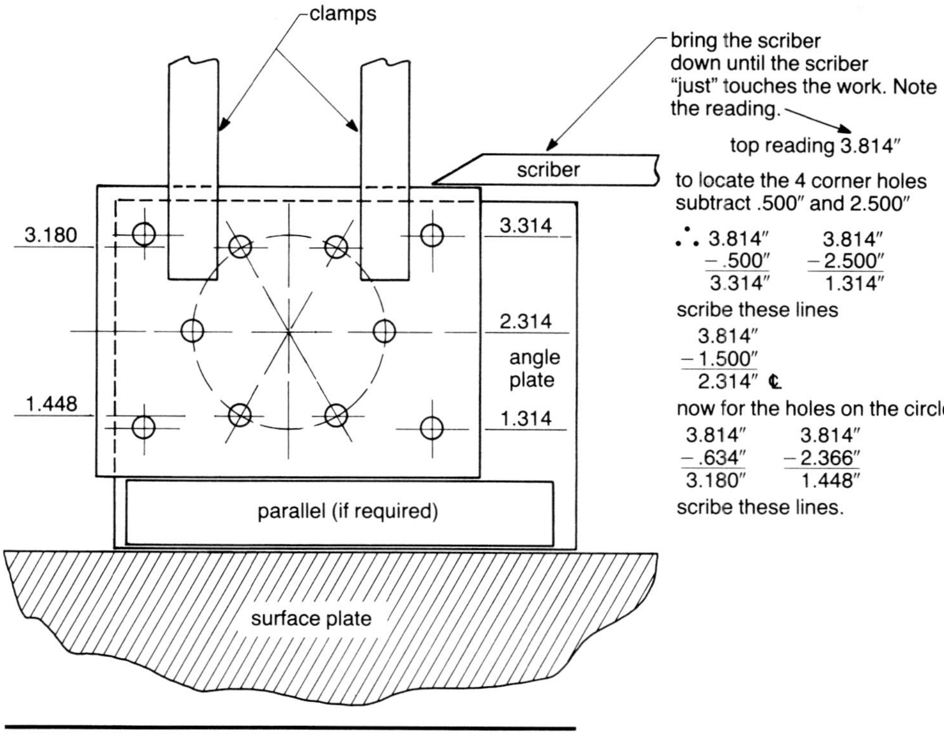

top reading 3.814″

to locate the 4 corner holes subtract .500″ and 2.500″

∴ 3.814″ − .500″ = 3.314″ 3.814″ − 2.500″ = 1.314″

scribe these lines

3.814″ − 1.500″ = 2.314″ ℄

now for the holes on the circle.

3.814″ − .634″ = 3.180″ 3.814″ − 2.366″ = 1.448″

scribe these lines.

to locate the corner holes:

4.987″ − .500″ = 4.487″ and 4.987″ − 3.500″ = 1.487″

centerline = 4.987″ − 2.000″ = 2.987″ and 4.987″ − 1.000″ = 3.987″ and 4.987″ − 3.000″ = 1.987″

now the holes at 60°:

2.987″ + .500″ = 3.487″ 2.987″ − .500″ = 2.487″

Turn the angle plate on end and note the top reading.

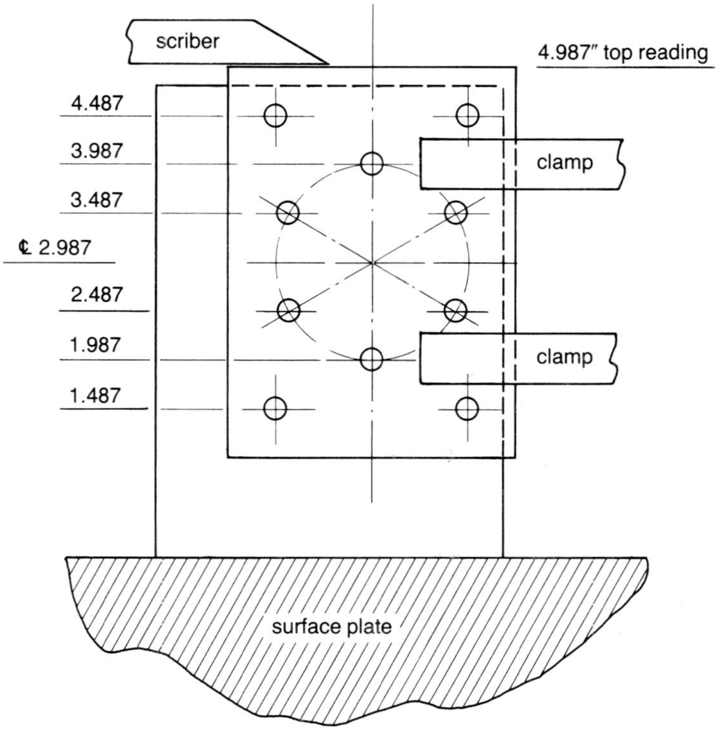

Sharpening Two-lip End Mills

Here is a simple method of sharpening two-lip end mills by hand using an ordinary bench grinder, a little care and common sense.

First, dress the grinding wheel so the wheel is straight and clean. Use the right-hand side of the wheel and raise the front of the cutter 7°-10° from the horizontal position (Figure 1). Just touch the wheel lightly and then turn the cutter over to grind the other cutting edge.

To check if both of the cutting edges are the same length, use a vee block on a surface plate. Place the shank of the cutter in the vee block with the cutting edges in a vertical position and test with a square (Figure 2). If one of the cutting edges is too long, grind and check again.

To check the relief angle of the land, turn the cutter in the vee block so the cutting edges are horizontal and use the sheet metal gage (Figures 3 and 4). To make this gage, no great accuracy is required. Simply mark out a piece of tin, cut with tin snips and bend the tab so the gage will stand by itself on the surface plate.

The secondary clearance is provided to give more chip room so the angles are not precise. Try to copy a new cutter when grinding these clearance angles.

This is a quick and easy method of sharpening cutters, so sharpen your cutters before they become very dull. I use this method for sharpening cutters in my school machine shop because I don't have time to sharpen the cutters

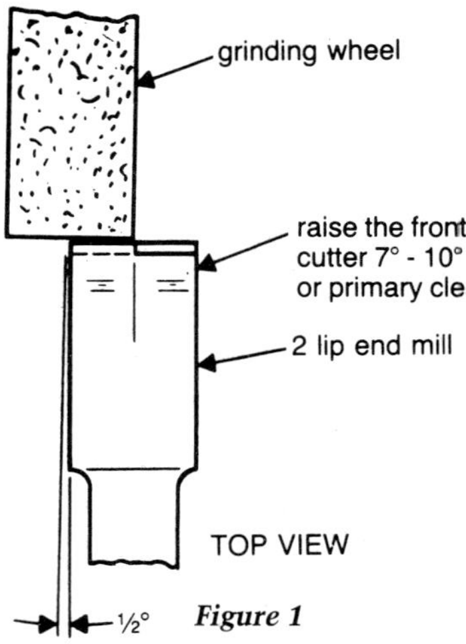

Figure 1

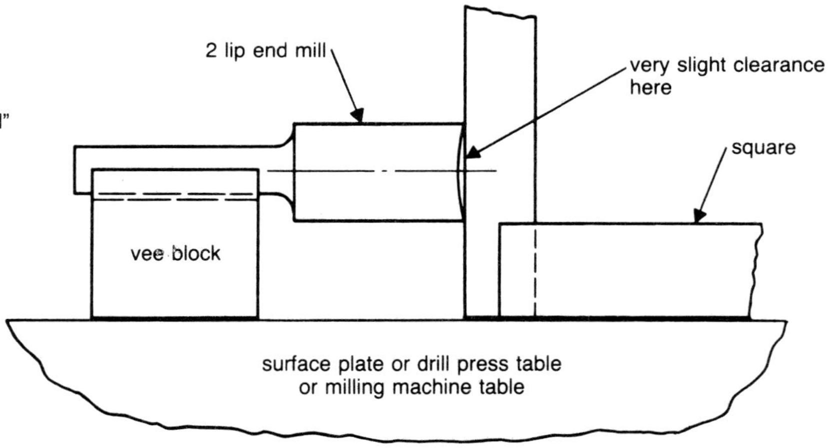

TEST WITH A SQUARE TO BE SURE BOTH CUTTING EDGES (LIPS) ARE THE SAME LENGTH

Figure 2

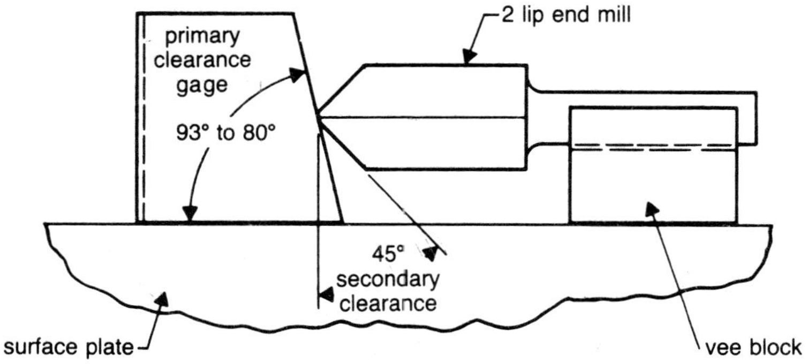

Figure 3 Checking the primary or "land" clearance the width of the land should be about 1/16°.

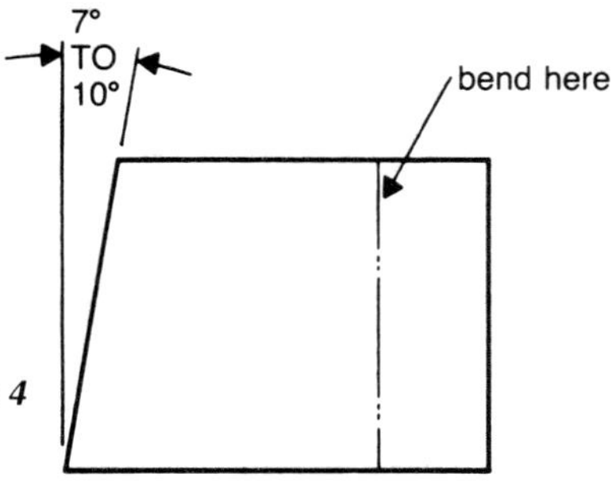

Figure 4

CUTTER CLEARANCE GAUGE

on a cutter grinder in the usual manner.

If the sides of the cutters become dull or chipped, they should be ground on a tool and cutter grinder. This will reduce the diameter of the cutter but continue its life. Sharpening by hand should only be attempted as a last resort. Start by coating the edges of the cutter with layout blue.

Now pass the cutter across the face of the grinding wheel and rotate the cutter at the same time. Try to maintain the same relief angle. Leave a blue line .002" to .005" to be finished by an oil stone. Stone inside the flutes as well.

To check the side or radial relief without a cutter clearance gage, place the cutter in a hole the same diameter as the cutter and visually check the relief.

A method commonly used in tool rooms employs a dial indicator to test the relief angle. Place the cutter in the vee block and hold the dial indicator in your universal surface gage or use a commercial stand or holder.

The operator on a cutter grinder checks the angle while the cutter is on the machine (Figures 5, 6, and 7).

The usual width of the primary land is 1/16" so at this distance each .001" is the same as 1° of relief angle. Example: 5° is .005" and 7° is .007". This is a quick and easy test, and does not require an expensive cutter clearance gage.

Figure 5

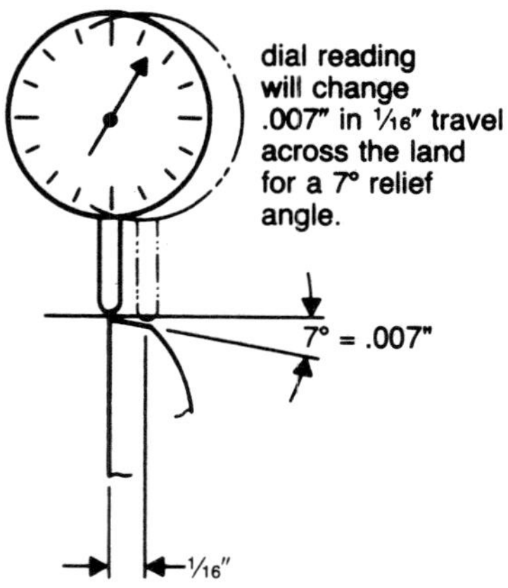

Figure 6

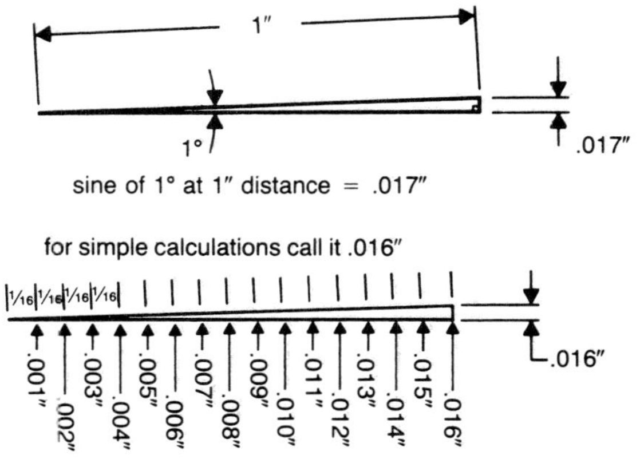

Figure 7

Sharpening Taps and Dies

Taps usually break because they are not correctly aligned in the hole to be tapped or are becoming dull. The misalignment can be corrected by good tapping techniques. The dull taps *will have to be sharpened.*

Taps are sharpened in factories on a tool and cutter grinder with tap grinding attachment. This attachment gives a radial relief to the tapered end of the tap. In a small shop taps can be sharpened with a round oilstone or a small hand grinder, such as a Dremel or a Foredom flexible shaft grinder. To sharpen a tap with an oilstone, you will need a round oilstone the same diameter as the flutes or slightly smaller. Hold the tap in a bench vise by the shank in a horizontal position, with one of the flutes on top. Move the oilstone back and forth along the flute while pressing the oilstone against the front face of the cutting edge (see drawings). Continue until the cutting edge is sharp when viewed with a magnifying glass. Repeat the procedure for each flute.

This method of sharpening can be done more quickly with a hand grinder. Select a mounted point (grinding wheel on a 1/8" shank) the same diameter as the flute or slightly smaller. If the mounted point is too large in diameter, reduce the size with the dressing stone provided in the kit. Hold the tap shank in the bench vise with one flute on top, and move the mounted point back and forth along the flute, pressing against the front face of the flute. Be careful not to grind too much, because the cutting action is much more rapid than using an oilstone.

To sharpen dies with an oilstone or a hand grinder, select an oilstone or a mounted point slightly smaller in diameter than the clearance holes in the die. Hold the die in a bench vise, and grind the front face as you did for sharpening a tap. You may find the dies easier to sharpen if the grinder is held in the bench vise on a stand and the die is held in your fingers or in a die stock. This makes it easier to see the grinding operation.

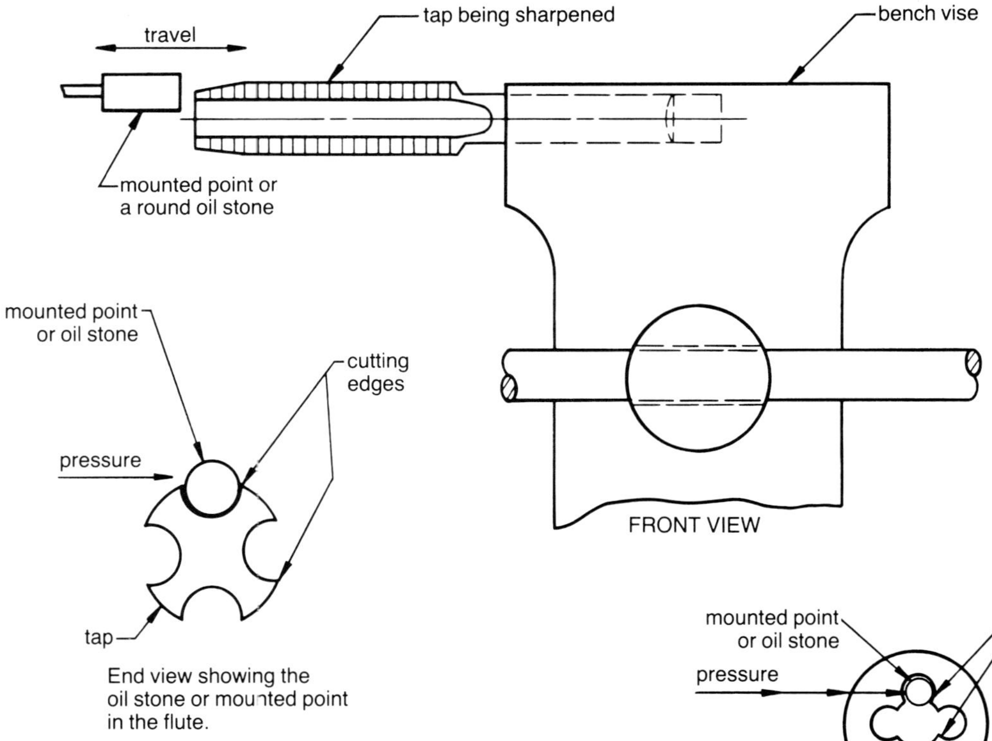

SHARPENING A TAP

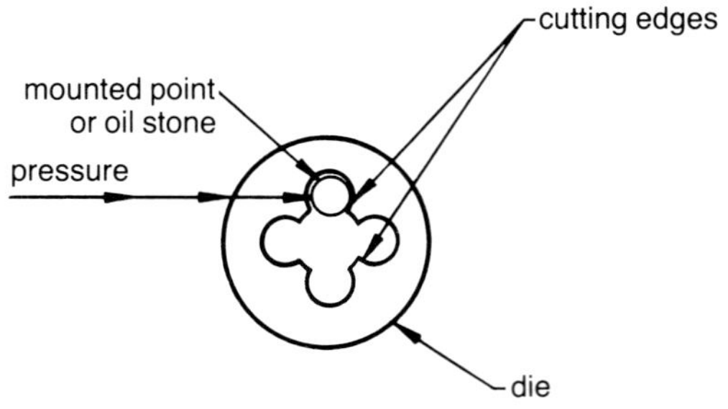

SHARPENING A DIAMETER

How to Install and Remove Ball Bearings

Ball bearings should be a snug or a very light press fit in the bearing housing and on the shaft. The manufacturers' catalogues will give the exact sizes, but an amateur will have difficulty working to these tolerances with the usual home shop equipment. Therefore, you should try for a snug fit. If the bearing is a bit loose on the shaft or in the housing, use a few drops of *Loctite* to hold the bearing in place. Be careful not to get any *Loctite* inside the bearing.

Mechanics often install bearings by tapping around the outer raceway with a ball peen hammer. A much better method is to make up a special tool from a short length of mild steel bar. This should be made to suit the bearing so the inner and outer raceways are being pushed at the same time. This will prevent the inner raceway from being displaced to one side, thereby damaging the bearing or reducing its life.

If the housing is not strong enough to resist the force of pressing the bearing, or if the work cannot be placed on an arbor press, then draw the bearing into place with a bolt and special washers. The same basic method can be used to remove bearings.

Special pullers are available to remove ball bearings but the cost is not warranted for occasional use.

A simple method is to drill three or four holes before assembly so a pin punch can be used to remove the worn bearings. The holes should be drilled so the punch will touch the outer raceway.

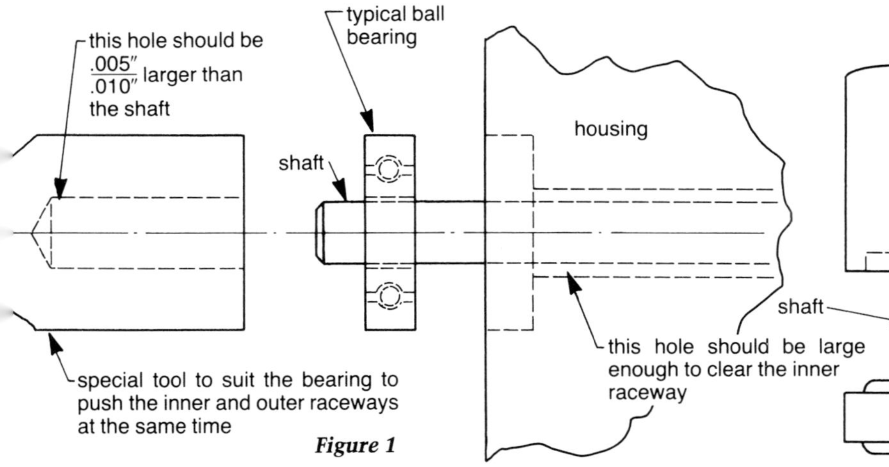

Figure 1

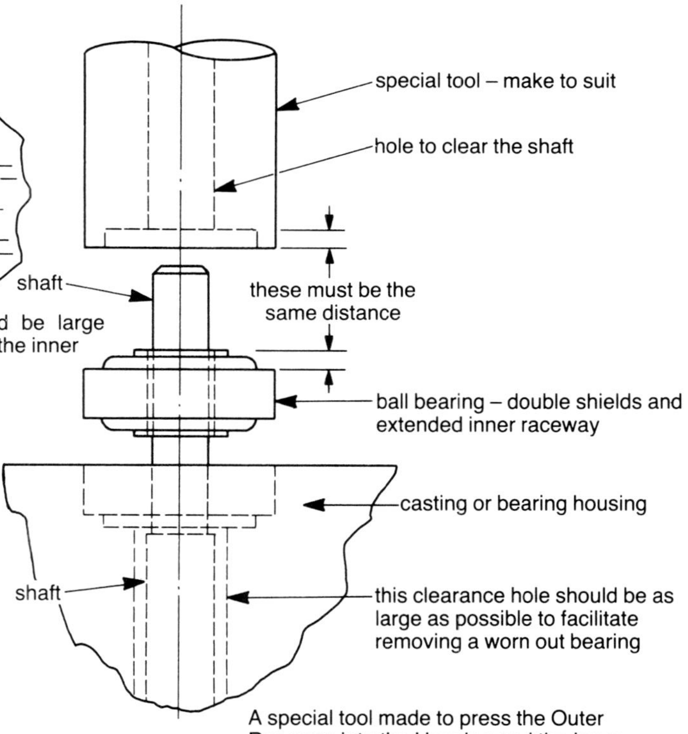

Figure 2

A special tool made to press the Outer Raceway into the Housing and the Inner Raceway on to the Shaft at the same time.

Improving a Milling Machine Vise

The proper type of vise to use on a milling machine is a Bridgeport or Kurt type with jaws 4" to 6" wide, depending upon the size of the milling machine. For a small mill/drill, the 4" size is best.

A Bridgeport type of milling machine will require a vise with 6" wide jaws. Unfortunately, these vises weigh about 100 lbs., so they will require some lifting arrangement to remove them from the table. I find that they are too heavy for me now, and I require some help.

Some readers may have to wait to buy a proper milling machine vise, having "licked the spoon" too much to buy their mill, so they will have to make do with the tilting drill press vise supplied with the machine.

Normally, the direction of the feed when milling will be toward the fixed jaw of the vise; however, there is a great danger of the vise tilting during the cut, so the work is ruined and the cutter is broken. This happened to me, so I altered the vise by placing two 10-32 NF Allen cap screws through the top of the vise, and threaded them into the bottom or base (Photo 3). The easiest way would have been to use flathead screws coming up from underneath. However, I chose to have the screws enter from the top so they could be easily removed for tilting the vise.

1 ***The baseplate is clamped to an angle plate resting on a surface plate.***

My vise had bolting lugs only on the ends, so the vise could be bolted to the mill table only by the center T-slot. I prefer to have the vise bolted across the table so the jaws are parallel

2 ***The vise is aligned on the baseplate with a precision square. Clamp the vise to the baseplate with bolts passing through the mounting lugs and threaded into the baseplate.***

3 ***Two 10-32 Allen cap screws are used to hold the vise from lifting during milling operations.***

4 *The vise is located with a precision square.*

5 *A small set square is used to locate the baseplate parallel to the front edge of the table.*

6 *Use a protractor to slew the vise to an angle. Clamp the baseplate to the table with straps.*

7 *The vise is raised for milling or drilling operations.*

to the T-slots (X-axis).

I mounted my vise on a baseplate 1/2 × 6 × 10". I used hot rolled mild steel, so I had to machine the edges, and surface grind the flats. A piece of cold rolled steel would be ideal.

Make sure the edges are true and free from burrs, since these edges will be used to locate the vise on the baseplate, and to align the baseplate on the milling machine table with a precision square. I will use a dial indicator to align the vise jaws only if the work is very precise.

The plate is drilled and tapped for 3/8-16 bolts along the center line to match the mounting lugs on the vise. The two holes on the sides are the same distance apart as the two outer T-slots. They are drilled 7/16" to clear 3/8" bolts. The vise can now be bolted to the table with the jaws parallel or across the table. The vise can also be set at an angle with a protractor (Photo 6).

Keys can be mounted under the baseplate to register with the T-slots. I find that aligning the baseplate with a square is so quick and easy, I haven't bothered with keys. If you want keys, then two shallow slots will have to be machined on the underside of the baseplate. One lengthwise and one crosswise will locate the keys accurately.

Broaching Square Holes

This method of making square holes will enable you to make your own boring bars and shaper extension tool holders for cutting keyways.

Use a tool bit the same size as the tool bits you intend to use in the boring bar to make the broach. This is done by grinding the ends of the tool bit square with the sides. I suggest you use the same size tool bits you use on your lathe. When the tool bits are too short to be used on the lathe, they can still be used in a boring bar. (I cannot deny my Scottish heritage.)

Locate and drill the hole in the bar by one of the methods previously described. Broaching and filing will be easier if the hole is a bit larger than the tool bit. The tool bit will still be held accurately in the bar by the corners. For example, drill a 9/32" hole for a 1/4" square tool bit.

Start the broaching operation by locating the broach over the hole and forcing the broach into the bar sufficiently to show the outline of the broach in the bar. This can be done with an arbor press or a large brass hammer.

Place the bar in the bench vise in a vertical position with the hole just above the vise jaws. Sit on a low stool so that your eyes will be near the same height as the hole. This will enable you to see more easily when you are filing to the outline.

Old-time tool makers used to hold the work in the vise in a horizontal position so the hole would be vertical with the outline showing on top. The file would be held under the work so the filing strokes were up and down. This requires some practice, but it is a good method because the file is coming up from the unbroached side of the hole on the cutting stroke. This means that there is less chance of the file enlarging the broached part of the hole.

The broach should penetrate the work about 1/32" each time. Do not try to force the broach more than this at one time, or the burr will fill the hole and make filing difficult. Cutting oil will help during the broaching operation, but be sure to wash off the oil after each broaching operation with Varsol or a similar cleaning solution.

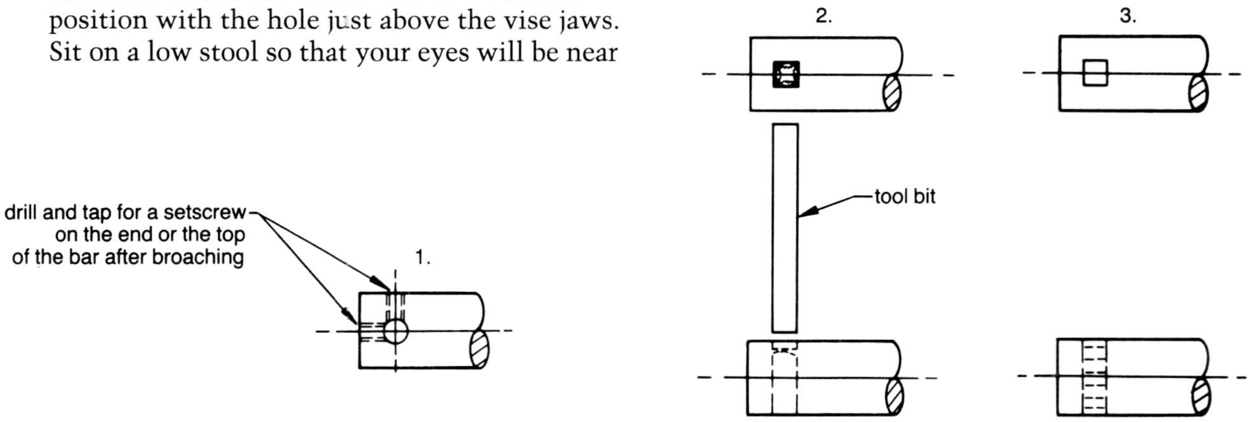

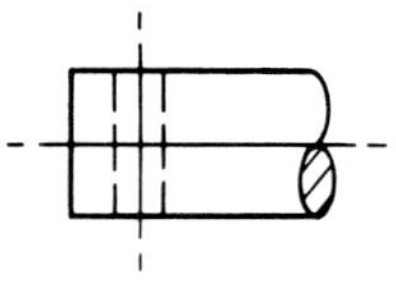

1. Drill a hole the same size as the square or slightly larger (1/32").
2. Grind the ends of a tool bit square. Place the tool bit over the hole and strike the tool bit with a brass hammer to drive the tool bit into the bar. Use a small square file to remove the excess material in the hole.
3. Force the broach into the bar a short distance. Punch the broach out with a small brass rod and then file the burrs in the hole. Repeat several times until the broach passes through the bar. Use cutting oil to make the broacing operation easier, but wash off the cutting oil with a solvent before each filing operation.

HOW TO BROACH A SQUARE HOLE

Drill Press Tools & Accessories

Homemade Drills

These drills are made as required when a regular twist drill of the correct size is not available. If the drill is to be used only once, then the purchase of a drill is not justified and a homemade drill will do.

Heat the rod of a suitable length of carbon tool steel or drill rod and flatten the end on an anvil. Permit the metal to cool slowly and then turn the end to the correct diameter. Turn the point to an included angle of 118°-120° or an angle of 59°-60° with the center line. Check to see how the graduations read on the compound slide rest set parallel with the lathe axis; see if the graduation reads 0° or 90°. Then set the compound to 60° or 30° as required.

Back off the sides and the cutting edges by filing. Harden and temper, then finish the cutting edges by grinding as for a twist drill.

When using these drills, remove the drill from the hole frequently to clear out the shavings.

Use a cutting speed of 20 to 30 feet per minute as the drill is carbon tool steel and not high speed steel.

These drills were once called flat drills and are often used for wood and brass. The earlier versions were called spear point drills and were used in bow drills. These drills are still used in India.

Another version of the flat drills are the spade drills used in industry for large diameter holes (up to 8"). They are held by a screw with tapered shank in a bar.

A variation of the spade drills may have several diameters with cutting edges at an angle to suit the work. There may also be several drills or cutters set along the bar to bore several diameters at the same time and, hopefully, all the diameters of the bores will

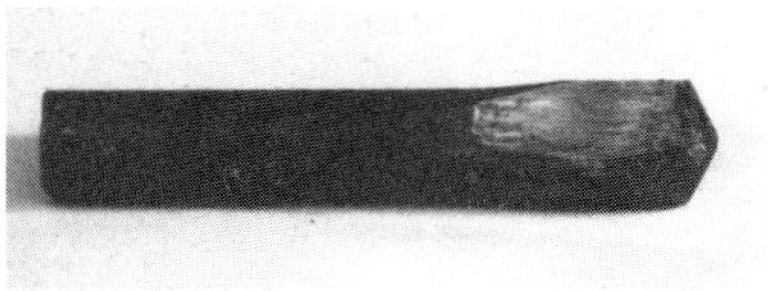

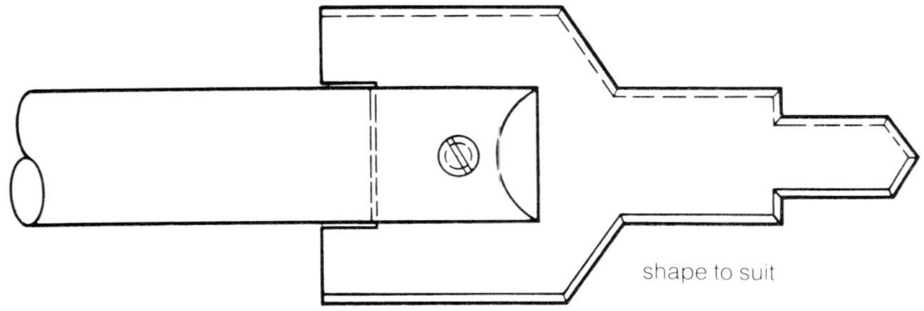

STEP DRILL

If a tool and cutter grinder is available a standard twist drill can be reground to a step drill.

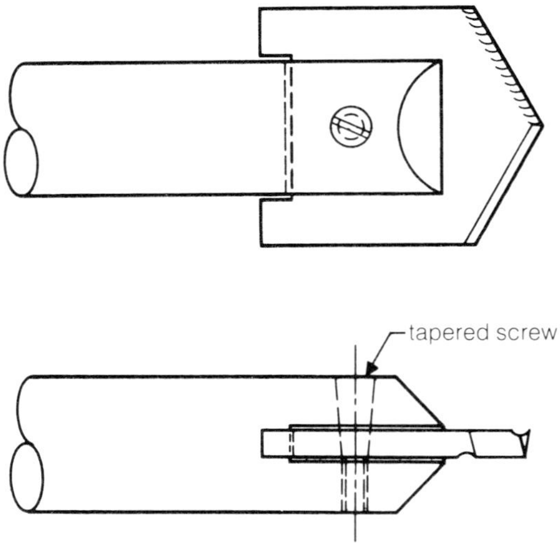

SPADE DRILL

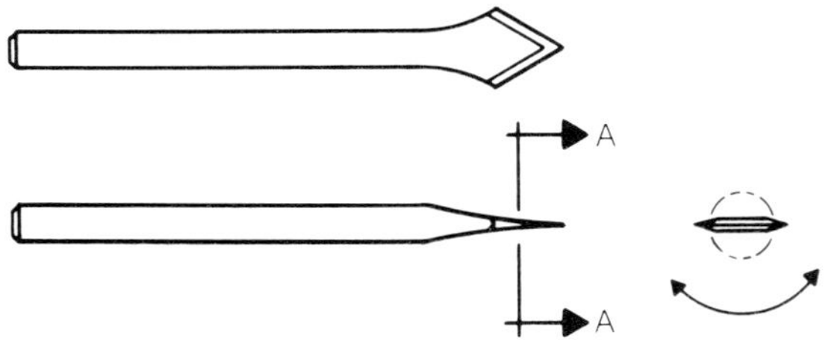

The drill rotates backwards and forwards and cuts by a scraping action.

Old fashioned spear point drill bit used in a bow drill or an archemedian type drill.

be concentric. These special drills or cutters are usually seen on large turret lathes and automatic lathes such as Warner & Swazey, Potter & Johnson, Herbert, Gisholt, etc. However, the idea may be useful in a small shop when the cost of making a special tool is mainly your own time.

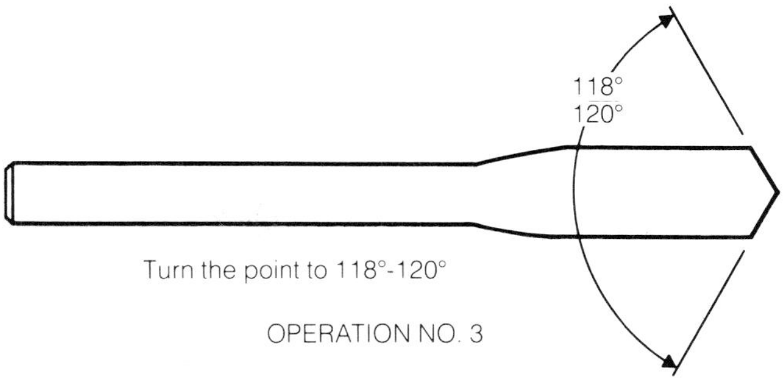

Turn the point to 118°-120°

OPERATION NO. 3

TOP VIEW

FRONT VIEW

Heat one end to bright red and then flatten the end with a heavy hammer on the anvil. Reheat and cool slowly.

Use carbon tool steel ½" diameter x 6" long or to suit. If ¾" or larger rod is used, turn down the shank to fit the drill chuck or turn a Morse taper.

OPERATION NO. 1

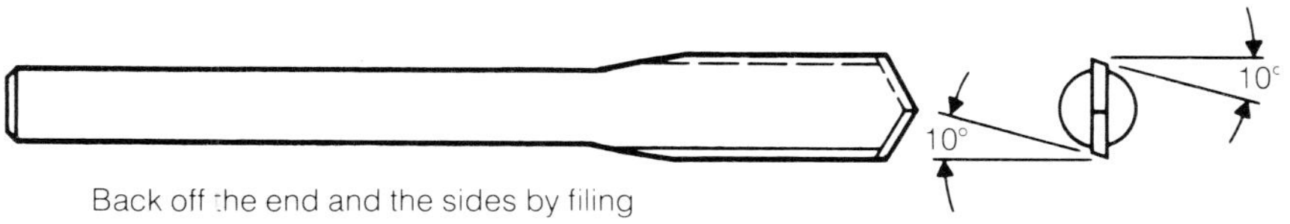

Back off the end and the sides by filing

OPERATION NO. 4

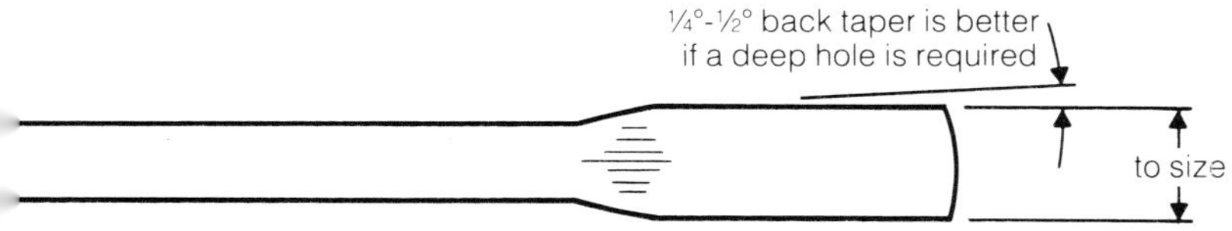

Turn to the correct diameter

OPERATION NO. 2

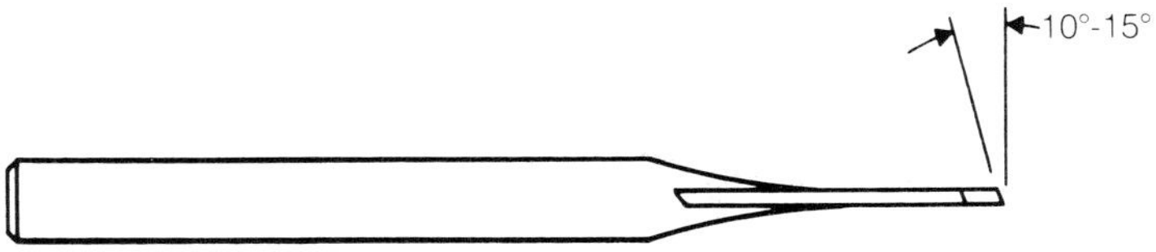

Harden and temper as usual.

OPERATION NO. 5

Reamers & D-Bits

MACHINE REAMERS

Machine reamers are used to finish a hole accurately to size: parallel, round, and with a smooth finish. They may be purchased in all the drill sizes: fraction, letter, wire and metric. They are also available in "over" and "under" sizes for "press" or "running fits" (Example: .249" dia., 1/4" dia., .251" dia.).

Reamers larger than 1/2" diameter usually have Morse taper shanks like the drills. The smaller sizes have straight (parallel) shanks to be held in a drill chuck and these are usually called "chucking reamers."

Most machine reamers have straight flues irregularly spaced to prevent "chattering." The flutes are divided or indexed by a method called "increment indexing" that gives a slightly irregular spacing but makes sure the flutes are diametrically opposite each other.

Some reamers have helical flutes to give a shearing cut and to ream holes with a keyway. The helical flutes will bridge the keyway. Unlike drills, the flutes on these reamers have a left-hand lead (viewed from the end). If a righthand lead was used, the helix would draw the reamer into the hole too rapidly and the reamer would probably break. These reamers cost more than those with straight flutes and are not as common.

TOOLMAKER'S REAMERS

These homemade reamers may be used when a proper machine reamer or a hand reamer is not available. These reamers work best on small holes for valve gears, oil pumps, etc., when the reamer is made from the same piece (length) of drill rod (silver steel) that will be used for the shaft. The reamer will cut slightly oversize to give a good running clearance.

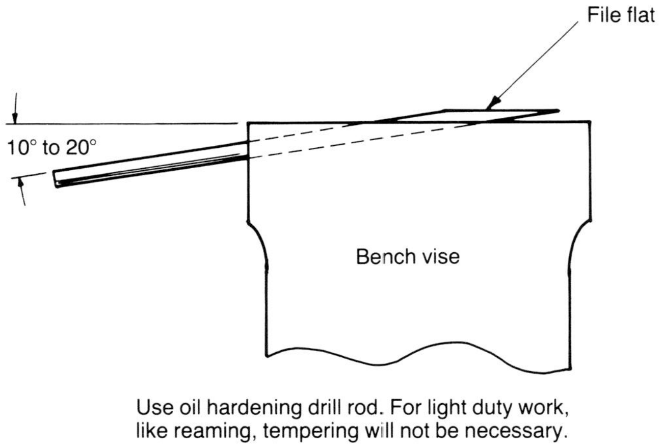

Figure 1

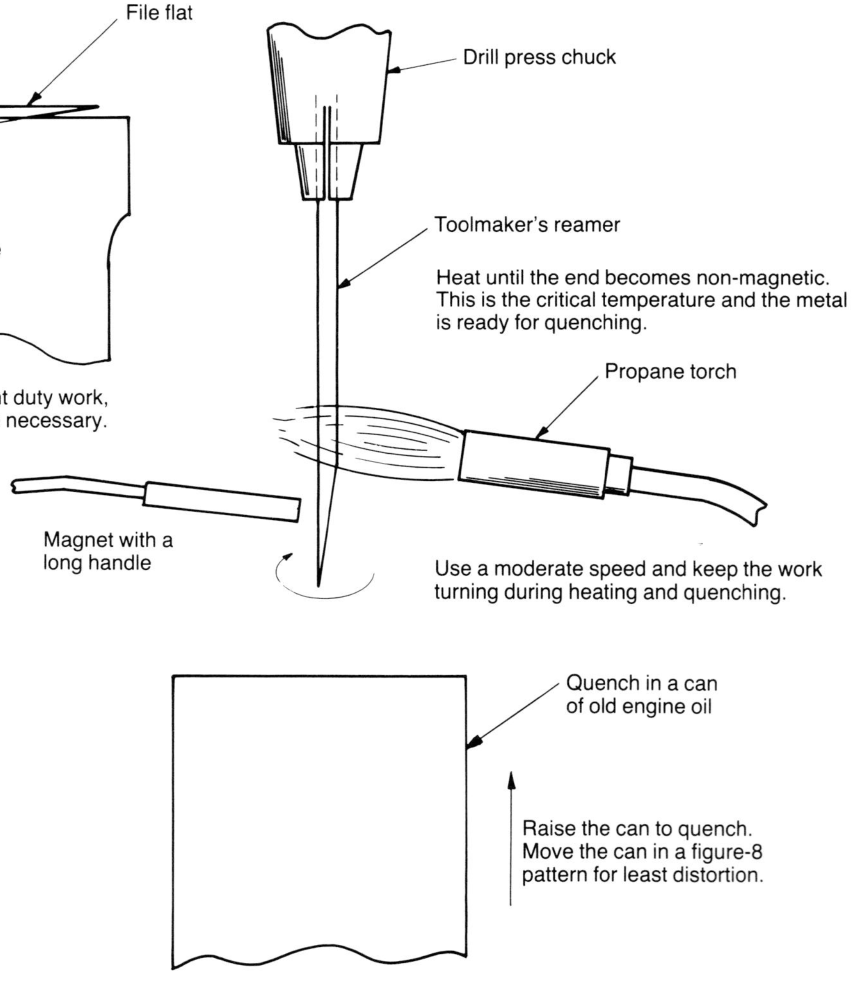

Figure 2

Use a piece of drill rod 4" to 6" long. File the flat to form the cutting edge as shown, harden and temper.

Use oil hardening drill rod instead of water hardening, if possible, for there will be less warping and the reamer will not have to be tempered.

To use these reamers, drill to the closest size and then ream. Use a speed about half that for drilling and use cutting oil.

D-BITS

The D-bit receives its name from its shape, as seen in an end view. It is a derivation of the old gun drills and is an end-cutting tool, rather than a side-cutting tool, such as an endmill.

While some amateur machinists use them as reamers, I think they are vastly overrated for this use, since after drilling, a hole must be started straight by boring before a D-bit can be used to finish the rest of the hole.

D-bits are most useful for finishing special-sized holes or the bottoms of blind holes. Years ago, I used them to finish the bottom of cavities in plastic molds where hemispherical or curved ends were required.

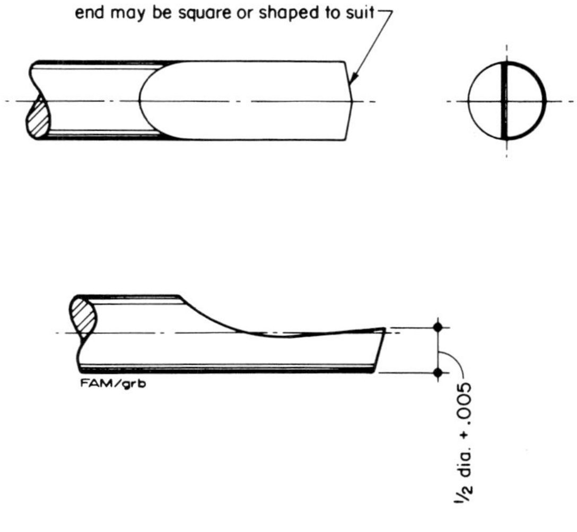

TYPICAL D-BIT

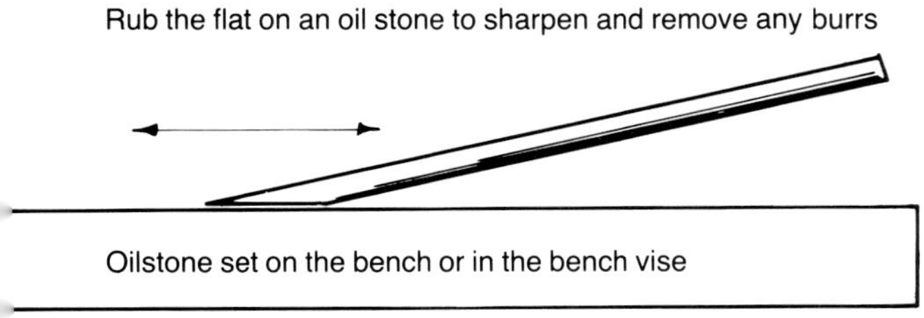

Figure 3

Sheet Metal Drills

These sheet metal drills will give neat round holes in thin or thick sheet copper, brass, aluminum and steel.

For holes up to 1/2" in diameter, grind an old or broken drill to the shape shown on the sketch.

To use: place the drill point in the hole left from the center punch and this will guide the drill as the two lips cut the metal, like the spurs on a wood auger.

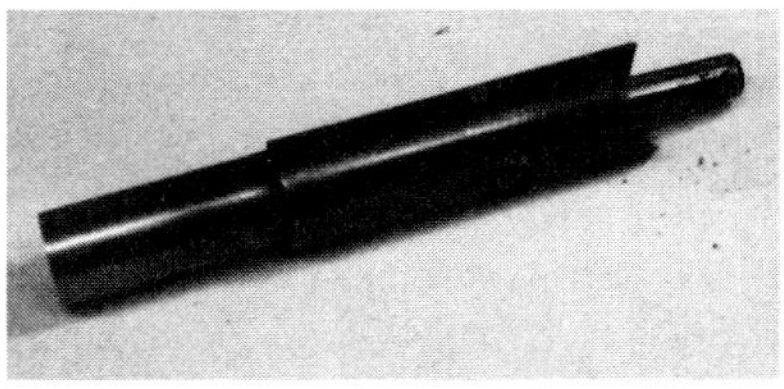

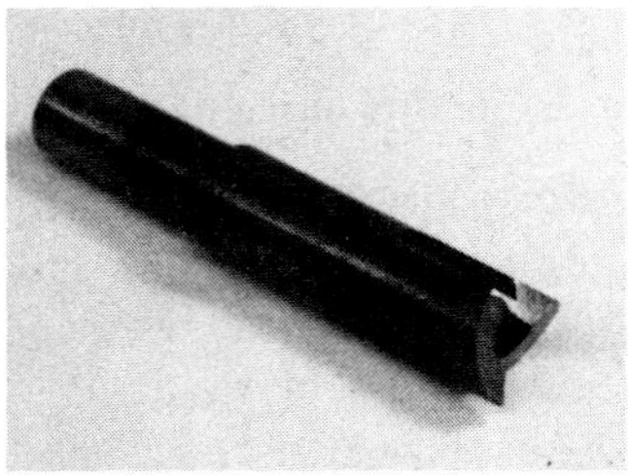

For holes larger than 1/2", the cost of a drill becomes prohibitive. Make up a cutter as shown on the drawing.

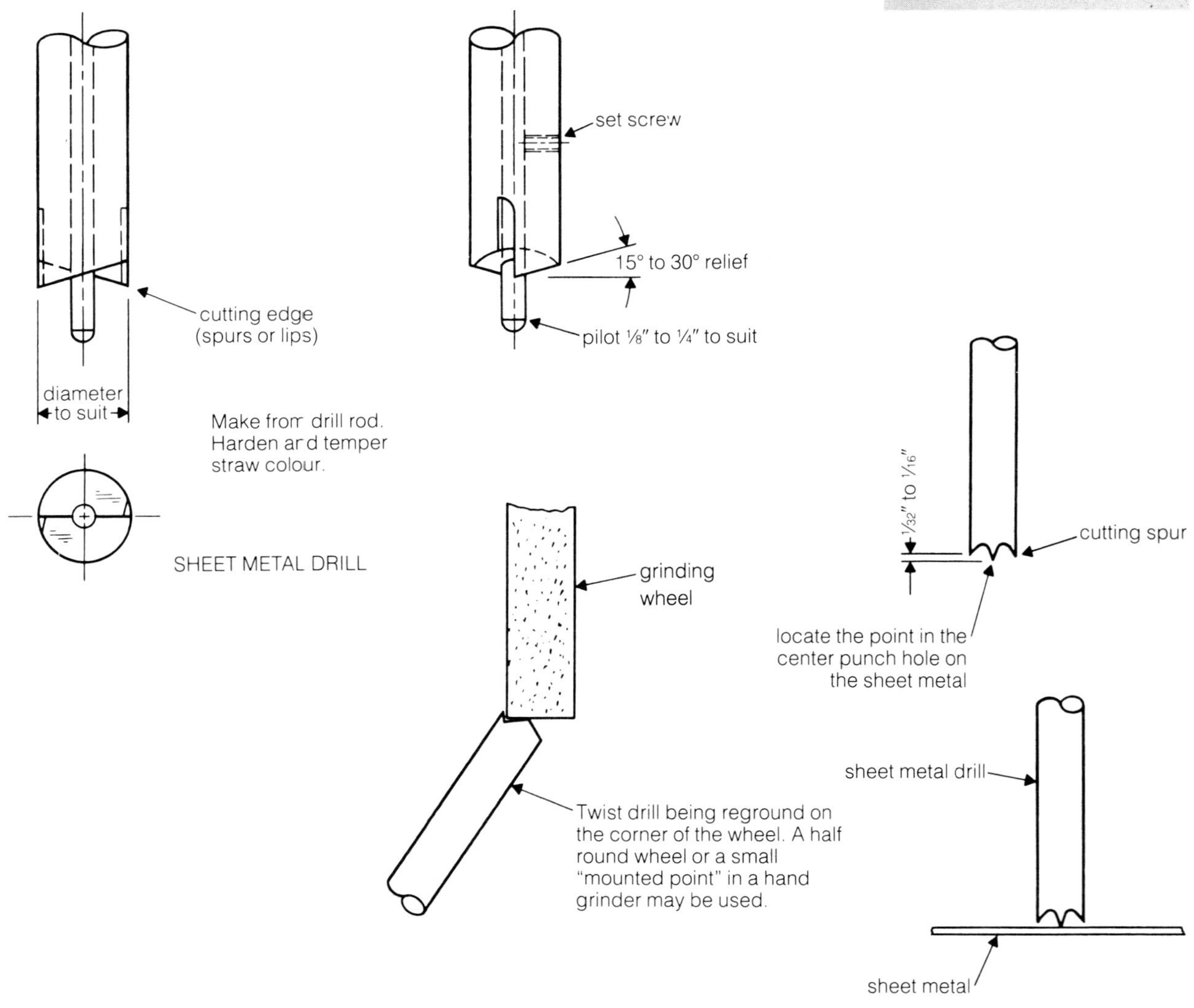

Lathe Accessories

Lathe Stands

A stand or table for a lathe must meet these requirements:

A. Must be Rigid
A cast iron cabinet or standard (legs) is the good old-fashioned method.

If angle iron is used to make a stand, the size of the angle must be heavy enough to avoid deflection and vibration. The joints may be bolted for easy assembly but they must be welded so they will not loosen from vibration.

B. Free from Atmospheric Changes
For those with a serious space problem in their shop, a lathe could be mounted on the left-hand end of the workbench.

An 1/8" metal plate on top of the bench under the lathe feet would distribute the load over a larger area and avoid crushing the wood fibers, and would also keep the wood from becoming saturated with oil. I would suggest that a large shallow pan (I use a cookie sheet) that just fits between the lathe feet be used to catch the shavings and the cutting oil. At the end of the evening, the pan can be removed and the shavings dumped into a garbage can.

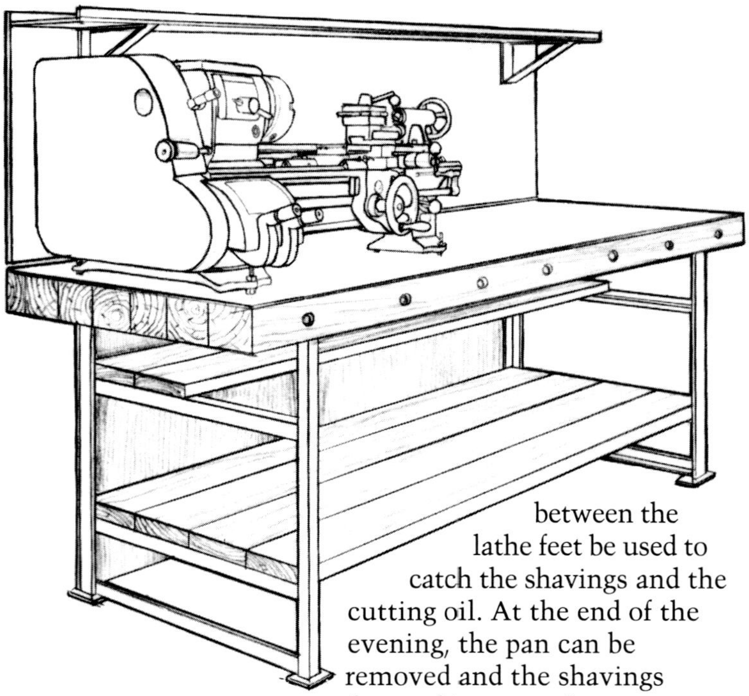

The lathe should be fastened down with bolts and sheet metal or brass shim stock placed between the metal plate and the feet.

When a wooden bench is used, the lathe will have to be checked from time to time with a precision level. The wood tends to move slightly from the changes in the humidity in the air.

C. Must be Level
Not only the bench must be level, the lathe itself must also be level. Many bench lathes have the bed twisted when bolted to the bench after leveling. This is one main reason we hear so many complaints about lathes boring tapers. Leveling the bench lathe is a most important function. Here are some tips on how to do it.

You will require the best machine level you can buy or borrow, and a dial indicator and some type of stand. Get yourself a 3/14" to 1" diameter bar of drill rod or cold rolled steel about three feet long (to reach the end of the bed). If you have a raised V-type bed, you'll need a pair of parallel bars. (Buy an 18" length of ground flat stock and saw it in half. Use 1/2 × 1" or a similar size.)

To level the lathe, first remove the tailstock if yours is a short bed lathe. Remove any burrs from the flatways of the bed with a very fine oilstone. Clean the bed thoroughly and wipe dry. Loosen the bolts holding the lathe to the bench.

Test the level of the bed along the length by placing the level along one of the flatways of the bed. While the bed should be level lengthwise, it is not critical.

The leveling of the lathe across the lathe axis is *very critical* if the bed is not to be twisted. Put the parallels in place on the flatways of the bed. Slide the parallels slightly to make sure there is no grit and to break through any oil film. Place the level on top of the parallels, again testing to make sure there is no grit or any oil. *Do not turn the level end for end.* Make this test in front of the headstock and at the end of the bed. Look directly down upon the graduations of the vial to have as accurate a reading as possible.

When the lathe is level, place the bar in the three-jaw chuck. Set the dial indicator on the bed or on the carriage with the contact point touching the side of the bar. (Do not take this reading from the cross slide.)

Gradually tighten the bolts holding the lathe to the bench. Shim the feet as necessary and retighten. On my lathe I used studs with nuts below and above the feet so that a minute adjustment is easily made.

With your lathe properly mounted, it will give you its best service for years to come and eliminate many frustrations.

D. Have Storage Space for all the Lathe Accessories

The old-fashioned cast iron lathe stands and the modern sheet metal cabinets usually have one shelf or one cupboard for all the lathe accessories and no place to hold a rule, micrometer or drills, ready for use. Shallow and deep drawers are a much better way to store lathe tool bits, boring bars, special cutters, knurling tools, tool holders, collets, etc. Shelves can be used for the heavier items such as lathe chucks, faceplates, vertical milling attachments, and the steady and follower rests.

A shelf mounted upon uprights, bolted to the back of the lathe stand, is ideal to hold fraction, wire and letter drills in stands. I also keep my taps in wooden blocks with the correct size tapping and clearing drills with each tap.

A shallow pan to hold drills and taps or reamers for use in the correct sequence will save a lot of time. I place a clean piece of paper towel on the shelf to hold my micrometer, telescoping gage or rule instead of placing them in the chip pan as commonly done.

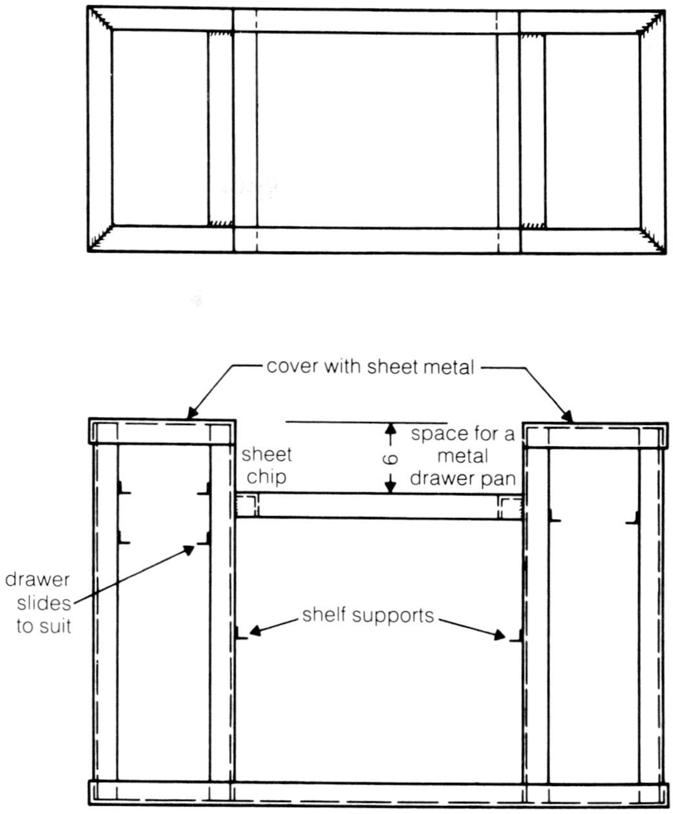

ALTERNATE TYPE OF LATHE BENCH
FOR A DRAWER TYPE CHIP PAN

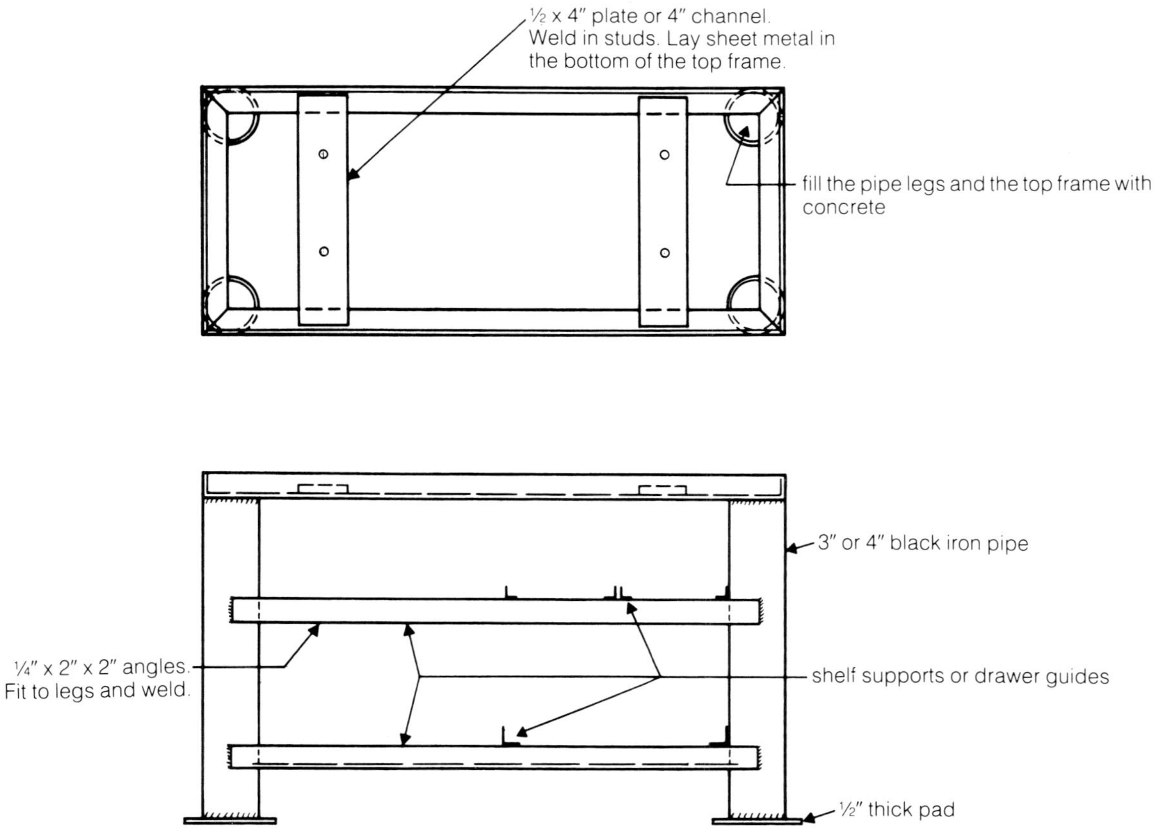

ALTERNATE TYPE VIBRATION FREE
LATHE BENCH

My lathe stand. Note the cross bracing (optional) and the wooden duck board.

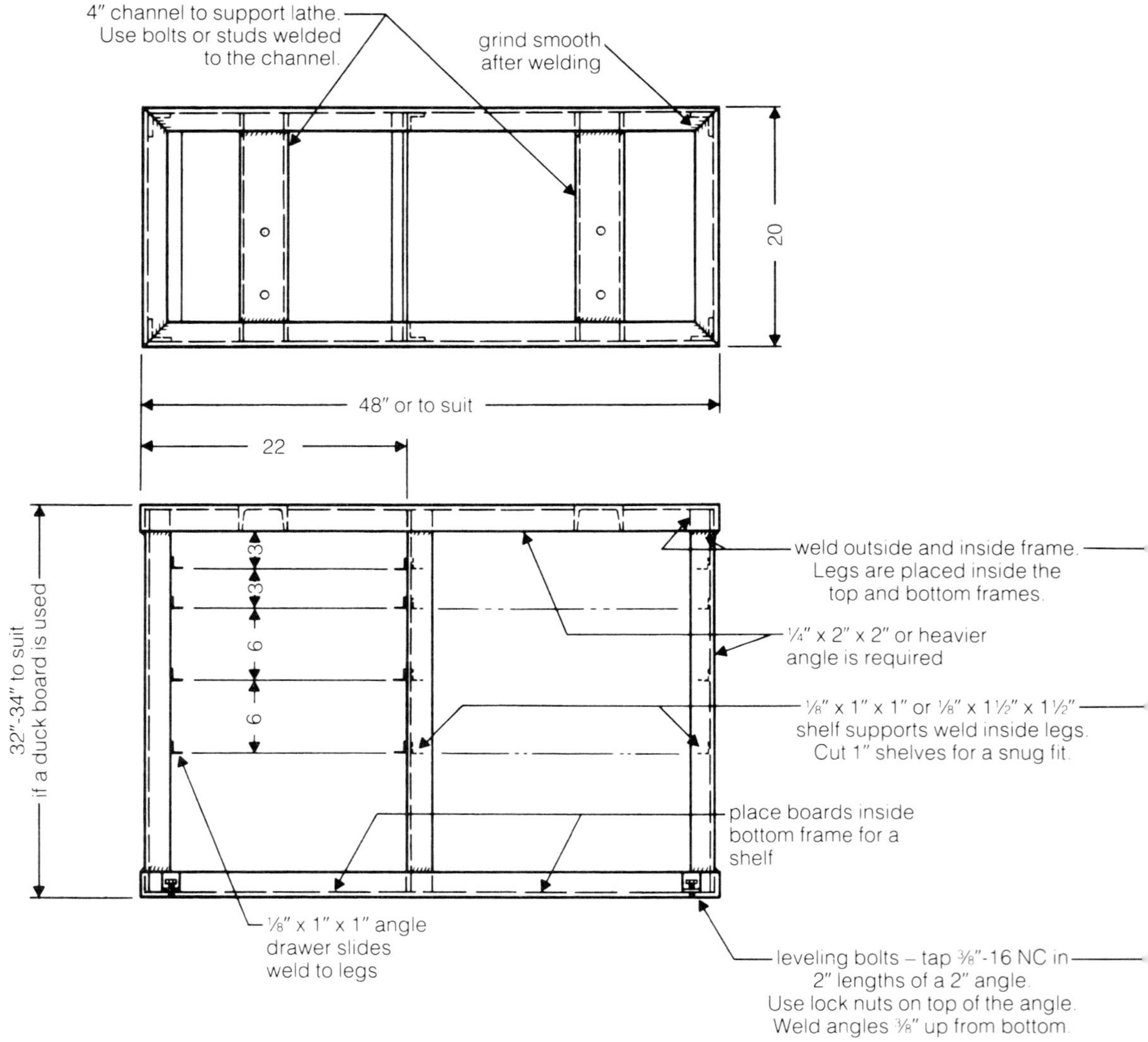

LATHE BENCH FOR A 9"-10" LATHE

The lathe chuck keys and lathe wrenches are placed on the front of the shelf either on nails or with the handle slipped into holes drilled into the edge of the shelf. The drill chucks rest in holes to suit the Morse taper shanks on the right-hand end of the shelf.

A piece of sheet metal is fastened to the underside of the shelf at the back and extends to the chip pan. This prevents the cutting oil from splashing on the wall behind.

E. Should be Portable

I have simple pipe handles welded to the ends of my lathe bench (visible in the photo) to facilitate sliding the bench along the floor when rearranging the shop or moving.

The lathe stand can be leveled on the floor with sheet metal shims, thin wooden wedges or with leveling screws; two in front at both corners and one in the rear in the middle. There are two other screws at the rear corners but they are adjusted to just touch the pads and not to bear any real weight. When the lathe stand has been leveled, the final operation of leveling the lathe is much easier.

Moving your workshop to a new location can be a formidable task. A 10" lathe on the bench, with all the accessories, weighs about 1,000 lbs. Unbolt the lathe and place it temporarily on another workbench or on the floor. Remove all the drawers with their contents in place. Remove all the chucks, etc., from the shelves. Now back off the leveling screws so the bench will slide easily on the cement floor. Place the lathe stand in the truck, replace the drawers and loosely bolt the lathe in position. Now rope the lathe stand securely in the truck and there will be no danger of the lathe falling, or having something else falling on it, as may be the case if the lathe is placed on the floor of the truck.

LEVELING PADS

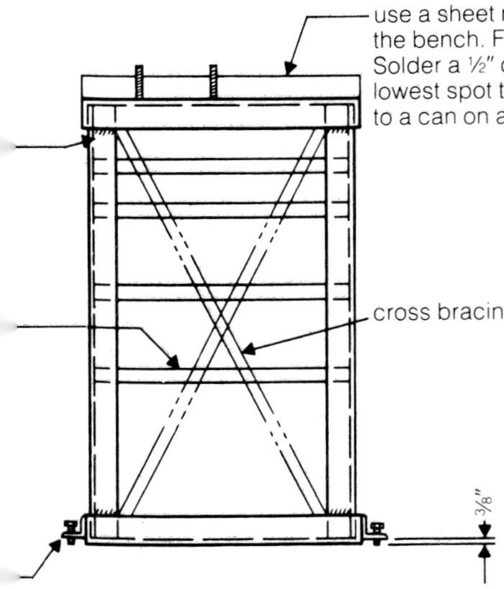

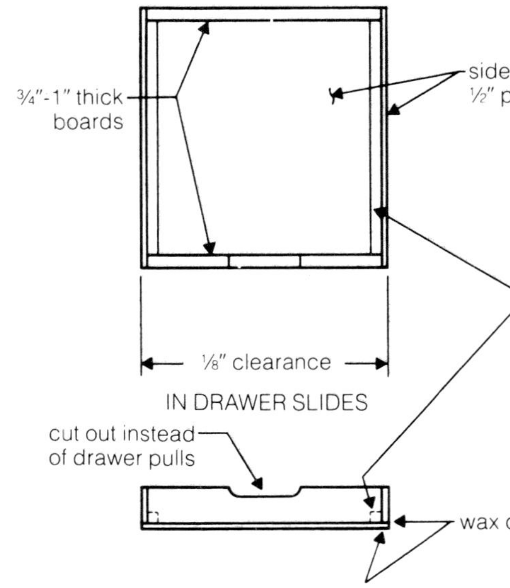

DRAWER CONSTRUCTION

Lathe Accessories

Four-tool turrets and other toolholders will extend the usefulness of your lathe and increase your pleasure in its use. The four-tool turret (Photo 1) will save a lot of time over the usual toolholders. The open-side toolholder (Figure 1, Photo 2) is better than the usual American or lantern-type toolpost for tungsten carbide-tipped tools, when heavy cuts are necessary and rigidity is essential. Otherwise the carbide tip may chip along its cutting edge. The English-type flat toolholder (Figures 2 and 3) can also be used for carbide tools and large tool bits. However, its special function is to hold boring bars in a vee block (Photo 3). Very large diameters can be turned with a square bar that is slotted to hold a tool bit.

The four-tool turrets (Figure 4 and 5) can be made entirely on the lathe, if a milling machine or shaper is not available. Face a block of hot- or cold-rolled mild steel on all six faces by holding the work in the four-jaw independent chuck. (These short ends or "off cuts" are often available from a metal merchant or a machine shop for a nominal price per pound.) Lay out and drill the bolt hole in the middle. Coat the four sides with layout blue (or use a felt tip marker). Place the block on top of the compound slide rest and scratch the lathe center line by sliding the block against the tailstock or headstock centers. Now use this center line as a reference line to lay out the width of the slot.

1

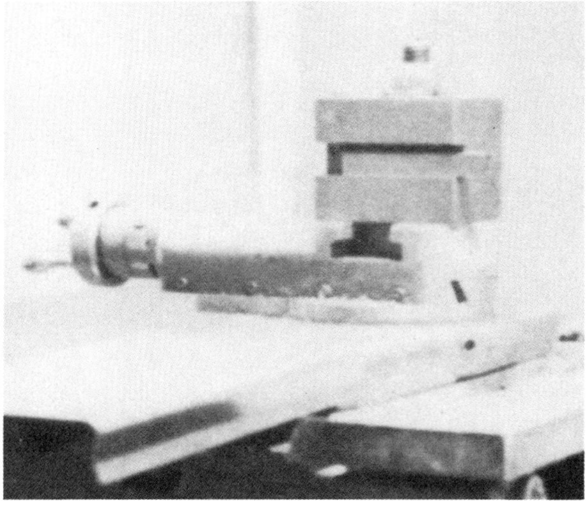

2

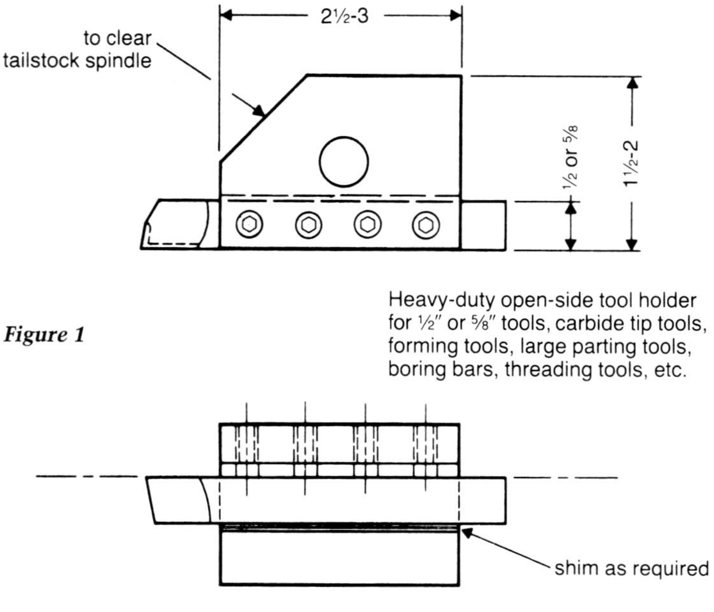

Figure 1

Heavy-duty open-side tool holder for ½″ or ⅝″ tools, carbide tip tools, forming tools, large parting tools, boring bars, threading tools, etc.

3

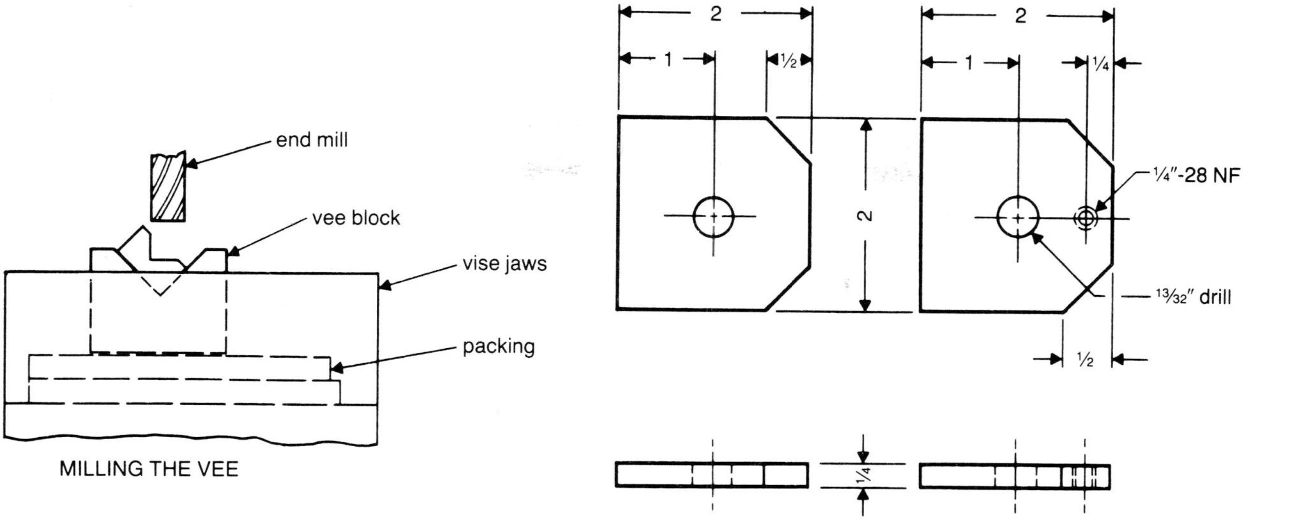

MILLING THE VEE

BOTTOM PLATE
CRS

TOP PLATE
CRS

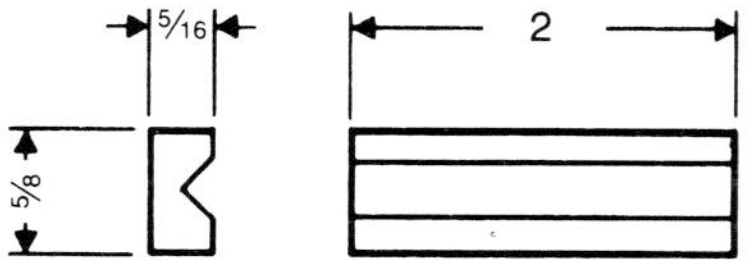

90° VEE BLOCK

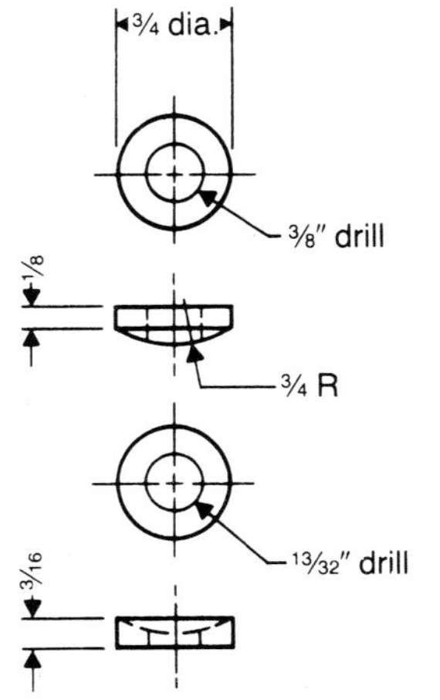

SPHERICAL WASHERS

Mark the lower line to suit the size of the tool bits you will be using. Do this carefully, for the top of the tool bits should be exactly at the center height. Mark the top line 1/16" to 1/8" above the center line so there will be enough clearance to pack the tool bit up with shims if required. This top line can be located to suit the size of milling cutters available in order to mill the slot, if the block has to be slotted on the lathe without a milling attachment.

Note: When grinding tool bits for a four-tool turret, try to grind the tool bits on the end, only, so the tool bit is not ground away on the side and top. This grinding method will give the maximum use of the tool bit until it is too short to clamp in the turret or a toolholder.

Those who have a milling machine will simply rough out the slots and then finish up to the layout lines as usual. A sturdy lathe milling attachment may be used to hold the block for milling the slots; however, the overhang will be substantial, so the milling operation will require light cuts (.010" depth of cut). Rough out the center of the slot and finish to the layout lines, remembering to feed against the direction of the cutter rotation (Figure 6).

Another method is to clamp the block upside down directly on top of the compound slide

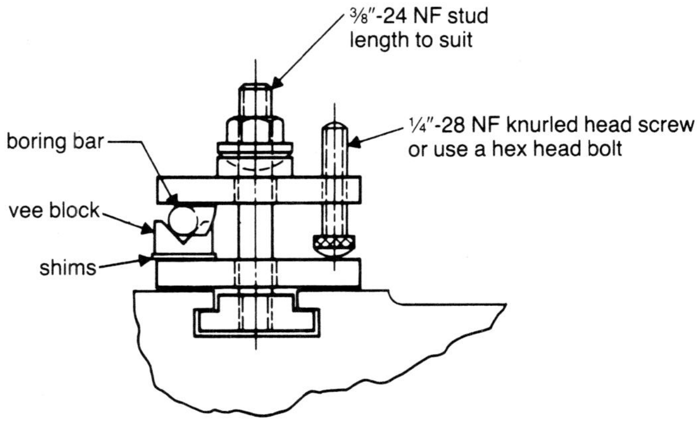

ENGLISH STYLE FLAT TOOL POST
FOR A 6" ATLAS LATHE

Figure 2

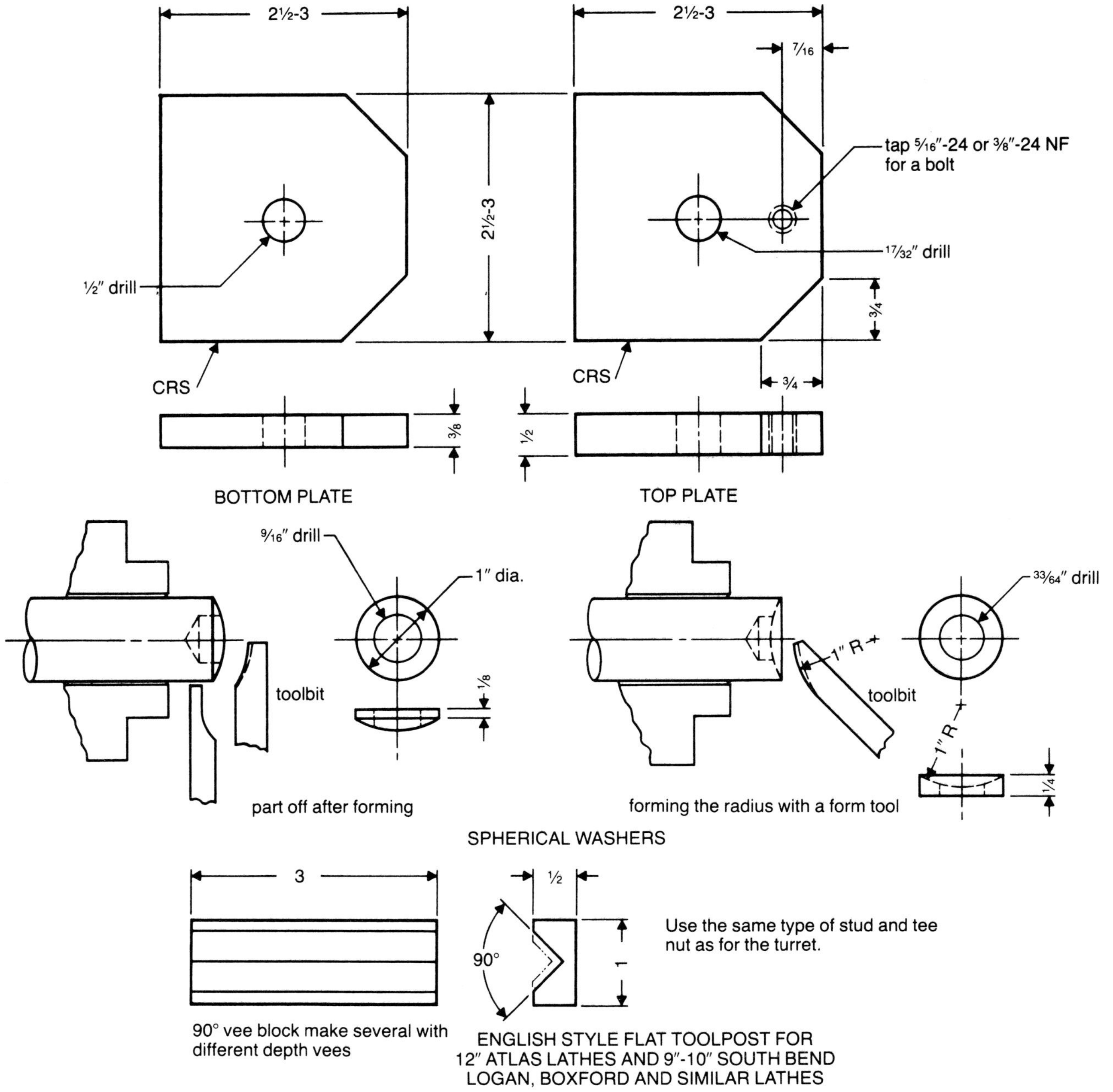

Figure 3

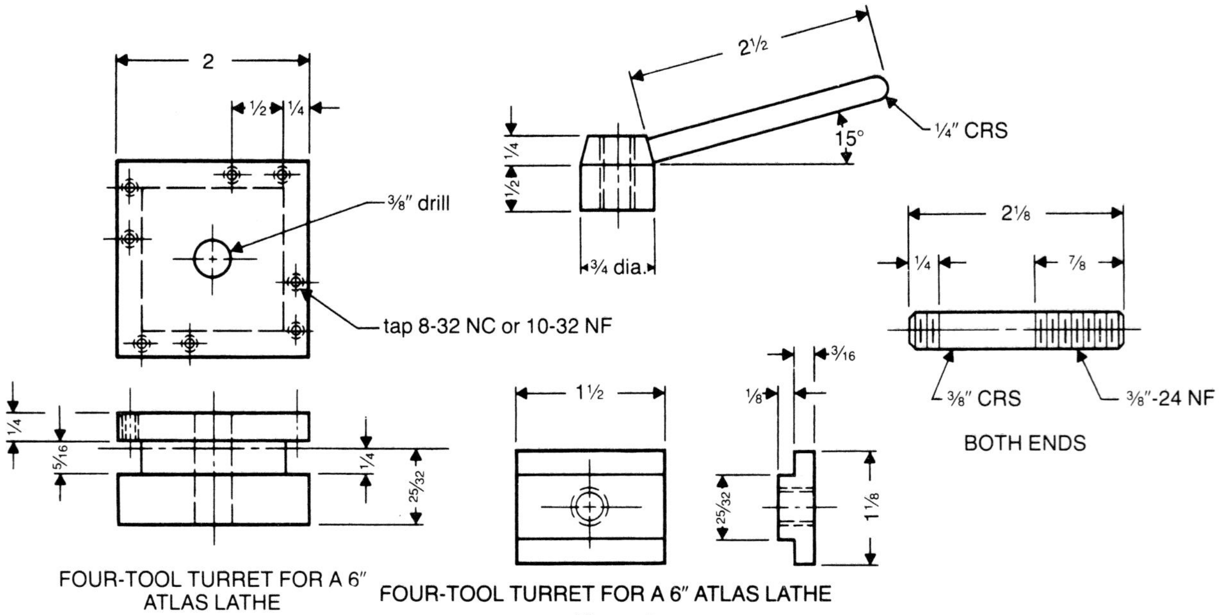

Figure 4

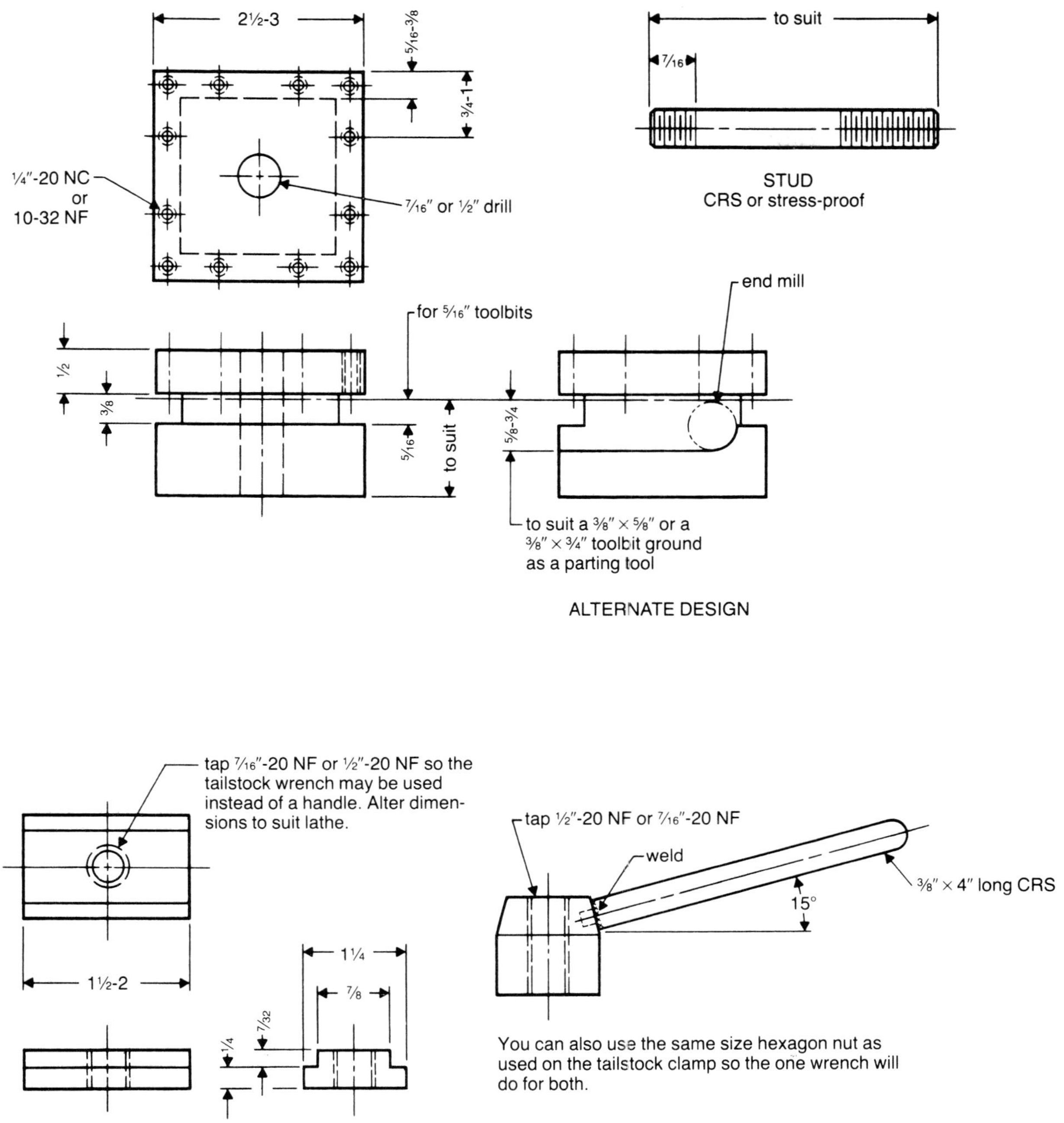

TURRET FOR 9"-12" LATHES –
ATLAS, SOUTH BEND, LOGAN, BOXFORD AND MYFORD

Figure 5

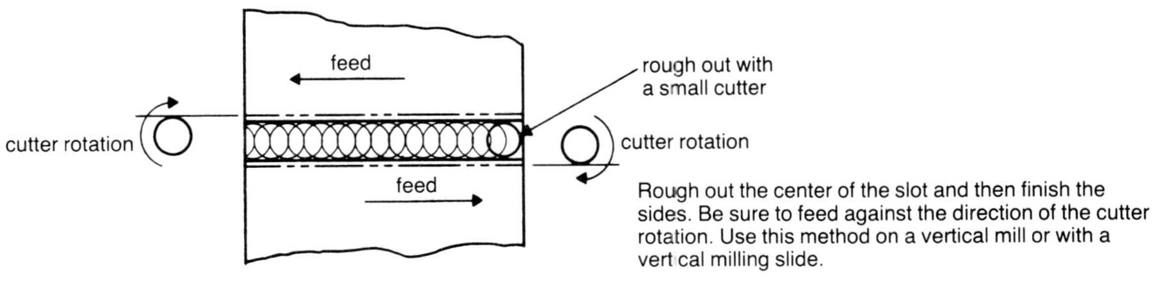

Figure 6

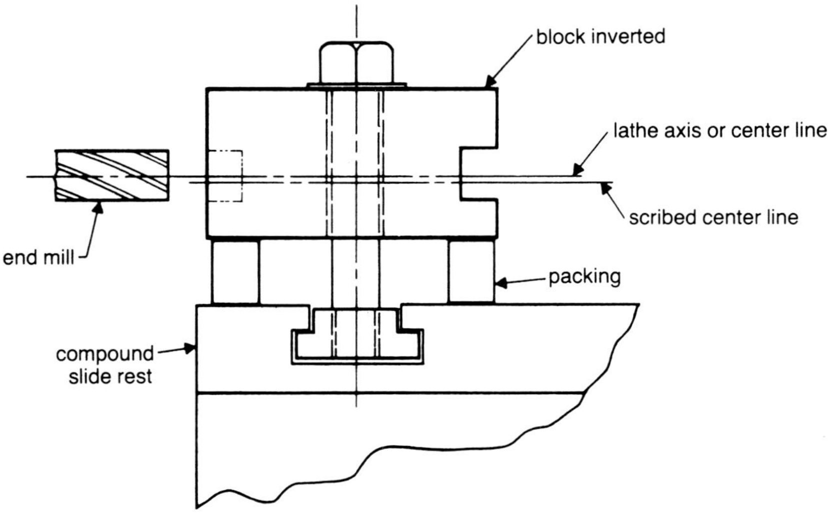

CLAMPING THE TURRET UPSIDE DOWN
ON THE COMPOUND SLIDE REST FOR MILLING
THE TOOL SLOTS

Figure 7

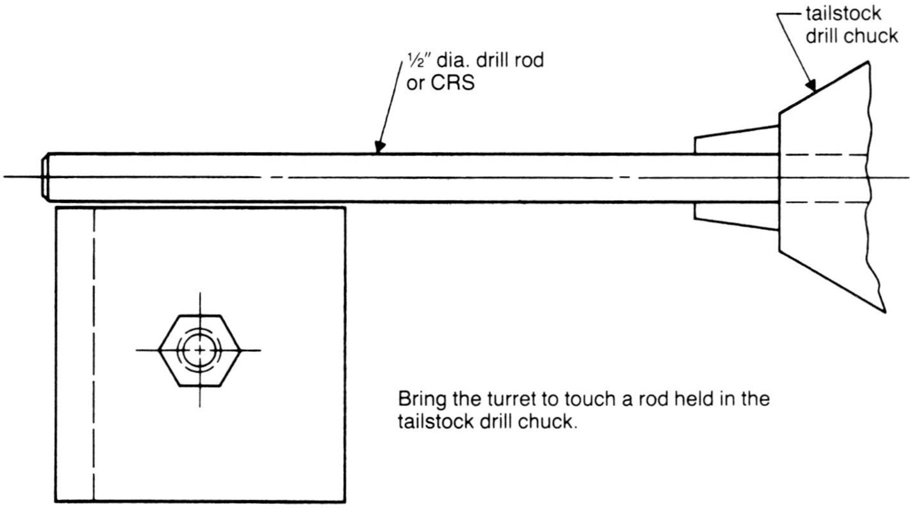

Figure 8

rest with packing underneath to bring the layout lines to touch the lips of the end mill. (Figure 7). Make up the tee nut first with the vertical milling attachment, for this will be required with the stud to clamp the turret in place for the milling operation. Set the compound slide rest parallel with the lathe axis. Align the block by bringing it up to touch a bar held in the tailstock drill chuck (Figure 8). Clamp the carriage to the lathe bed after moving the carriage so the block just touches the end mill. Set the depth of cut with the compound slide rest and traverse the block past the cutter with the cross slide. Take cuts of .010" to .030" until the full depth of the slot is reached. Repeat for each slot. A micrometer carriage stop could also be used to set the depth of cut.

A back tool post (Figure 9) for a cutoff blade or a forming tool will be a time saver when a number of similar or identical parts are to be made. If your lathe has tee slots at the back of the cross slide (or tapped holes), one of these can be used to fasten the back tool post in place. Lacking a tee slot, a hole can be drilled and tapped into the cross slide, if you don't mind altering the lathe. The tapped hole should be plugged with a short screw when the tool holder is not used so the hole will not fill with shavings or allow shavings to fall onto the cross slide screw.

The cutoff blade may be set horizontally for brass or set with a 15° top rake for steel as shown in Figure 10. If you don't want to use

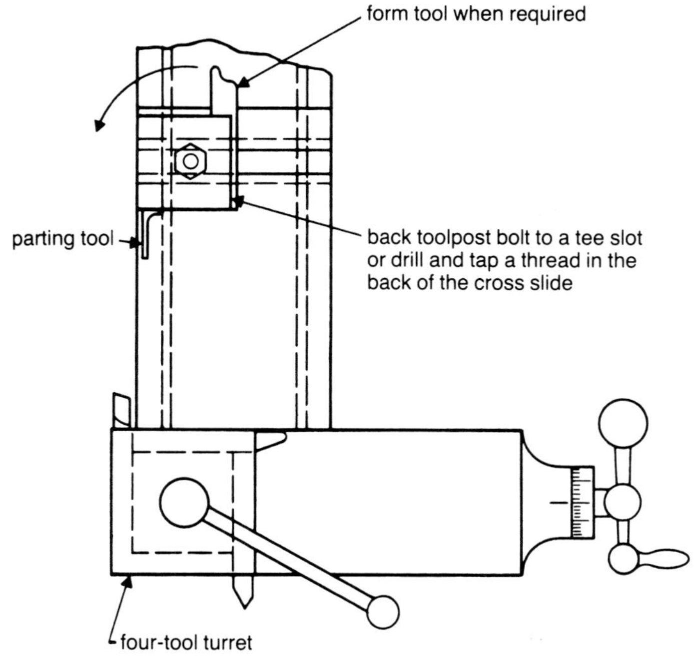

BACK TOOLPOST FOR REPETITIVE TURNING FOR SMALL PARTS

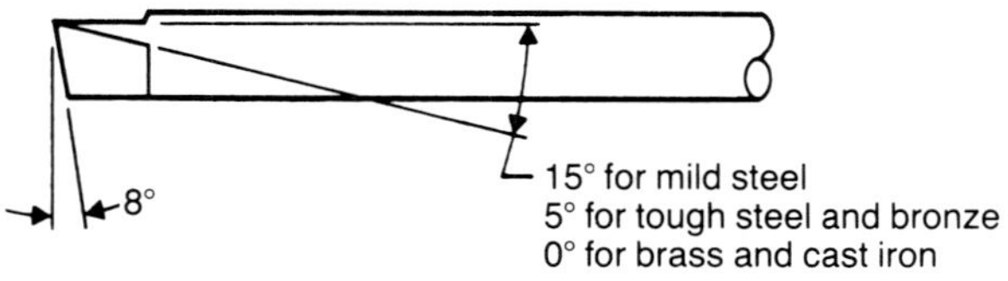

Figure 10

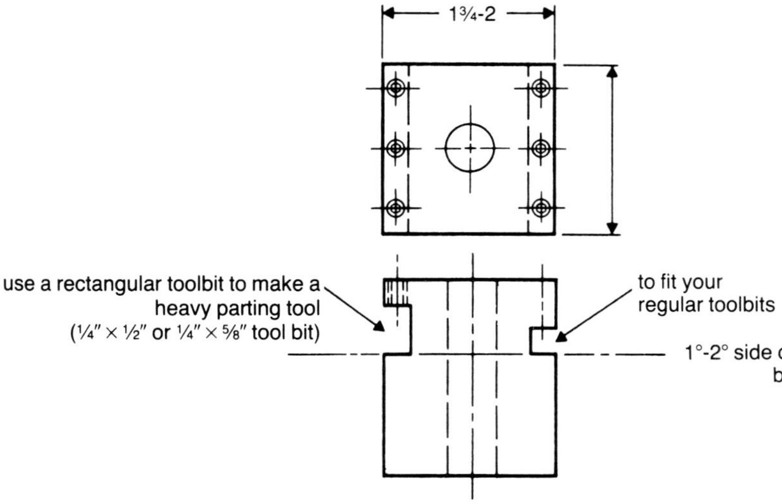

BACK TOOLPOST

Figure 9

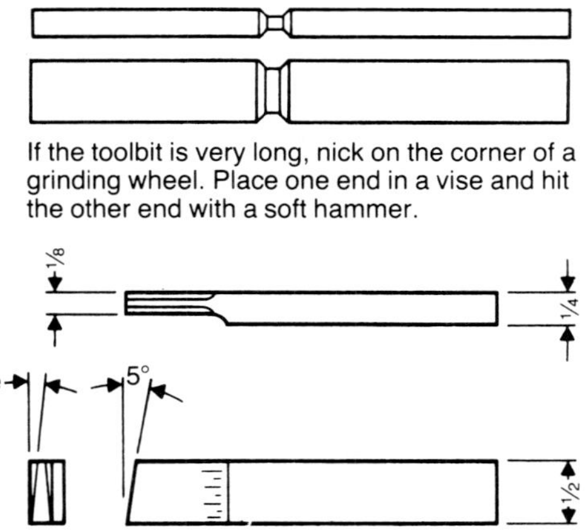

heavy parting tool or cutoff tool

Figure 11

a cutoff blade, a parting tool (Figure 11) can be ground from a square or rectangular tool bit. I still have one that I used on an automatic lathe over thirty years ago.

Parting or cutting off from the rear of the cross slide seems to be easier than from the front of the lathe. I've heard several explanations for this; however, my own feeling is that there are three principal reasons: First, the parting tool being inverted, the shavings fall clear and do not jam in the kerf (Figure 12). The second reason is the coolant can reach the cutting edge instead of being restricted by the shavings on top of the tool. The third reason is that the cross slide is drawn up against the dovetail vee ways so that any wear or looseness

4 ***Heavy forming operation to make the curved washer for an American or Lantern Tool Post.***

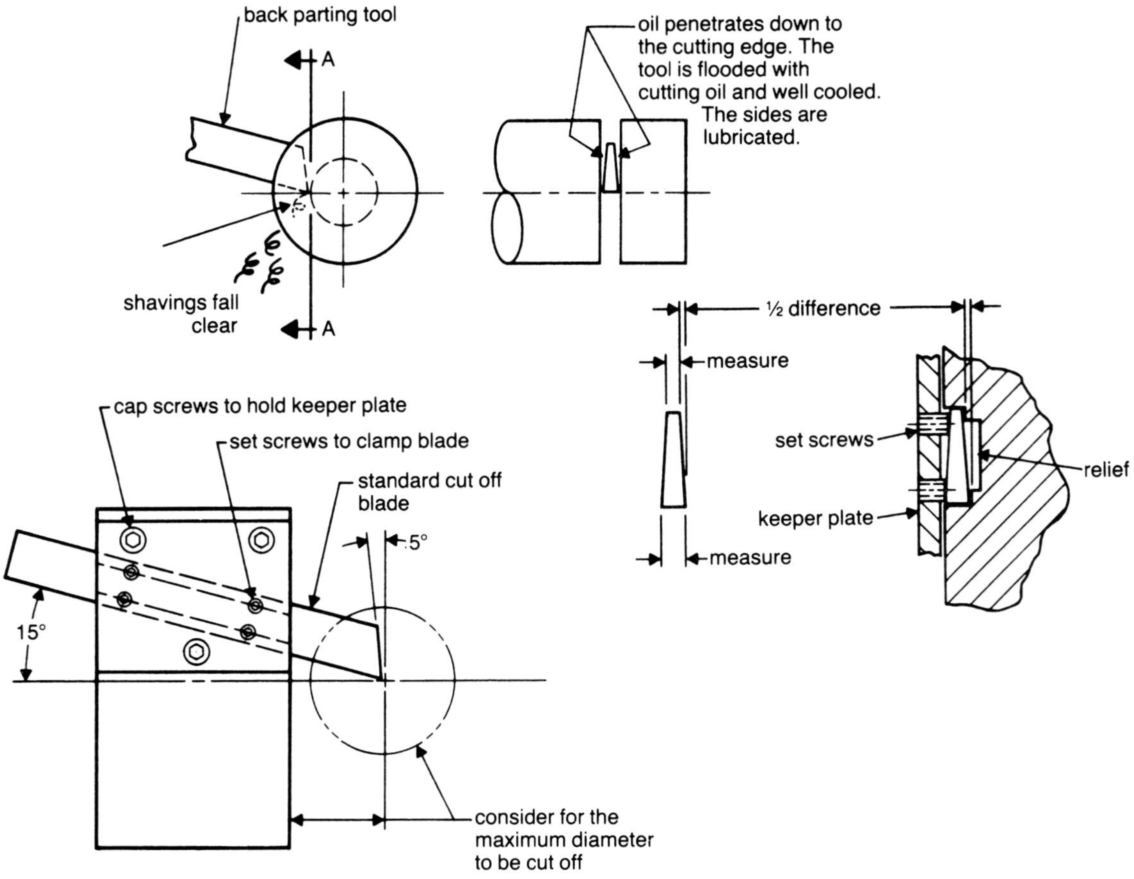

Make another block with the blade set horizontal for brass, bronze, cast iron, etc. if the production warrants a special cutoff tool.

BACK TOOLPOST WITH A STANDARD CUTOFF BLADE SET WITH 15° BACK RAKE FOR FREE CUTTING MILD STEEL

Figure 12

5 ***Knurling tool in an open side tool holder.***

6 ***Two tool turret.***

is taken up. There will be a slight amount of friction between the vee ways instead of having the cross slide floating on a thin film of oil on the flat ways. This resistance reduces the chances of a "dig in." Of course, a substantial lathe spindle and a minimum overhang beyond the chuck are desirable, just as for parting off from the front of the lathe.

The photographs (4 through 7) show some of my turrets and other tool holders. I have

7 ***Bar set to turn very large diameters.***

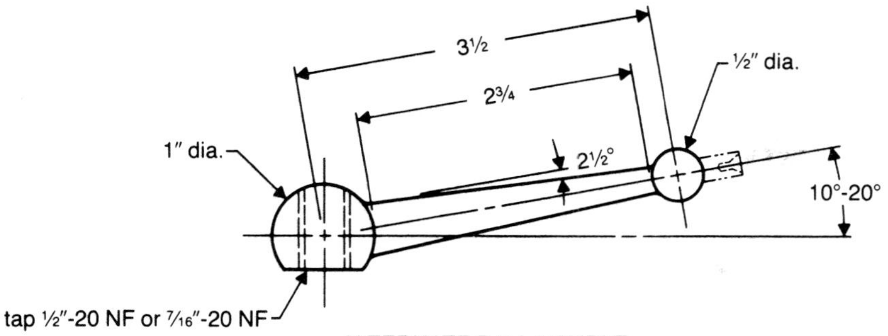

ALTERNATE BALL HANDLE

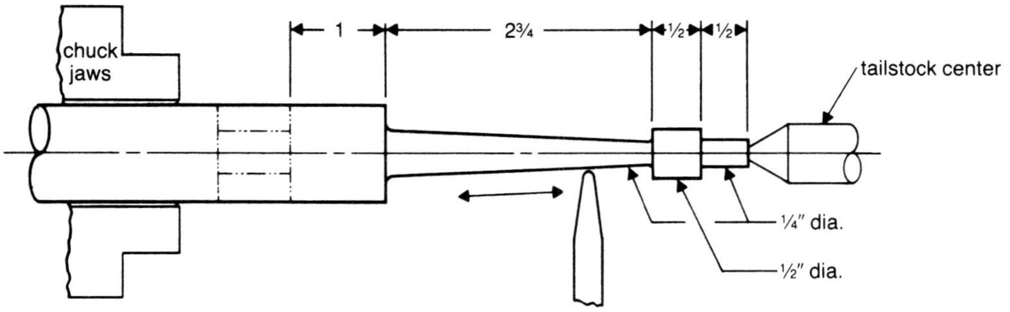

OPERATION NUMBER 1
Place a 1″ diameter bar in the chuck with 6″ projecting beyond the jaws. Face and center drill. Turn to ⅝″ dia. × 3¾″ long. Starting 1″ from the end, turn the taper with a round-nose toolbit. Then turn the ½″ and ¼″ diameters at the end.

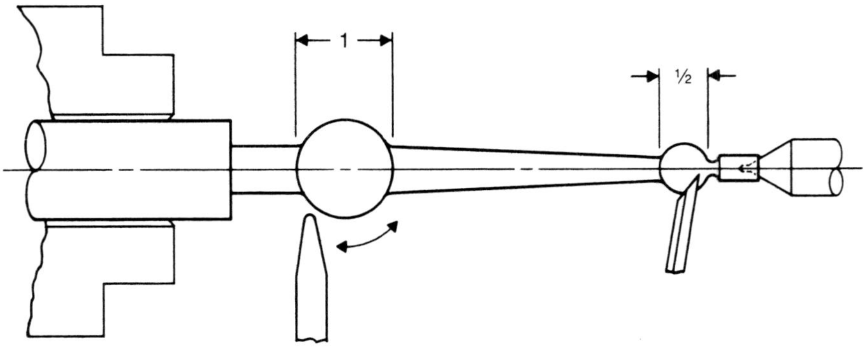

OPERATIONS NUMBER 2 AND 3
Rough out the balls with a round toolbit. Finish turning with a graver, another hand tool. Saw off the waste from the balls with a hacksaw.

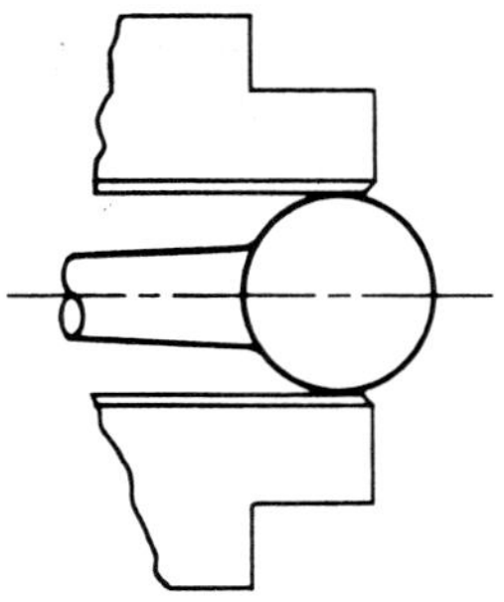

OPERATION NUMBER 4
Place the handle inside the chuck to finish the ends.

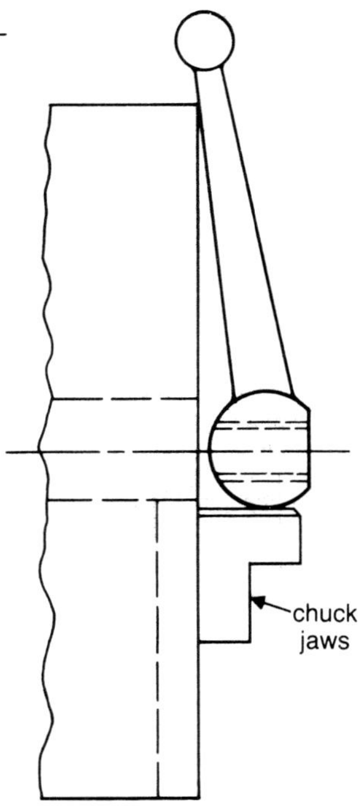

OPERATION NUMBER 5
Place the handle in the 3-jaw chuck to face. Drill and tap.

Figure 13

four turrets set with tools for mild steel, brass, cast iron, etc., to save time changing tools. I intend to make up a back tool post when time permits. I used one for many years on a Myford Super 7 lathe.

The four-tool turrets do not have a method of locating each face; I prefer to adjust the angle slightly for different operations, such as facing, turning, etc.

Also included here is an alternate design for a ball handle (Figure 13) you can use on your turret, as well as some hand turning tools (Figure 14), boring bars (Figure 15) and a sheetmetal radius gage (Figure 16) – all very useful tools that will increase the versatility of your lathe.

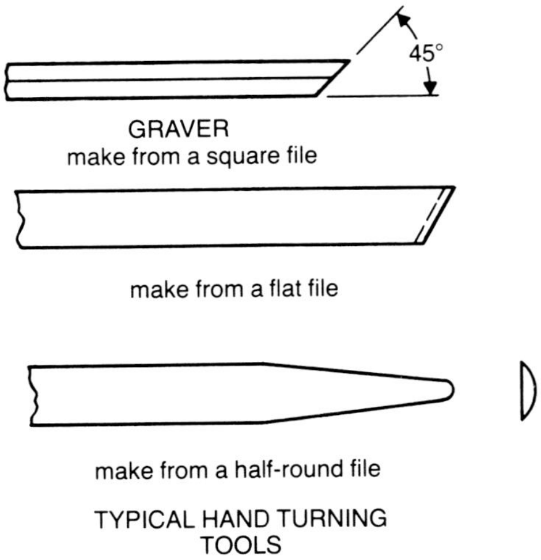

Figure 14

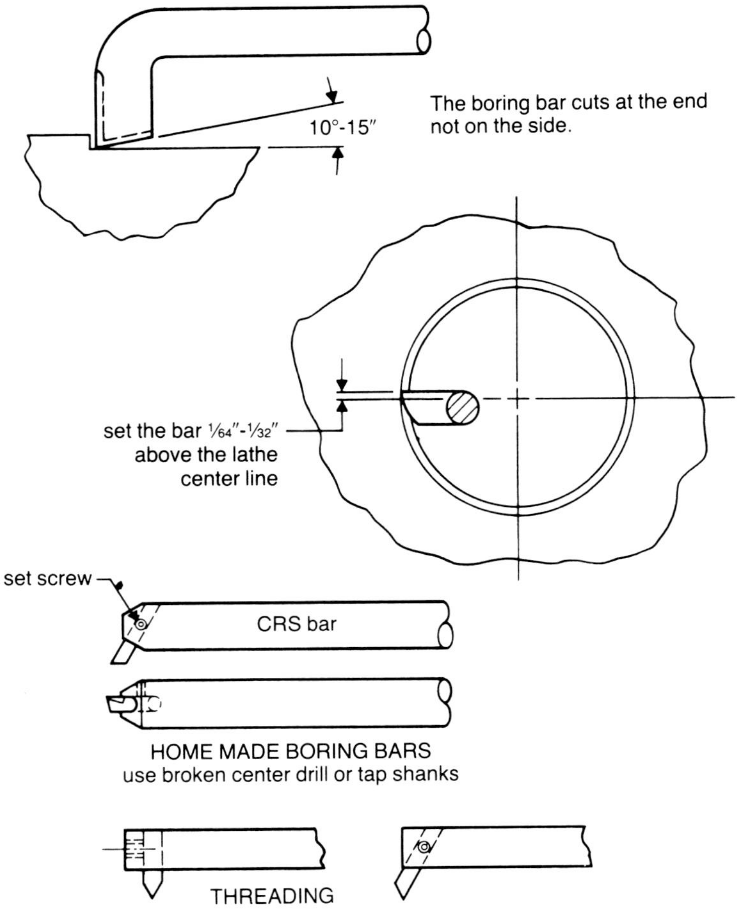

Figure 15

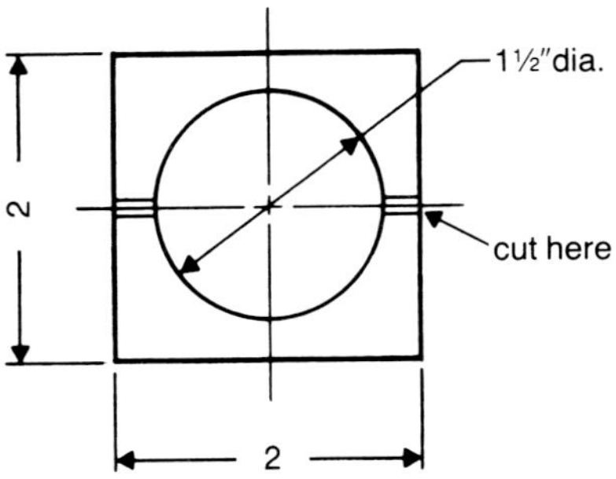

Figure 16

Lathe End Mill Adapters

For those who must do their milling operations on their lathe with the work clamped on a vertical slide, these adapters have less overhang and will be more accurate than holding end mills in a universal three-jaw chuck.

The Atlas Press Company used to supply such adapters for their 6" lathe (I had one), but they are easily made with a No. 2 or a No. 3 Morse taper to fit your lathe.

Turn the taper as described the article "Turning Tapers." However, the hole will have to be drilled and the tap started before removing the work from the universal chuck.

Start the tap by holding the tap shank in the tailstock drill chuck or by supporting the tap with a tailstock center. Rest the tap wrench (or a lathe dog) on top of the compound slide rest. Pull the driving belt by hand and feed the tap into the work with the tailstock handwheel. When the tap has made two or three turns, remove the work with the tap still in place to the bench vise and continue the tapping operation. (Don't forget the cutting oil or the tap will seize in the work.)

Now place the adapter in the headstock spindle taper (remove the universal chuck).

To hold the adapter in place so it cannot loosen or twist during the milling work; you will need a piece of threaded rod long enough to screw into the adapter and project beyond the end of the spindle by about 1". A large washer or a nut will complete the "draw bolt."

Now face, drill and bore the adapter to size for your end mill shanks. Measure the diameter of the shanks of the end mills you use and try to finish the bore no more than .001" larger. Make one for each size of shank that you use.

Next, drill and tap the setscrew hole. Check from your end mills where the setscrew should be located.

When placing an end mill in these adapters, remove the setscrew to make sure that the flat on the end mill shank is correctly located under the setscrew hole. Replace the setscrew and turn the screw to just touch the end mill shank lightly. Wrap a rag around the end mill flutes and pull the end mill out until the back end of the flat rests against the setscrew. Now tighten the setscrew. This will prevent the end mill from being pulled out of the adapter by the force of the cutting on the helical flutes, thereby spoiling an expensive casting or other work that you have spent many hours of careful work on.

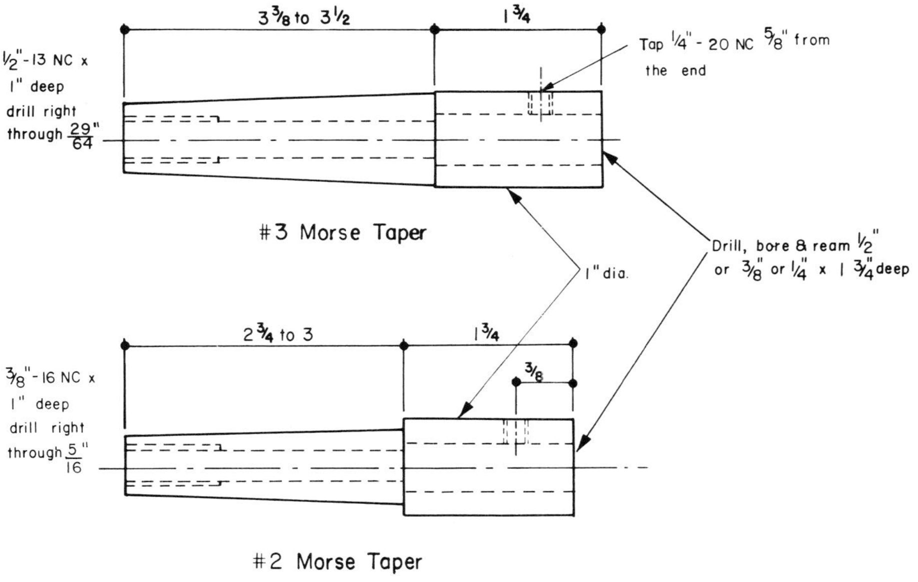

END MILL ADAPTERS

Tailstock Die Holders

A tailstock die holder should be among the first accessories an amateur makes for his lathe, since threading with a hand-held die stock is seldom satisfactory.

To make the die holder, chuck a 4" length of bar stock of a suitable diameter in the universal chuck with about 3-1/4" projecting beyond the jaws. Face and center drill. Support the work with the tailstock center for the turning and knurling operations. Reverse the work in the chuck with a single layer of cardboard or emery cloth wrapped around the work to protect the knurling. Face and center drill. Drill through with a small drill and open up the hole in stages until the hole is about 9/16" diameter. Now bore the hole to size or, if you have a 5/8" machine reamer, leave about .015" to .020" for reaming. Note: These reamers are available with Morse Taper Shanks or use a chucking reamer. The shanks are left soft so they may be turned to a smaller diameter to fit the drill chuck if required.

To cut a long bore like this, you will need as large a diameter boring bar as will pass through the hole mounted in a vee block or other suitable holder so that the bar is parallel with the hole and the cutting edge is about 1/64" above the center height to allow for the spring (deflection) in the bar (see the drawings).

When the 5/8" hole is finished, reduce the overhang of the boring bar to about 1" to bore the recess for the die. Face the bottom of the recess flat so that the die will seat properly.

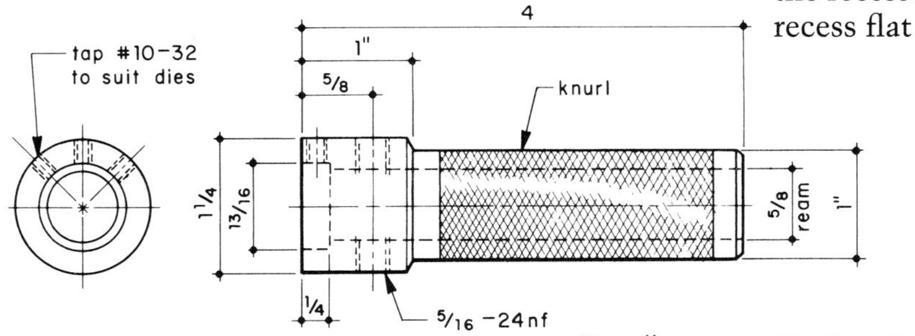

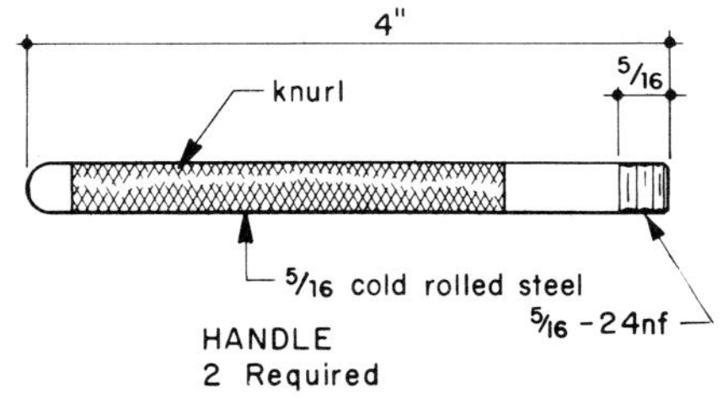

13/16" TAILSTOCK DIE HOLDER

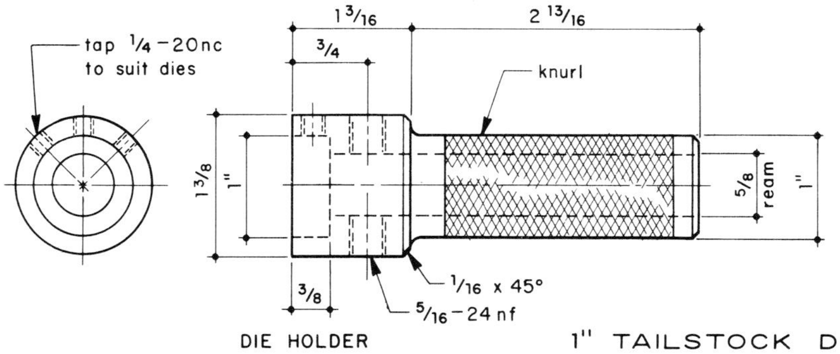

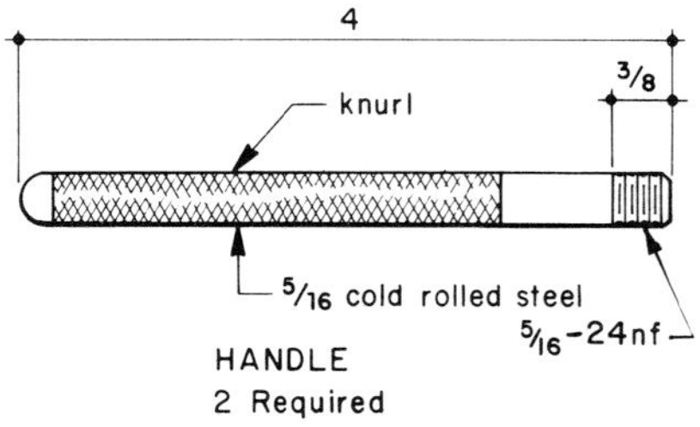

1" TAILSTOCK DIE HOLDER

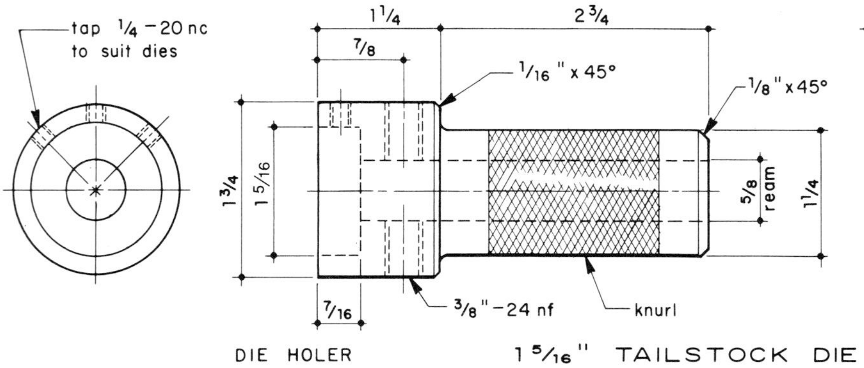

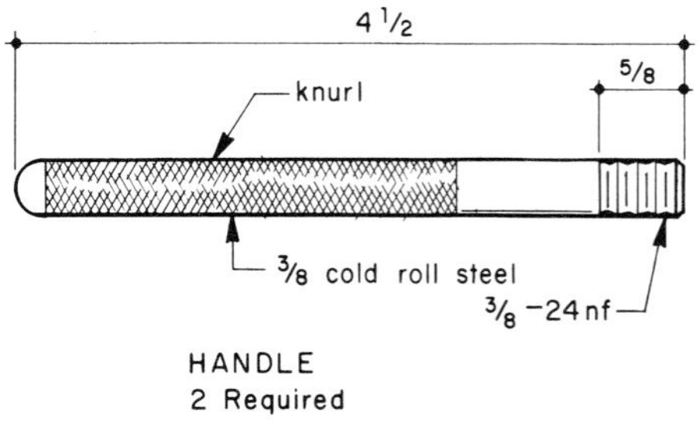

1 5/16" TAILSTOCK DIE HOLDER

Scratch a line at the center height along both sides and across the face. This line will be set vertical with a small square when the holder is in the drill press vise for drilling and tapping the holes for the two handles.

Drill and tap the holes for the setscrews to suit the dies you have. The two handles are so simple that they don't need any explanation.

The taper on the shank is made the same way as the soft center previously described. Use the headstock spindle sleeve as a gage and be sure to make a good fit.

Set the shank in the sleeve in the headstock spindle for turning the .623"-.624" diameter and drill to suit.

When using the tailstock die holder, don't forget to oil the shank where the die holder slides and to clean out the shavings frequently.

Another method of finishing a long hole like this accurately is to drill out the hole to about 1/16" undersize. Then bore about 1" depth to the correct diameter. Then finish the remainder at the hole to size using a 5/8" "D" bit.

The D-bit will have to be backed out frequently to remove the shavings. However, the D-bit will make a true hole.

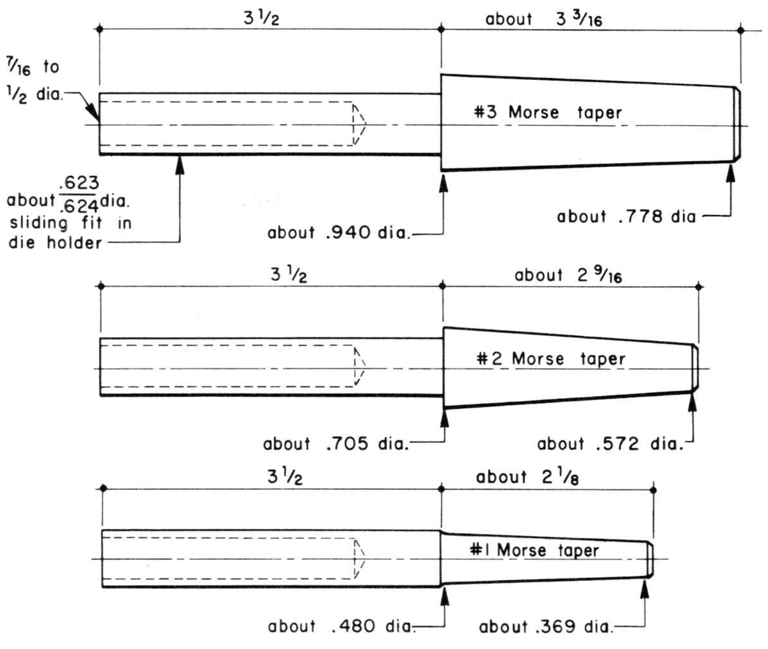

MORSE TAPER SHANKS

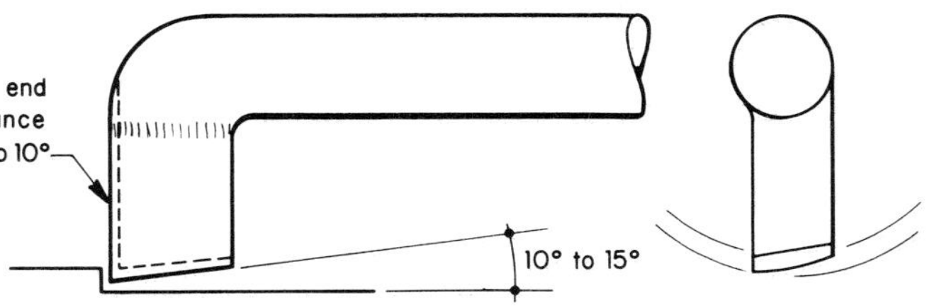

the boring bar cuts on the front edge only and not on the side

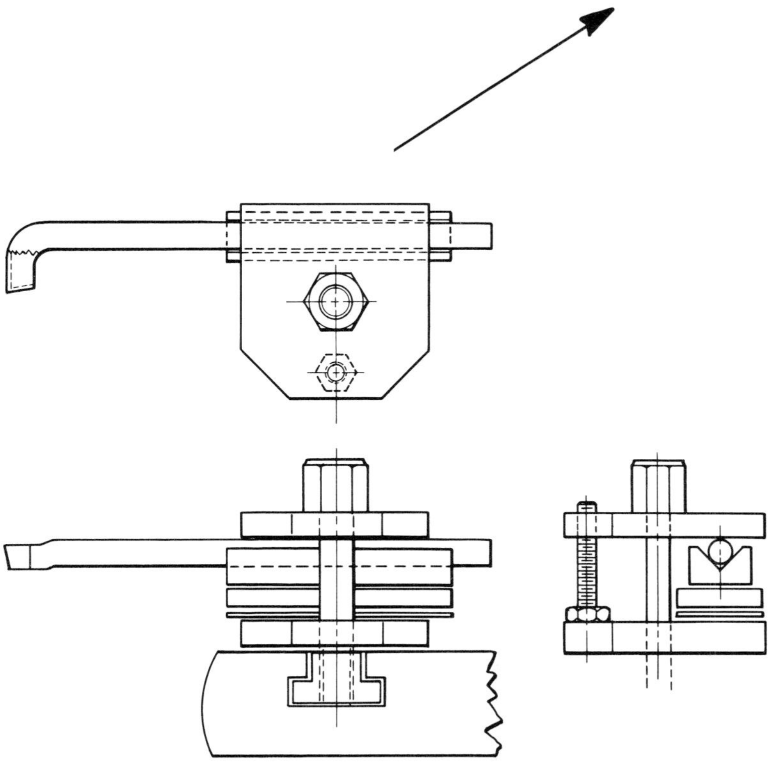

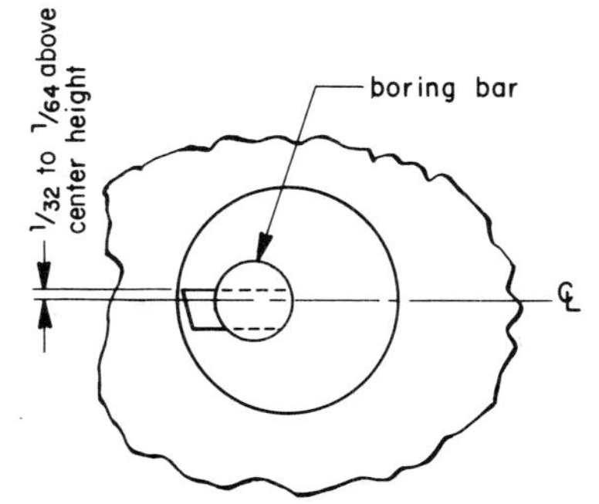

BORING BAR MOUNTED IN VEE BLOCK

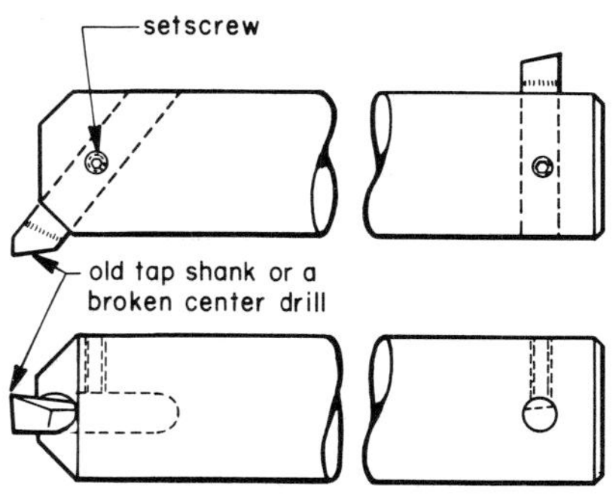

HOMEMADE BORING BARS

Roller Tool for the Lathe

This is an old-fashioned tool used by the old-time turners for several purposes. Mine came from an old character called MacIntyre who wore steel-rimmed glasses and an old peaked cap, and had two tobacco streaks from the corners of his mouth in his beard.

This old-timer was a first-rate machinist and he helped me several times when I was just starting to work in a tool room. He had a 0-1/4" micrometer on his watch chain that he had made. He had also forged his own tin snips and hacksaw frame and had made almost all of his tools. There was a hook made from 1/4" rod hanging on the outside of his tool chest. The tool chest was the size of a large steamer trunk with some deep drawers, so when he wanted a tool at the bottom of the drawers he went "fishing" with the hook until he pulled out the correct tool.

1

The old fellow was still working in his late seventies when he died. His relatives removed the micrometers and other saleable tools and threw the rest into the garbage. I managed to rescue the roller tool and a few others. The foreman was in a panic the next week when he realized all the special tools for the odd jobs that old MacIntyre had been doing had been thrown out.

This tool has several uses: (1) When chucking a disk or a wheel casting, and the work is usually "wobbling" in the chuck, in spite of all your efforts to have it running "flat" (Photo 1).

Set the lathe to run at its lowest speed. The chuck jaws should be tightened just enough to hold the work.

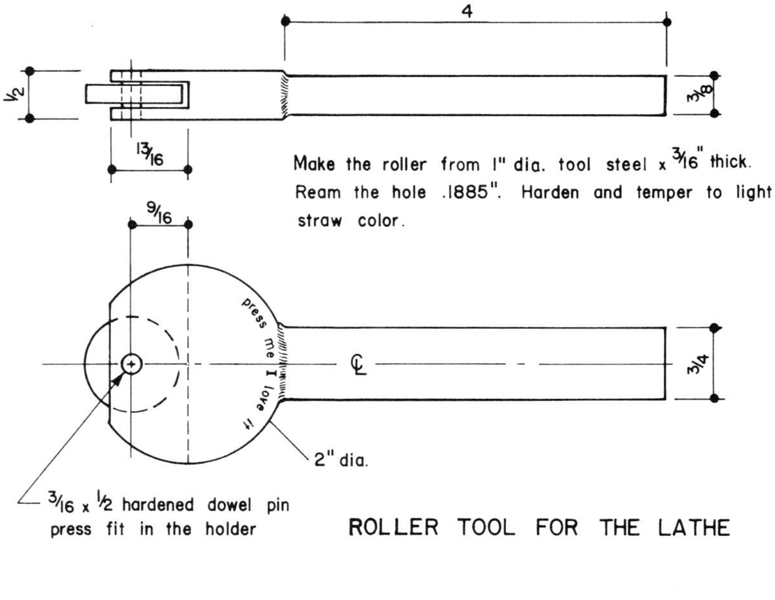

ROLLER TOOL FOR THE LATHE

Set the compound slide rest parallel to the lathe axis (Photo 2) and bring the roller tool up to touch the work. Clamp the lathe carriage to the bed and then apply pressure to the work with the roller tool from the compound slide rest. Continue the pressure until the work is running true. Stop the lathe and tighten the chuck jaws.

(2) When a large bar or pipe is being placed in the lathe (Photo 3). The chucked end is running true but the outer ends running eccentric or out of true.

Set the lathe to run at the lowest speed or turn it over by hand and apply pressure to the roller tool with the cross slide until the work is running true.

(3) For swaging work.

(4) We used this tool to straighten long reamers that had warped badly during the heat treating process. These were made of high-speed steel so they could stand some heating on the long shank without "drawing the temper."

The reamers were placed between centers on an old lathe that had seen better days. A dog was attached and the lathe set for a low speed. The warped shank was heated with an acetylene torch and the tailstock backed off as the work expanded. When the shank was hot enough, pressure was applied by the roller tool and the lathe continued turning with the roller pressing against the work until the work was cool. Most of the warpage was removed by this method.

2

3

A Wiggler

This is a simple straightforward job that should take only one evening to make. The shank and the cap don't require any explanation.

The balls are old ball bearings that have been heated red hot and then cooled slowly to soften them. Drill them in the lathe with the ball held in the three-jaw chuck.

The center finder is tapered in the lathe with a single cut file.

Use any convenient spring or else wind one up from piano wire.

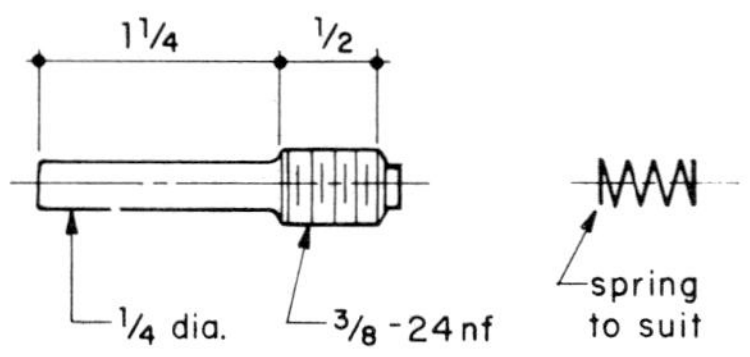

SHANK

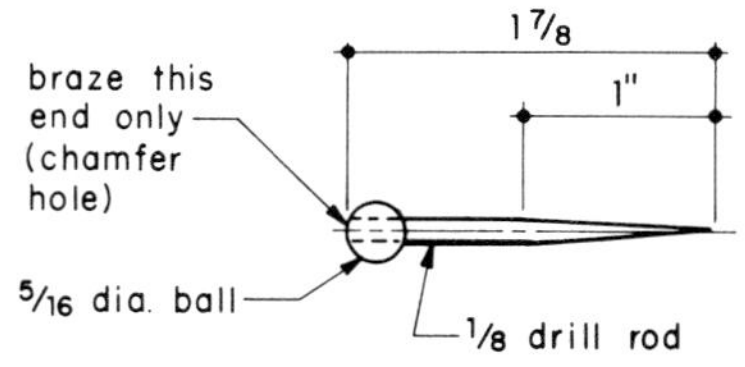

CENTER FINDER

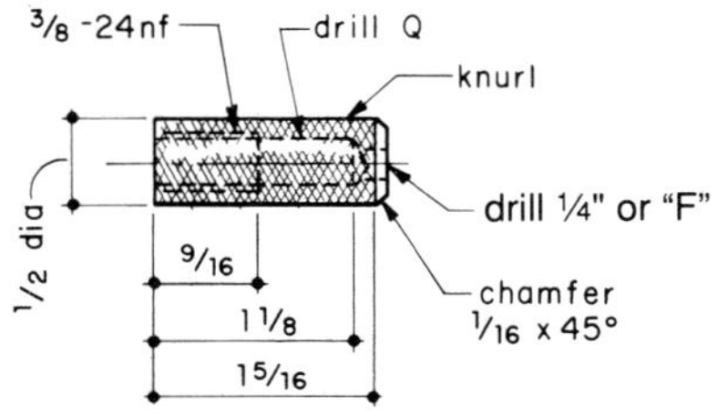

CAP
1/2" Cold Rolled Steel

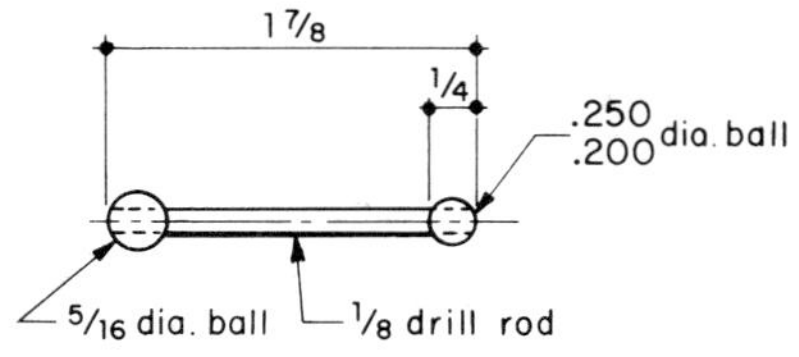

EDGE FINDER

Milling Machine Accessories

A Baseplate for a Dividing Head

The table on my small vertical mill is too short to hold the dividing head and the footstock and still have room enough between them for the work. I solved this problem with a piece of ½ × 6 × 21″ hot rolled mild steel plate. Do not use cold rolled steel; it tends to warp when the slot is milled on one side.

My plate was ground on the two edges and both sides on the surface grinder at the school. You will have to use the surface grinder at your local high school shop or send the plate to a commercial machine shop to be ground on a 8 × 24″ capacity grinder.

When the plate has been ground, mark out the location of all the holes and the edges of the slot. If the bolt holes down the center of the slot are marked out on the bottom of the plate, they can be drilled and tapped after the slot is machined. If the bolt holes are marked out on the same side as the lines for the slot, drill the 21/64″ holes prior to the milling operation. The tapping can be done, however, after the slot has been machined to size. Be sure that the ½″ holes for fastening the plate to the milling machine tube are located directly over the table tee slots.

I checked my plate for flatness by resting it on a surface plate. I tried to place a .002″ feeler gage between the surface plate and the edges of the baseplate. I found that the plate had not warped after the machining operations, so a second grinding operation was not necessary.

I had the use of a large milling machine to mill the slot. However, the slot can be machined on a small milling machine by milling the slot in "installments" and the result will be satisfactory.

Use the bolting holes to clamp the plate to the milling machine table. The plate will probably be narrower than the milling machine table, so you can use one of two methods to align the plate from the front edge of the table: 1) Use a depth micrometer or, if this is not available, a combination square (Figure 2) will suffice.

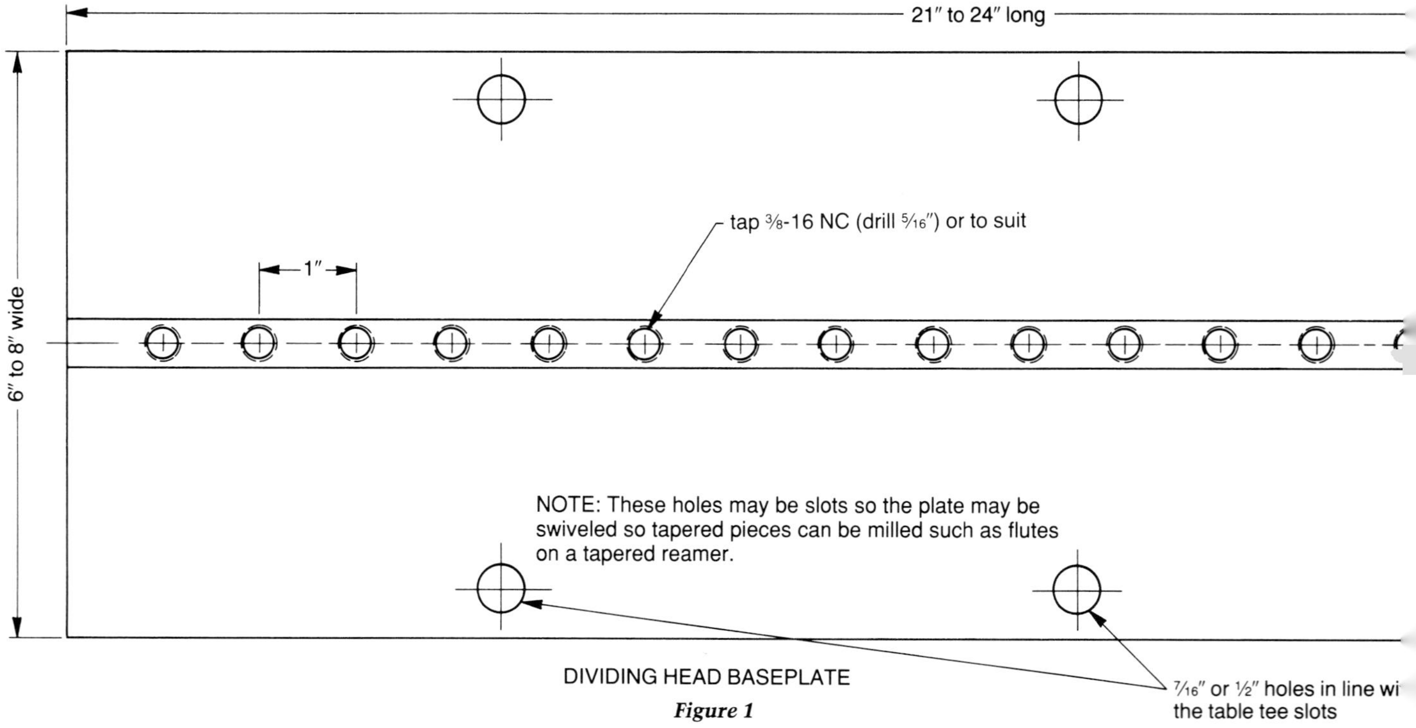

DIVIDING HEAD BASEPLATE

Figure 1

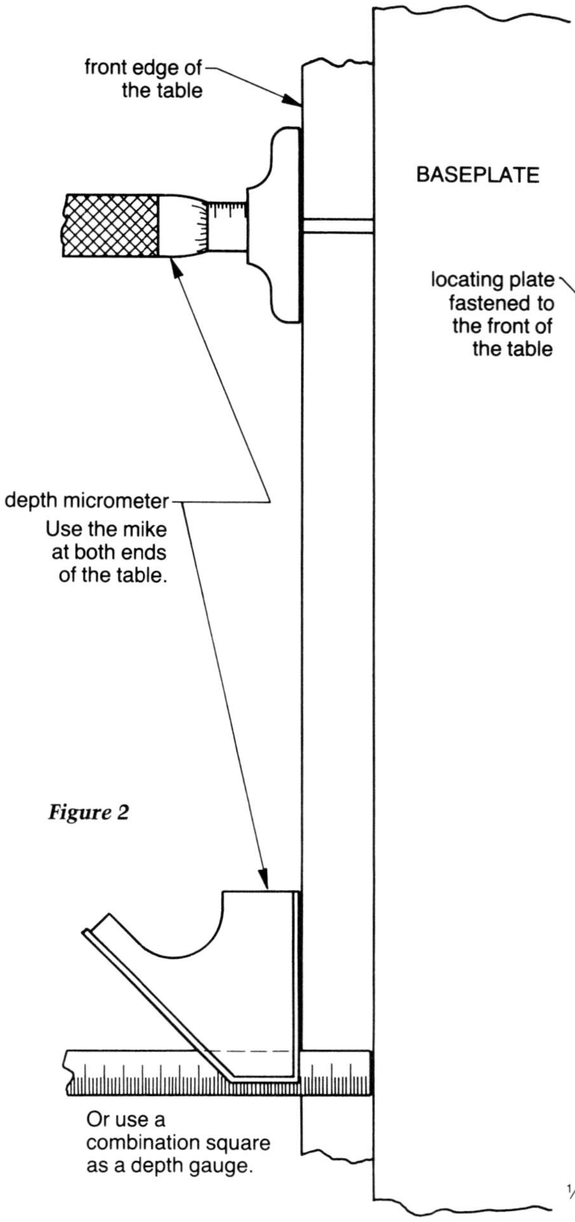

Figure 2

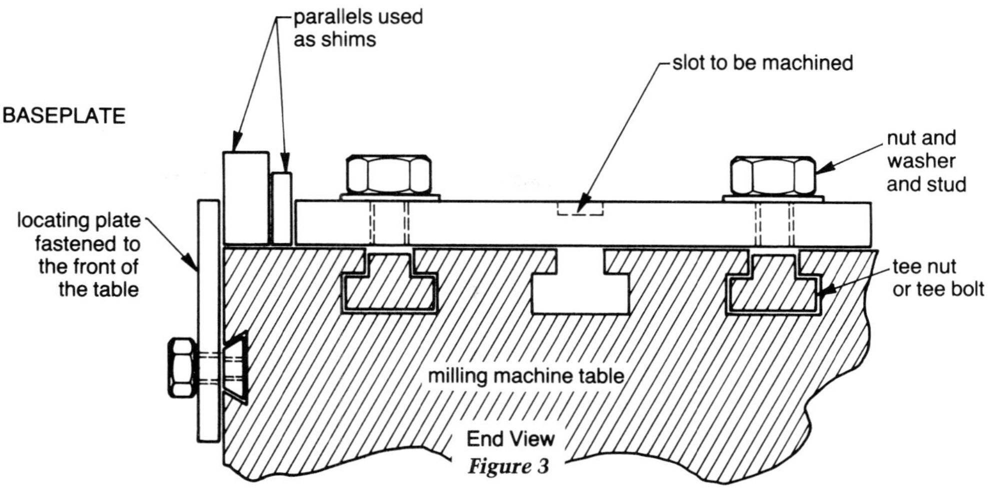

Figure 3

2) Clamp a piece of flat bar stock to the front of the table with the machine screws and dovetail nuts from the table stops. The bar should project ¼″ to ½″ above the table. Now place a parallel between the baseplate and the bar in front of the table to align the baseplate with the edge of the table (Figure 3).

Now run the table to one extremity of the table travel and clamp the baseplate to the table. Align the milling cutter with the layout lines and set the depth of cut to the full depth, if possible. Proceed to mill the slot until the table has reached the full length of its travel. Loosen the clamping bolts and return the table to the starting position. The baseplate will remain in the same position and the cutter will remain in the slot. Reclamp the plate and continue milling the slot.

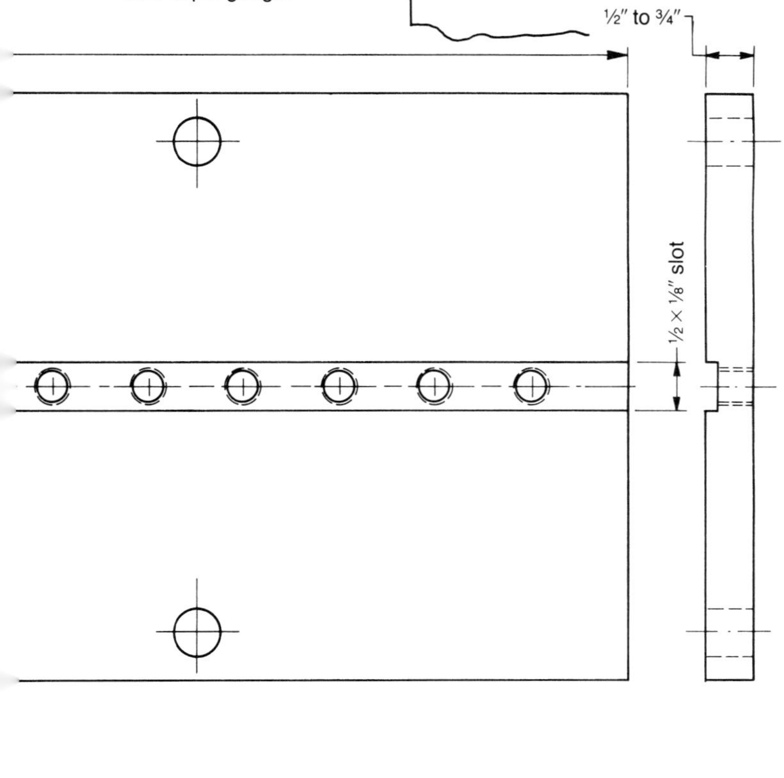

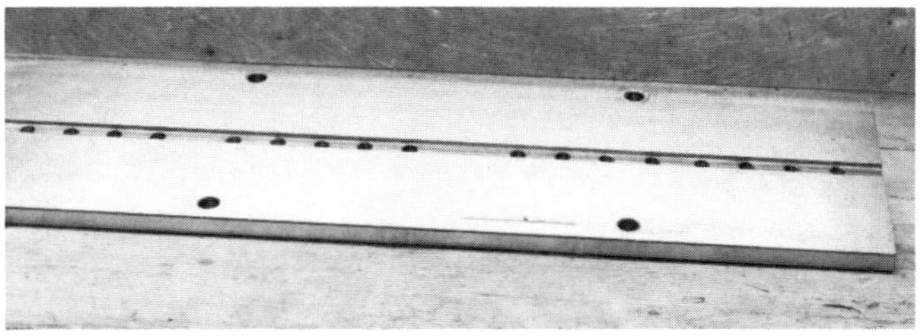

General view of the baseplate for the dividing head.

Dividing head and footstock resting on the baseplate.

A Finger Stop for the Vertical Milling Machine Vise

This is a simple locating device used in jobbing machine shops when tooling costs must be kept to the minimum. It will very accurately position identical parts for drilling, boring and reaming in your drill press vise.

Drill and tap a hole on one side of the fixed jaw of the vise to take a 1/4" bolt.

Make the body of the stop as you see it in the drawing to suit the vise. The rod is 1/4" diameter drill rod faced flat on one end. The rod is clamped in place with a No. 10-32 NF setscrew with a nylon tip. A brass screw could be used as an alternative.

For jobs like locomotive axle boxes, this stop will save a lot of time. Make a mark on each axle box so they can be located in the vise in the same position they will occupy in the chassis. Lay out one axle box as accurately as possible. Set this box in place in the vise and clamp all the table slides when the block has been centered with a wiggler. Now center drill each block in turn, pilot drill, drill out nearly to size, bore with a micrometer boring head if available, and ream to size. Change the blocks instead of the tools for each operation. I made four axle boxes for a locomotive in an hour by this method.

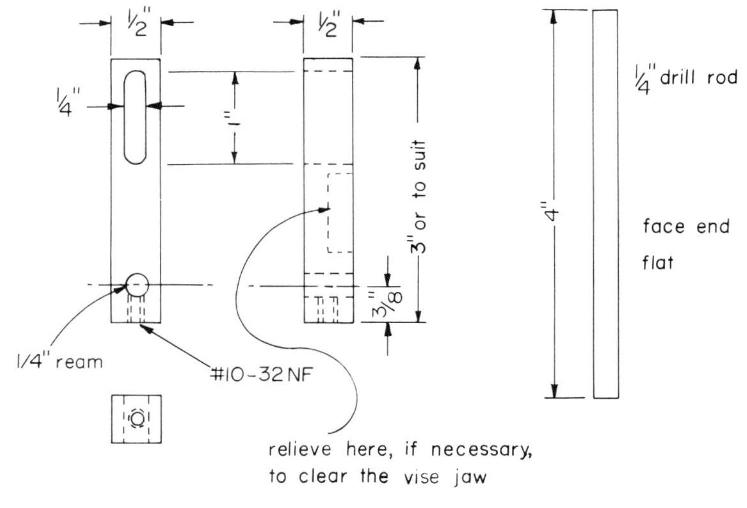

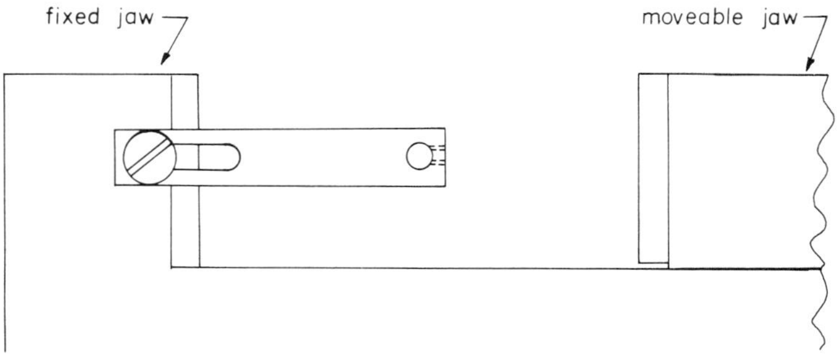

Lighting Your Vertical Mill

The photographs show how I adapted a cheap lamp from a local store into an ideal milling machine light. I bought the cheapest model with a plastic clamp (which I threw in the garbage can where it belonged). I then made up the small block shown in the photographs from a piece of aluminum from the scrapbox. Most any material will suffice – even hardwood.

There is a ½″ diameter hole for the lamp with a ¼″ screw in the side to prevent the lamp from turning too freely. I used 10-32 screws threaded into the cap on the column; however, practically any size can be used.

The lamp can be adjusted so you can easily work to a layout line. It makes a good one-evening project.

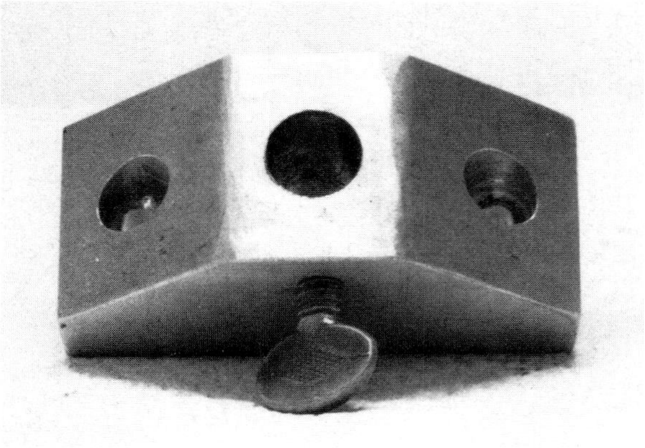

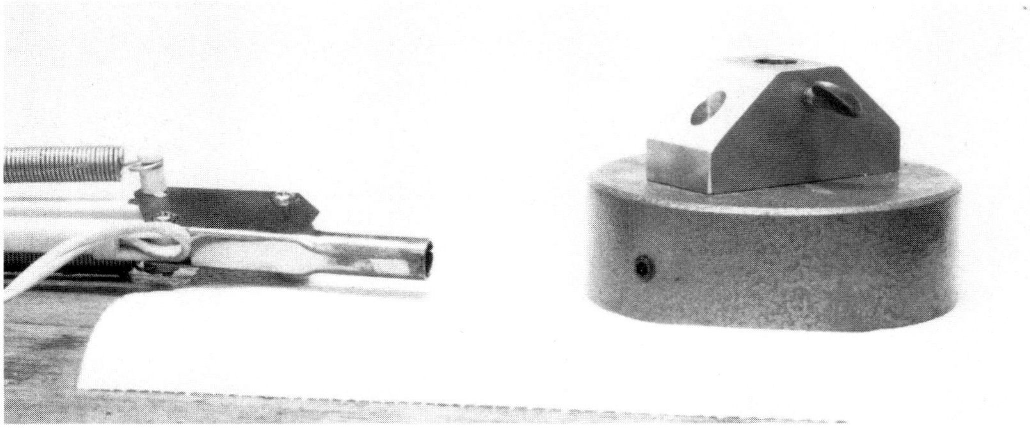

A Micrometer Stop Nut

I have never found that the twin nuts on the threaded depth gage at the front of my milling machine were satisfactory for accurate depth control. Recently, I made a micrometer stop nut similar to the type used on the Bridgeport milling machines. My stop nut is graduated in increments of .005". The wider spacing between the graduations makes them easier to read, and I can estimate in between; however, I find that .005" is accurate enough for most of my work.

Unlike factory production work, most tasks done in home workshops will be satisfactory if they meet the personal standards set by an old friend of mine who said, as he became old and his eyesight diminished, "I am satisfied with work that is pleasing to the eye and can be measured with a rule."

Now I bring the cutter down to just touch the work. Then I screw down the stop nut until it quits turning on the lug of the head casting. The reading is noted. Then the nut is "backed off" the required depth of cut, and the quill is brought down until the nut is firmly against the lug. Lock the quill in place and mill with confidence. Some vertical milling machines do not have a fine feed handwheel for the quill but only the feed for drilling. The micrometer stop nut will enable them to set the depth of cut accurately.

The first time I used my micrometer stop nut, I noticed that the stop nut was pressing so hard on the lug after the work had been machined that I had to use the fine feed handwheel to raise the quill slightly before I could reset the stop nut for another cut. Obviously, the quill

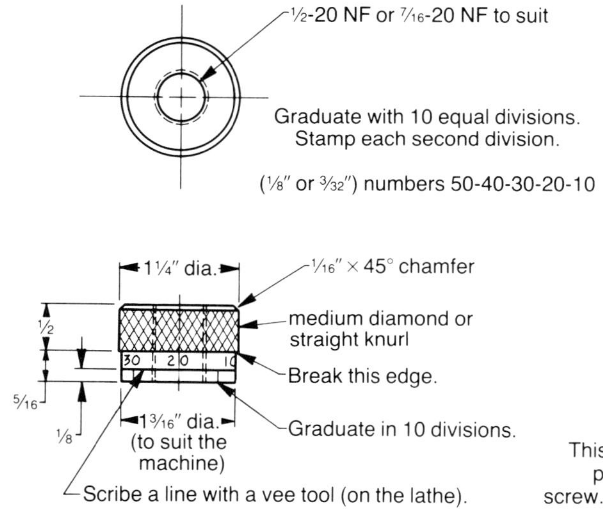

Figure 1

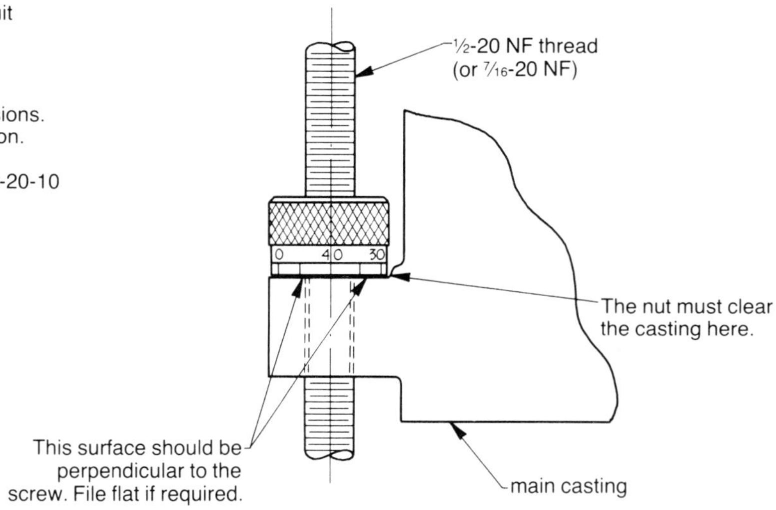

Figure 2

had been moving down slightly during the milling operation. This was caused by the downward force from the helical flutes on the milling cutter, even though the quill had been locked in position.

My micrometer stop nut is made from a bar of C12L15 leaded screw stock. However, any low carbon steel or brass or aluminum may be used. If you prefer aluminum, try to use 6061-T6 or another hard aluminum intended for machining, rather than one of the soft aluminum alloys that do not machine well.

Face the end of the bar flat and then take a light cut along the bar to be sure the work is running true. If the work is running eccentric, a satisfactory knurling job is almost impossible. Medium sized diamond or straight knurls may be used. I have straight knurls in a squeeze-type knurling tool that was made from instructions in *Model Engineer* magazine about twenty years ago. Knurling is a very severe operation and puts great strain on the spindle bearings and the cross slide screw and nut when a standard knurling tool is used. Guy Lautard sells drawings and the knurls for this type of knurling tool. To knurl a bar this diameter, a very slow speed is required – about 50 rpm – so the knurls will have a chance to form the metal, since this is a swaging operation rather than a cutting operation. Be sure to use some oil and keep your fingers, sleeves and rags clear of the work.

Turn the part where the graduations will be made to a diameter to suit the space available on the lug casting. Scribe a line with a vee threading tool about ⅛″ from the end of the

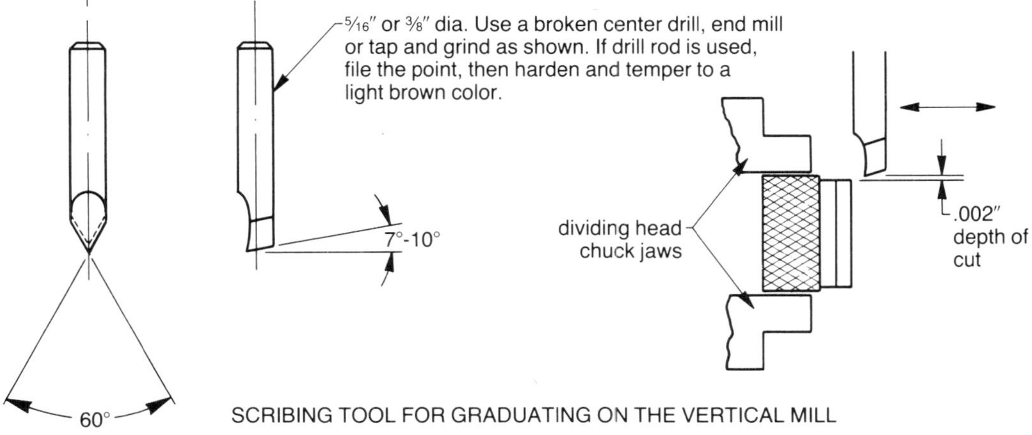

Figure 3

work to indicate where the graduations should stop. The 3/16″ wide space above this line is where the numbers will be stamped (Figures 1 and 2).

Center drill, then drill a pilot hole of 1/4″ or 5/16″ diameter. Then drill the tapping size hole (25/64″ for 7/16-20NF and 29/64″ for 1/2-20 NF). Countersink 1/2″ diameter for the 7/16″ thread and 9/16″ diameter for the 1/2″ thread to be sure the nut will be flat after tapping.

The tapping operation should be done on the lathe to be sure the thread is perpendicular to the bottom face of the nut.

Hold the shank of the tap in the tailstock drill chuck; slide the tailstock along the lathe bed while the lathe spindle is turning at its lowest speed. Reverse the spindle rotation to back out the tap. If you are using a spiral flute tap, the cuttings will be coming out along the flutes and you can go to the full depth without removing the tap. However, if you are using a standard four-flute plug tap, you should reverse the spindle rotation once or twice to withdraw the tap so the shavings can be removed with a jet of compressed air. If you don't have a small air compressor, put this on the list of planned purchases for your shop.

The torque required to tap these fine threads is so slight that there should be no trouble with the tool turning in the drill chuck or having the drill chuck rotate in the tailstock spindle socket. (If this occurs, see the paragraph at the end of this article). Now part off and reverse the nut in the three-jaw chuck to face and chamfer the top of the nut.

My micrometer stop nut was graduated on my milling machine using a dividing head. The graduating tool was an old broken center drill, ground to an angle of 60° as shown in Figure 3. The depth of cut was .002″.

For those people without a dividing head, the graduating can be done on the lathe before the parting off operation. The bull gear on the

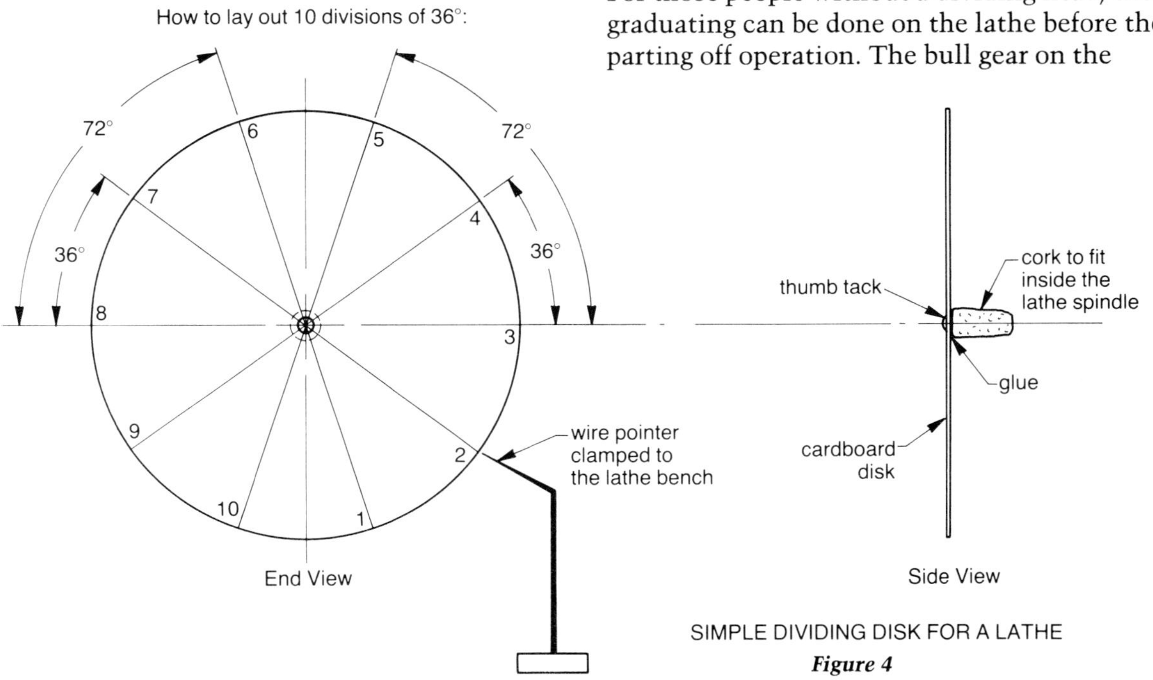

SIMPLE DIVIDING DISK FOR A LATHE

Figure 4

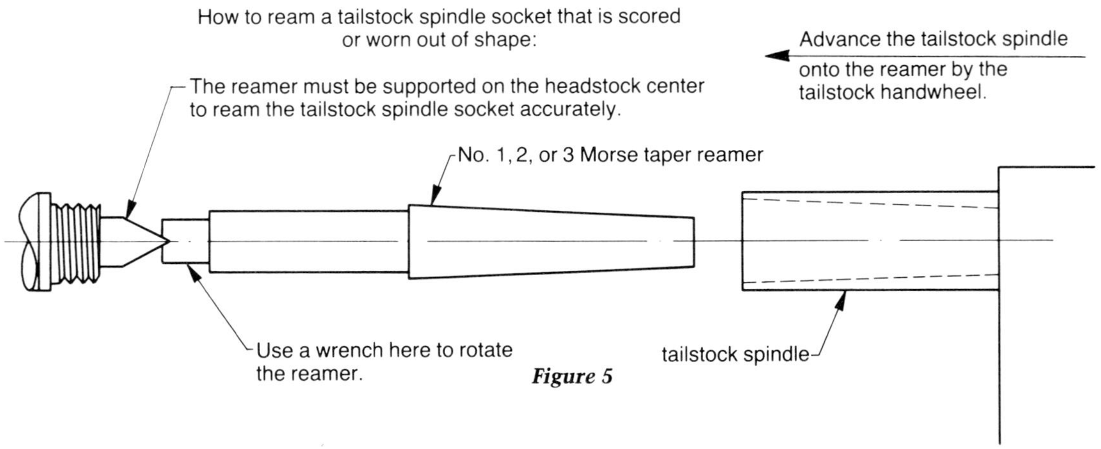

Figure 5

lathe spindle is often used for dividing; however, most of these gears are not divisible by ten, so another method will be used. Make up a cardboard disk 8" to 12" in diameter and divide this into ten divisions of 36° with a protractor (Figure 4).

Now glue the disk to a cork that fits into the back end of the lathe spindle. Arrange a wire clamped to the lathe bench as a pointer. Use a vee threading tool clamped in the tool holder on its side and traverse the carriage by the handwheel. A more elaborate dividing disk can be made of metal with several circles of holes and a spring-mounted locating pin. This method of "indexing" was used by the early clockmakers and on the ornamental turning lathes made by Holtzapffel, Evans, and others.

Clamp the nut in the bench vise by the parallel surfaces (top and bottom) for stamping the numbers. Use 1/8" or 3/32" size number stamps.

Run the nut down the threaded rod until the bottom of the nut touches the top surface of the lug. If the top surface of the lug is not parallel with the nut, this will have to be corrected by filing. For those not proficient with a file, this will be a good chance to learn. Make a final test with a thin coating of mechanic's blue on the bottom of the nut. File off the blue spots until the nut bears evenly all around.

SLIPPING DRILL CHUCK

If the tap rotates in the drill chuck: there may be oil present in between the tap shank and the jaws; the chuck has not been tightened enough; or the chuck may be old and worn out.

If the chuck arbor rotates in the tailstock spindle socket: the arbor may not have been driven into the socket firmly enough; or there may be oil or dirt present. If the chuck has been slipping frequently, this can be remedied. Buy or borrow a No. 2 or 3 Morse taper "Finished Reamer." The roughing reamer has notches along its cutting edges and will not be suitable for this operation.

Locate the back of the reamer on the headstock center and slide the tailstock along the bed until the reamer enters the spindle socket. *Note that the spindle should be extended about one inch so the ejector bar will not touch the reamer.*

Clamp the tailstock to the lathe bed (Figure 5). Use a wrench on the square end of the reamer and advance the tailstock spindle by the handwheel. The reamer should make only two or three rotations, for this will be a "light" scraping only,to renew the accuracy of the Morse taper.

Now make three or four lines with chalk along the arbor of the drill chuck. Insert the arbor into the Morse taper and rotate the arbor. If the chalk is removed all along the arbor, the fit is correct. If the chalk is removed only at one end, this will have to be corrected by careful filing. Hold the drill chuck in the three-jaw chuck by the cylindrical portion below the ring. Use a single cut file to file the large end of the arbor. Test frequently until you are satisfied.

SHOP NOTE

There are times when I want to use a vise with jaws in a horizontal position for filing some small part that cannot be clamped satisfactorily in the ordinary bench vise. The photos show how I converted a hand vise by brazing a rectangular bar of steel along one side and how this is clamped in the bench vise.

The Humble Angle Plate

The proud owner of a new milling machine will soon discover that the vise supplied will have to be supplemented with several sizes of angle plates.

These can be purchased from the advertisers in *HSM*. Be sure to test the plate with a precision square. If you don't have a precision square, here's a cylinder square that can be easily made in the home shop, and will be as accurate as the usual precision squares sold for shop use. (I'm excluding the hardened squares sold by Starrett and others from this category, since they are too expensive for most home shop machinists.)

Most angle plates are made of iron, and usually warp after machining. Try to buy one large plate, such as is shown in the photos. Clamp the plate to the table and add shims as required, to have the face absolutely vertical. Now clamp the other angle plates to this for machining to compensate for any warping that might have taken place.

Once one plate is true, use it to machine all the other plates. I keep one angle plate for this purpose.

Some suppliers sell rough castings so you can make your own angle plates. The machining is easy. Photo 1 shows a rough casting being machined on a shaper instead of a milling machine. This was done because I have a shaper, and tool bits are cheaper and easier to sharpen than milling cutters. Note the use of two bolts and nuts as jacks to compensate for the uneven thickness of the casting, and to accommodate the center rib on the casting. This plate has a center rib instead of the usual ribs at the ends to make clamping the work to the angle plate easier. The rib helps to prevent the casting from warping, and stiffens the plate in use.

Structural steel supply houses usually buy their material in long lengths (60 feet), and then cut to the customer's requirements. They will often have pieces of angle iron left over in 1-, 3- or 4-foot lengths. This will make very good angle plates with a little machining. Useful sizes are 4 × 4" by 1/2" or 5/8" thick, and 6 × 6" by 3/4"

The first cut on a rough casting. Here, a shaper is used because tool bits are easier to sharpen than milling cutters. Note the two bolts used between the moving jaw of the vise and the rough casting.

Use a precision square to align the casting vertically.

A substantial cut. This cut has increased in depth due to the draft on the casting.

Another view of maching the ends of the casting.

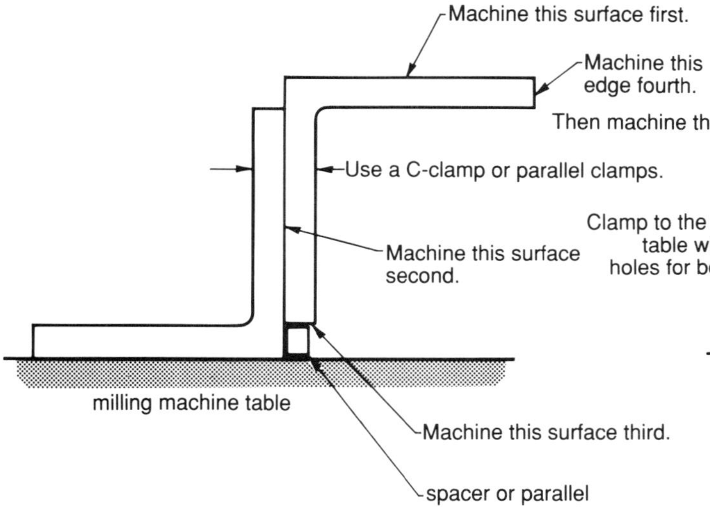

MACHINING PROCEDURE
FOR CASTING OR STRUCTURAL ANGLE IRON

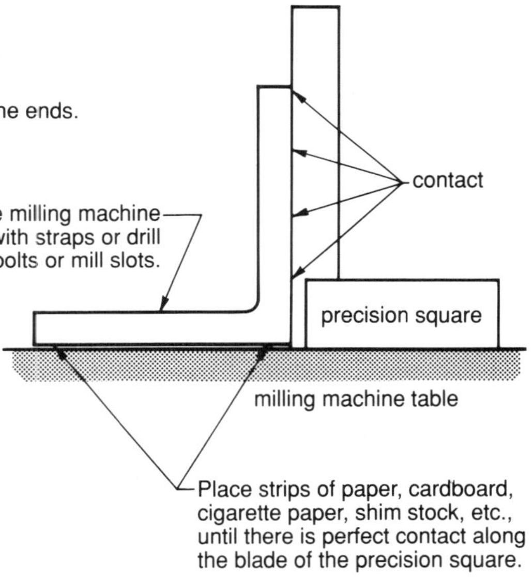

Use this method to set one angle plate so the vertical surface is 90° to the table surface.

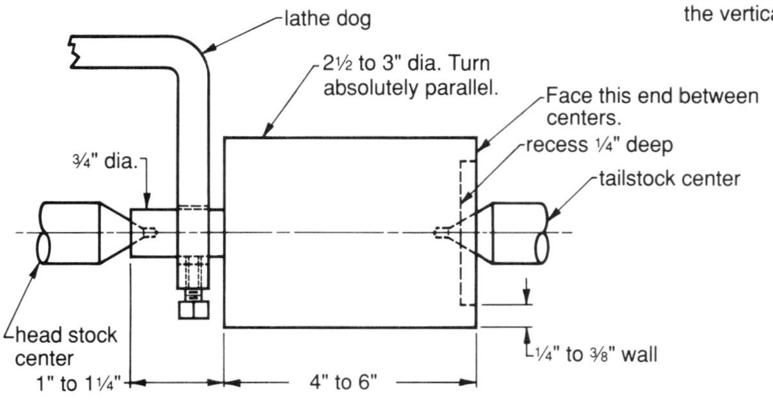

Rough turn both ends and center drill with the work in a chuck. Set up between centers to turn the cylindrical portion, then face the bottom also between centers. The bottom will be square with the cylinder.

HOW TO MAKE A PRECISION SQUARE
called a Cylinder or Bottle Square

Testing on suface plate after all the machining has been completed.

thick, or whatever's available. They may not want to make out a bill for these short pieces. A donation to the coffee fund will usually suffice.

I recently made three angle plates 15" long for a double spindle wood shaper to be used to hold wooden fences. Structural angle iron was ideal. Machining five slots 5/8" wide in each angle plate took eight hours, but was certainly preferable to making a special pattern, having castings made, and so forth.

If you are making castings, you will probably have to take a cut all over the faces, edges, and ends to remove the scale. Then machine any slots, drill the holes, and so on. Let the casting sit for a few weeks. Now take a light cut with a fly cutter to finish the plate. You will find that the plate warped when the stresses were released in the first machining operation.

I've shown two special angle plates. One has an end plate so the plate can be turned upside down on the drill press, and can be used to clamp the adjacent side and the end for making a box or drill jig, for instance.

HOW TO MACHINE AN ANGLE PLATE

Let us presume you are using an iron casting with a center stiffening rib. First, grind a 45° bevel 1/8" wide along all the edges. This will permit the cutter to start in soft metal under the hard surface scale. Now place the angle plate in the vise with a piece of cardboard between the casting and the fixed jaw to protect the vise from the rough casting and to give a better grip, for the cardboard will compensate for the irregular surface of the casting. The casting must be held firmly against the fixed jaw. Use two blocks of wood or two machinists' screw jacks. Now machine the surface marked No. 1 on the drawing. Next, place this surface No. 1 against the fixed jaw and machine the surface No. 2. To machine the edges No. 3 and No. 4, place the No. 1 and No. 2 surfaces down on the vise seat.

To machine the ends No. 5 and No. 6, set the angle plate vertically in the vise and align with a square as shown on the drawings. Don't forget to remove the burrs after each machining operation or the angle plate will not be "square."

A LONG ANGLE PLATE

This long angle plate is made with two ordinary angle plates and a piece of 1/2 × 3" cold rolled steel plate about 20" long. My two angle plates are 3 × 3 × 6" with a stiffening rib in the middle. These plates are

The pattern is shown on the left and the finished angle plate is on the right. The pattern was made by my father over fifty years ago.

Another modification of a standard angle plate. The end is useful for clamping the sides and end of a box for drilling and tapping holes. The end also permits the angle plate to be turned upside down.

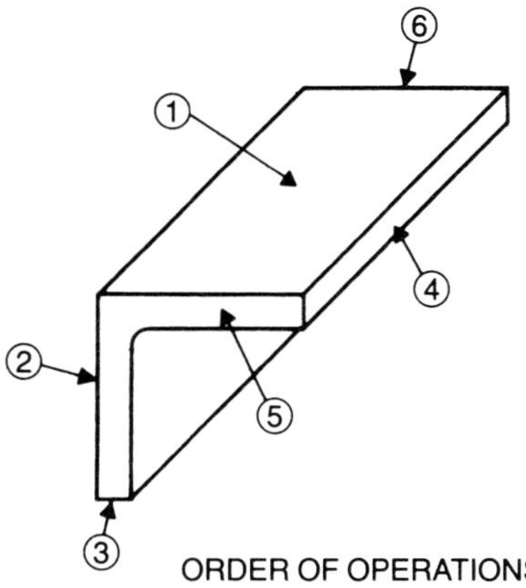

ORDER OF OPERATIONS

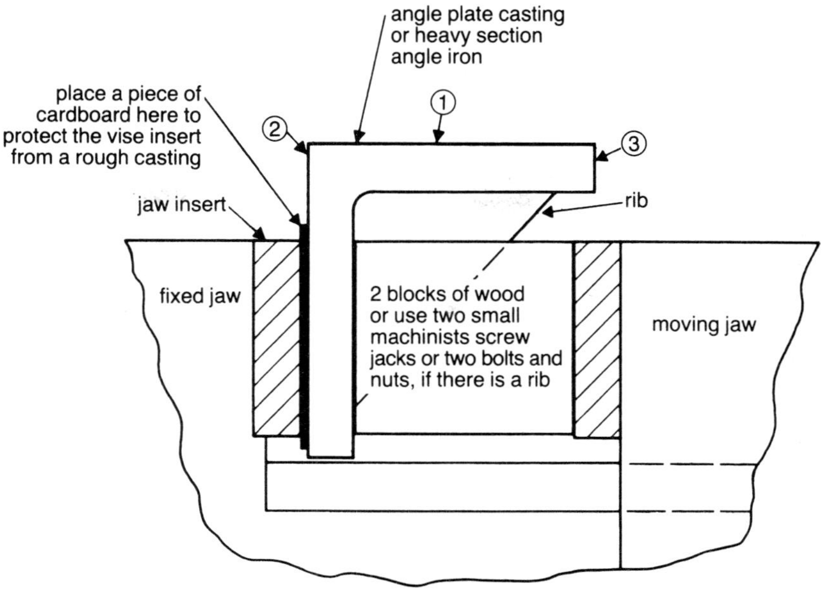

SET UP TO MACHINE THE TWO MAIN SURFACES OF AN ANGLE PLATE ON A VERTICAL MILLING MACHINE OR A SHAPER
If the angle plate is cast iron try a fly cutter if you are using a vertical mill to avoid damaging an expensive end mill.

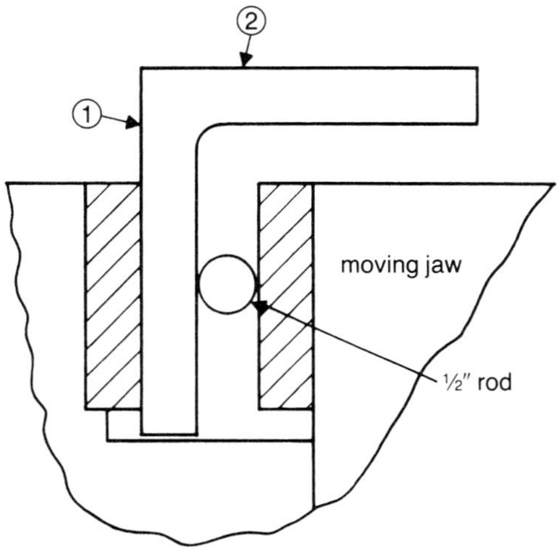

ALTERNATE METHOD WHEN ANGLE IRON IS USED OR THERE IS NO STIFFENING RIB

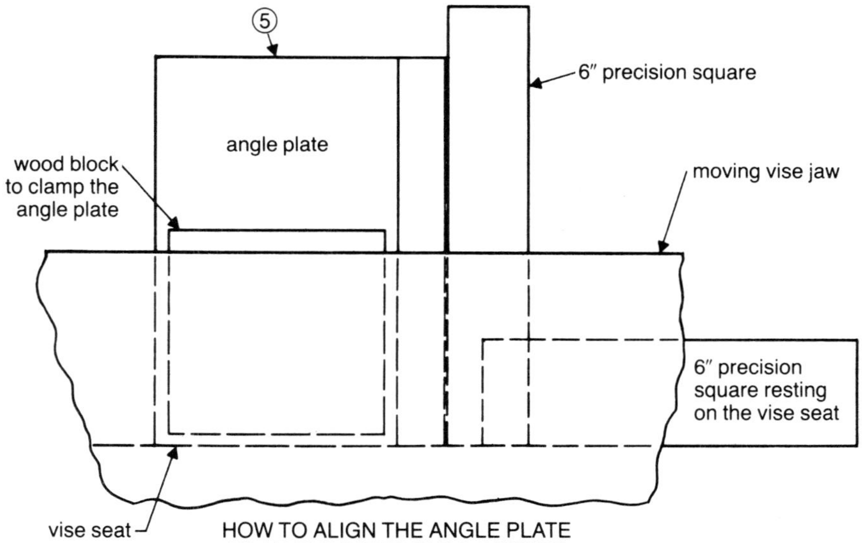

HOW TO ALIGN THE ANGLE PLATE TO MACHINE THE ENDS SQUARE WITH THE SIDES

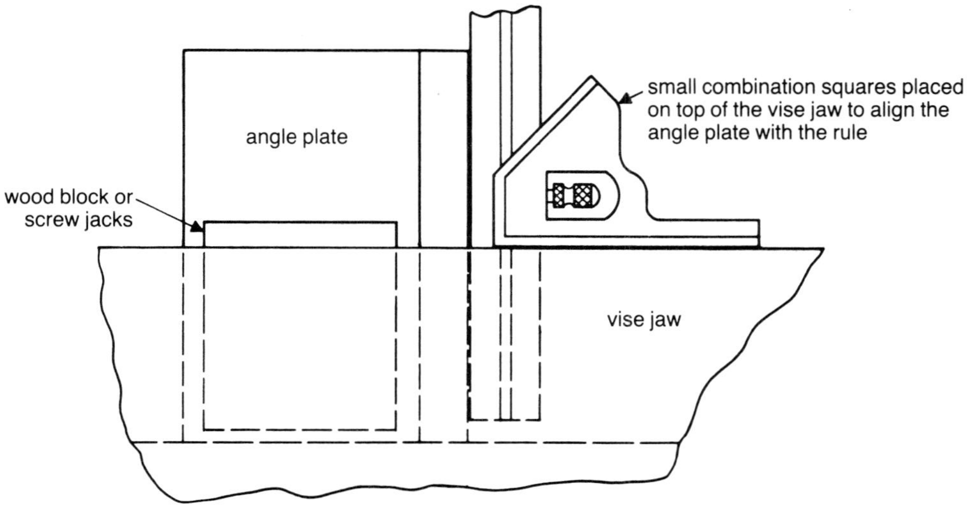

ALTERNATE METHOD TO ALIGN THE ANGLE PLATE TO MACHINE THE ENDS

homemade from a pattern designed and fabricated by my father more than fifty years ago.

When the long angle plate is not needed, the two angle plates can be removed from the cold rolled steel plate and used for other operations. I used 1/4-20 NC flathead Allen screws to fasten the cold rolled plate to the angle plates because I had them. However, I could have used Allen cap screws countersunk into the cold rolled plate, or fastened from the back through the angle plates and threaded into the cold rolled plate. It doesn't really matter which method you use.

To use the angle plate, remove the milling machine vise and align the cold rolled plate parallel with the table. A dial indicator is the most accurate method; however, a quick method I use is to measure from the front of the table to the ends of the cold rolled plate with a combination square.

To mill the edges of plates or bars, rest the work directly on the table or on parallels and clamp the work to the cold rolled plate with large C-clamps. These clamps can be moved as the cutter moves along the work. Very long plates can be machined in stages by returning the table to the starting position, sliding the work along the table, and then reclamping the work to take another cut. The edges of long locomotive frames can be machined in this manner.

To machine the axle box openings and the pedestal keys, either mill out a clearance in the plate before machining the frames or simply mill the axle box opening in the frame and pass through the plate, as well. This will give maximum support with a large clamp on each side of the opening. The cold rolled plate can be put aside until a similar job comes along and a new plate can be fastened to the angle plates. To machine connecting rods, first remove the plate from the angle plates and drill a hole in each end to bolt the plate flat on the milling machine table. The rod will be fastened to the plate with bolts passing through the holes for the bearings at the ends of the rods. Drill and tap two or three holes in line with these two holes for screws to touch the back of the rod. This prevents springing when fluting the rods, and compensates for the taper when reducing the thickness of the rods. The connecting rod will be fastened to the plate with bolts and thick washers under the bolt heads. Place a thick washer at each end between the plate and the rod so the cutter will not touch the plate.

Clamp the plate flat on the table and line up the layout lines with one of the cutting edges of the milling cutter (or use a wiggler). Don't

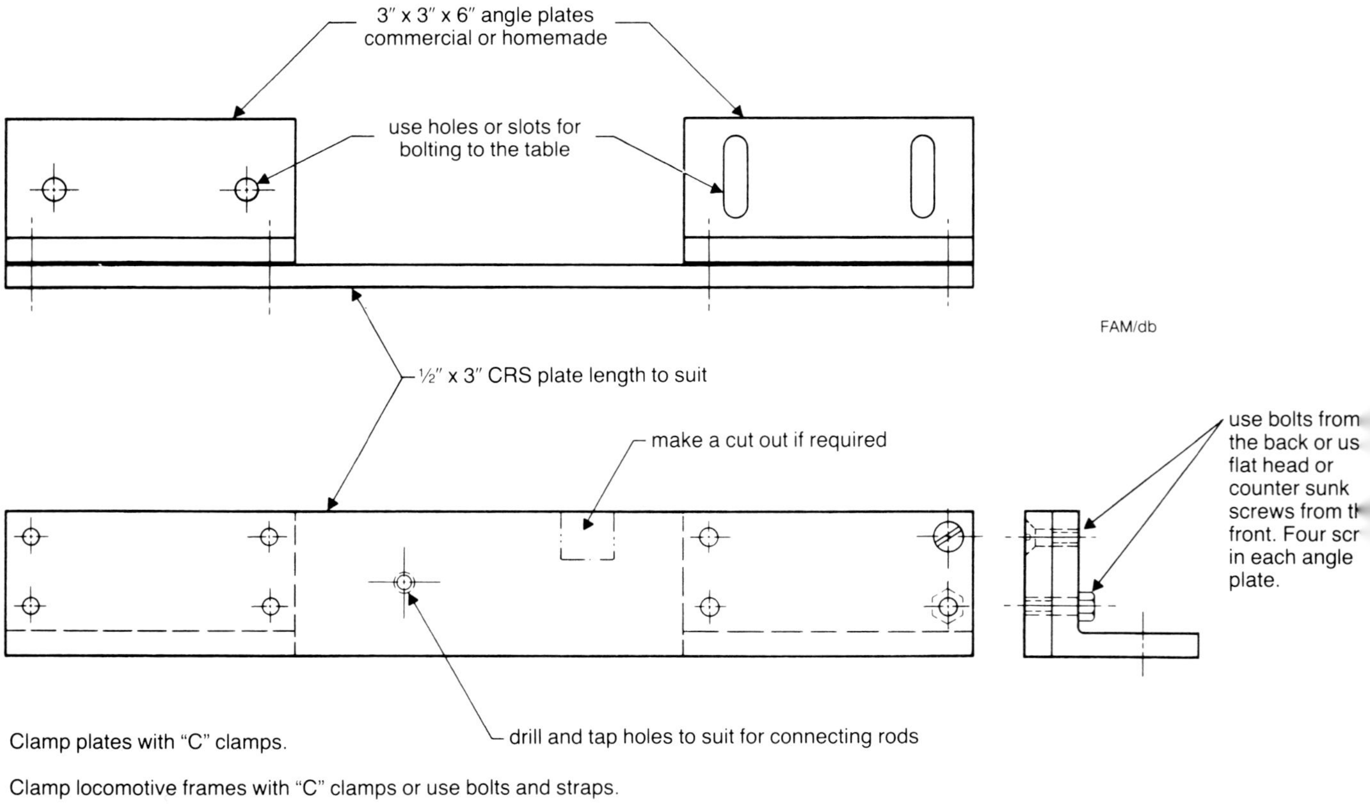

HOMEMADE LONG ANGLE PLATE

forget that the direction of the table feed must be *against* the direction of the cutter rotation.

To machine the sides (reduce the thickness) of the rods, the cold rolled plate will have to be fastened to the angle plates. Use the washers of suitable thickness between the plate and the rod to bring the surface to be machined parallel with the plate. Adjust the screws in the plate to just touch the back of the rod so that you prevent springing during the machining operation.

To flute the rods, use a Woodruff-type cutter and adjust the big end of the rod up and down to bring the edge of the flute to be milled parallel with the table. Check with a surface gage. A tapered flute will require two cuts: one for the top edge and then the rod will have to be adjusted to locate the bottom edge of the rod parallel with the table for the second cut.

The long angle plate is a milling fixture that will greatly increase the versatility of your vertical milling machine. Originally, I intended to use a 20" length of 4 × 4" angle iron. This would have had to be machined over its length as most angle iron, as rolled, is not accurate enough to be used as an angle plate. By using a cold rolled plate there was no need for a machining operation since the finish is good enough for a fixture. The length of the plate can be made to suit the work, and plates can be drilled and tapped as required to suit various jobs.

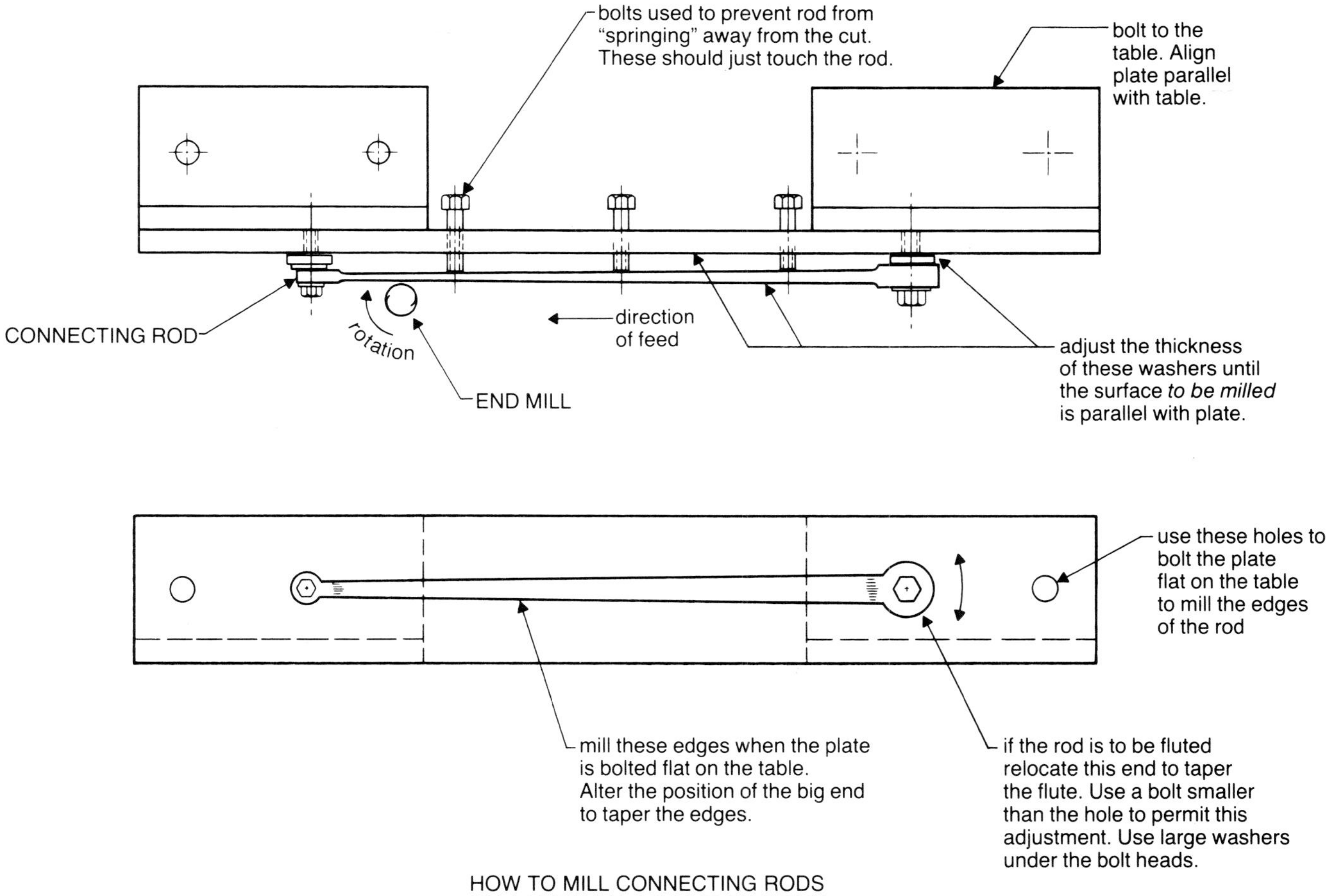

HOW TO MILL CONNECTING RODS ON A VERTICAL MILLING MACHINE

1 ***Showing the 7/8" diameter hole with the 1 9/32" diameter counterbore and the two slots for removing the center bushing.***

A New Rotary Table

Last year, I bought a good 8″ diameter rotary table from Busy Bee. This table has an index plate for easier and faster dividing or indexing than is possible with rotary tables using a dial for dividing.

The worm-to-worm gear ratio on my plate is 1:72. That is, the crank must be turned 72 times to move the table one revolution. The index plate has circles of holes on both sides. (Some rotary tables have sets of two or three plates.) The plate has 49, 41, 33, 27, 20, and 17 holes on one side and 47, 39, 31, 23, 19 and 16 holes on the other side.

Use this formula to calculate the number of turns of the crank for the required number of divisions:

$$\text{Number of turns of the crank} = \frac{72}{\text{number of divisions required}}$$

$$\text{To find the indexing required for 36 divisions} = \frac{72}{N} = \frac{72}{36} = \text{2 turns of the crank}$$

$$\text{To find the indexing for 21 divisions} = \frac{72}{N} = \frac{72}{21} = 21\overline{)72}\ \ 3,\quad 63,\quad 9$$

$= 3\frac{9}{21}$ Reduce fraction to lowest common denominator.

$= 3\frac{3}{7}$ Now the only circle of holes suitable for the denominator 7 is 49.

$\therefore 3\frac{3\times7}{7\times7} = 3\frac{21}{49}$ Set the locating pin on the crank to fit the 49 hole circle and the brass arms on the index plate will be set for 21 spaces.

When setting the arms, do not count the hole the pin is resting in, but rather count the spaces. Then set the arms so the pin will drop into the correct hole. Advance the arms to the right, or clockwise, after each indexing. Note: Indexing is always done by turning the crank clockwise to avoid errors.

My table came with a No. 2 Morse taper hole in the center. This is standard; however, I don't find it convenient, so I converted it to a 7/8″ diameter straight bore with a counterbore

1⁹⁄₃₂" diameter by ⁵⁄₃₂" deep to accommodate the flange on the center bushing. Two slots were milled ½" wide by ¼" deep to permit two screwdrivers to lift out the center bushing by the flange.

I was in a great hurry, as usual, and did not align the tee slot correctly, so the slot came out at an angle. Later, I aligned the tee slots with a dial indicator and remilled the slots correctly, as I should have done the first time.

HOW TO USE THE ROTARY TABLE

The work will be more easily located if a suitable hole can be drilled in the center of the work. Then the work can be easily located over one of the centering pins. The work is usually supported on parallels, or a piece of scrap steel or aluminum can be placed between the work and the table to avoid running a drill or cutter into the table. The work can be clamped by the usual straps and bolts into tee nuts.

However, there usually is not enough room to use straps effectively, so it is easier to drill four holes at 90° and pass bolts directly through the work into the tee nuts. Extra material can be left on the work for this purpose, to be machined off later. Incidentally, you will probably have to make up special tee nuts to fit the tee slots in the table.

My table came from India, and was a great value for the money. However, I had to remove a few burrs and "ease" a few tight spots to have the table working to my satisfaction. The machinery and equipment from India and Taiwan can usually be improved by a little "tuning up" when you receive it. This is because the assembly has been done quickly to bring the price down to where most of us can afford to buy the tools.

2 ***Milling out the slots correctly, this time.***

4 ***The ½" centering pin or plug in place.***

3 ***Removing the center bushing.***

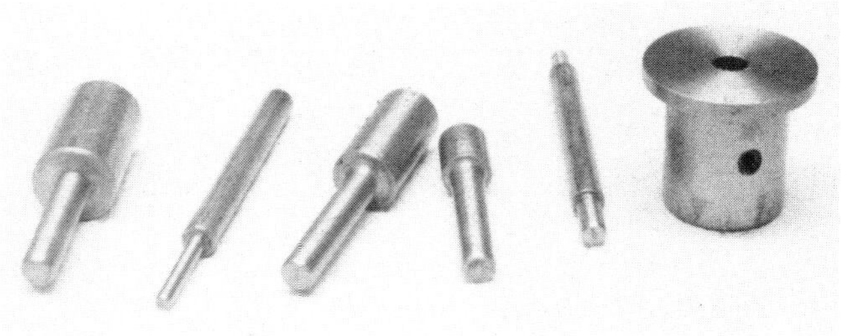

5 ***The center bushing and an assortment of centering pins.***

6 ***The 1/8″ centering pin in place.***

7 ***The center bushing showing the setscrew for holding the centering pins.***

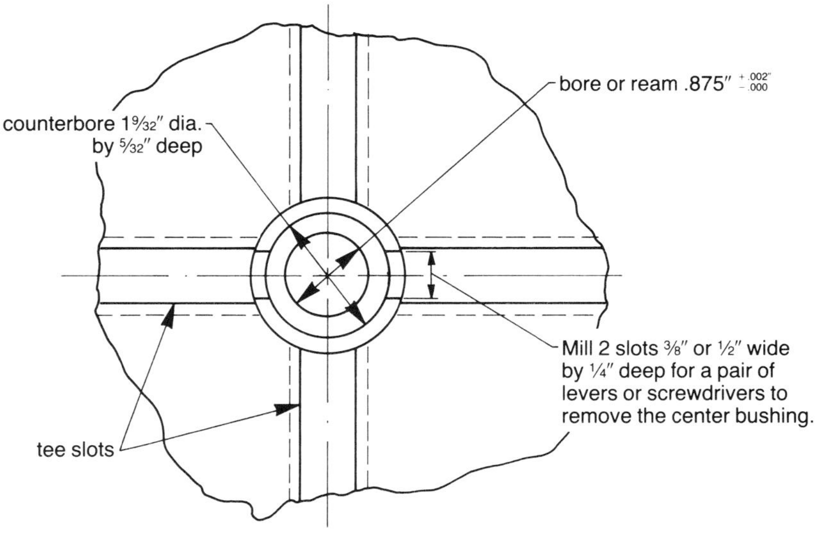

MY ROTARY TABLE
Top View

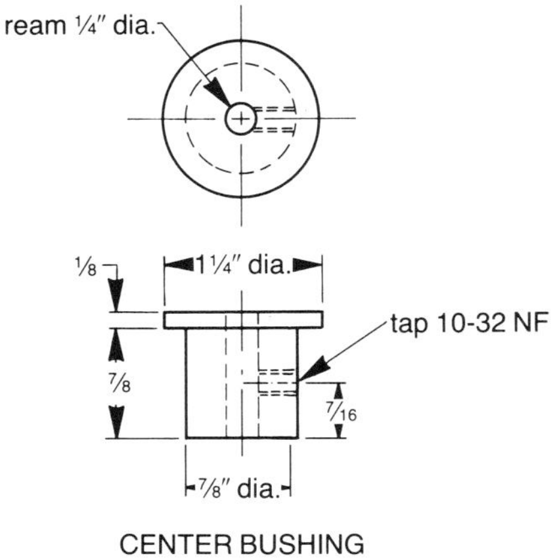

CENTER BUSHING

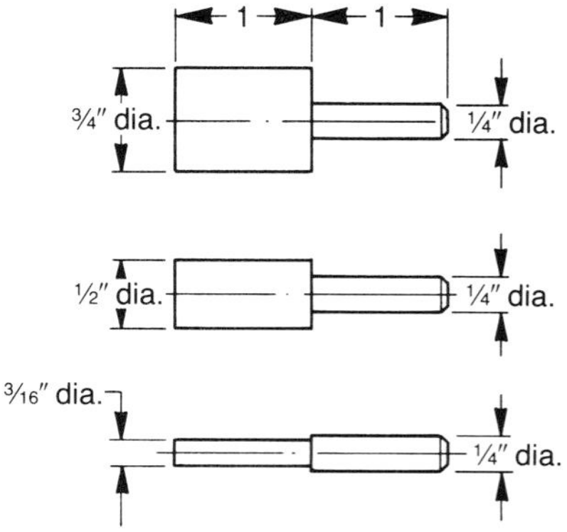

TYPICAL CENTERING PINS

Note: To locate the rotary table exactly under the milling machine spindle axis, use a 1/4″ diameter pin in the collet or drill chuck and adjust the table until the pin slips into the center bushing. This is much quicker than using a dial indicator and nearly as accurate.

Homemade Milling Cutters

In the home shop, the cost of end mills and other standard milling cutters can be a major factor. Special cutters, such as tapered end mills, ball end mills, corner rounding cutters and dovetail cutters, etc., are too expensive to be purchased if they are to be used only once for a special job.

Small cutters can be made from drill rod. For larger size cutters, use oil- or water-hardening tool steel instead of drill rod, as drill rod is very expensive in the larger sizes.

Make the cutters by turning to the shape required and then cut the flutes by milling or by hacksawing and filing. The cutters described later in this article are made in the same manner as milling cutters. Harden as usual and sharpen with a small oilstone.

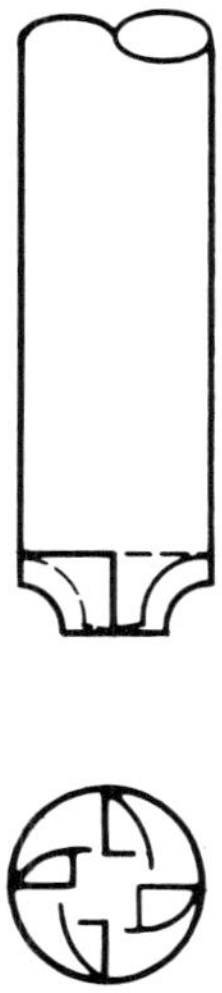

CORNER ROUNDING CUTTER

CUTTER TO FORM A FILET

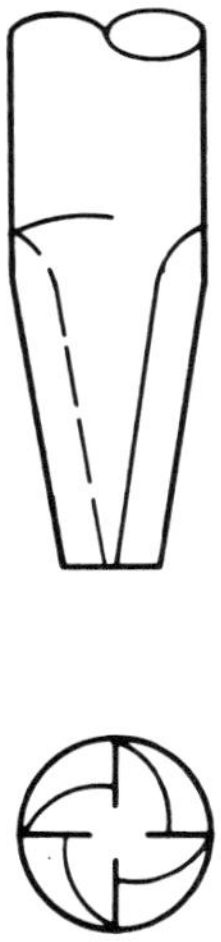

OUR LIP TAPERED END MILL

BALL END MILL

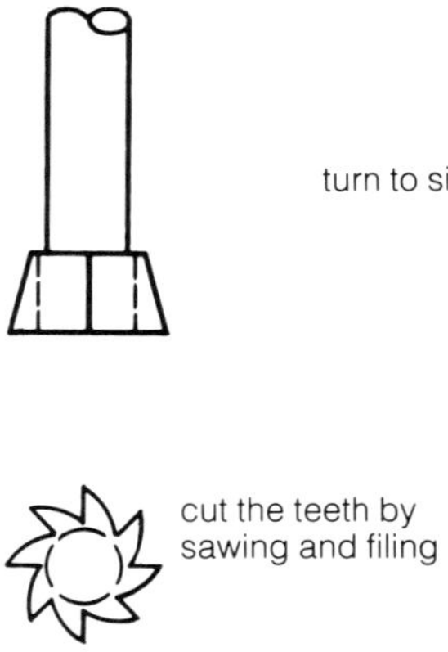

DOVE TAIL CUTTER

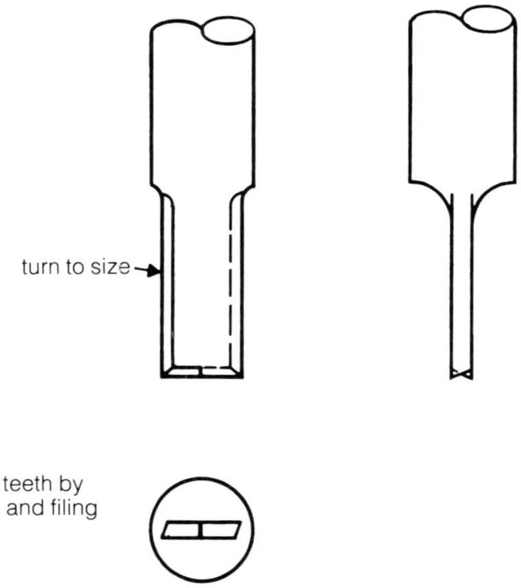

OLD STYLE HOME-MADE SLOT DRILL

FAM/db

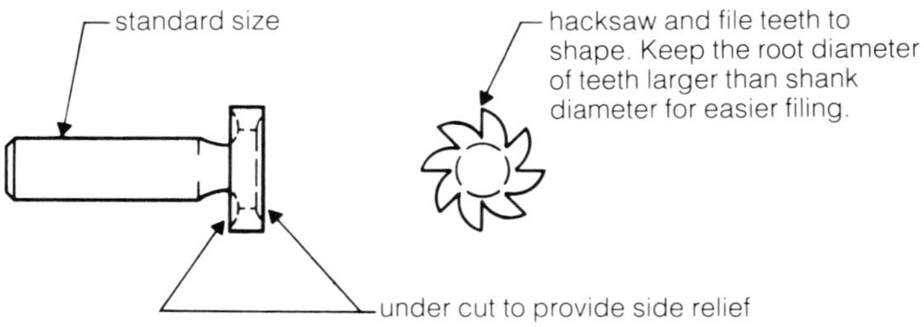

WOODRUFF TYPE CUTTER

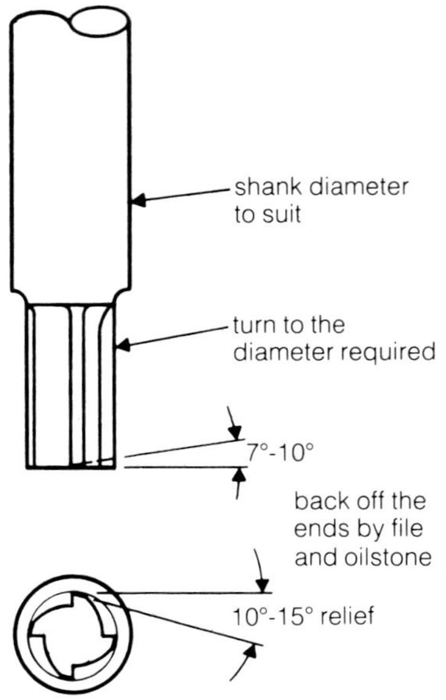

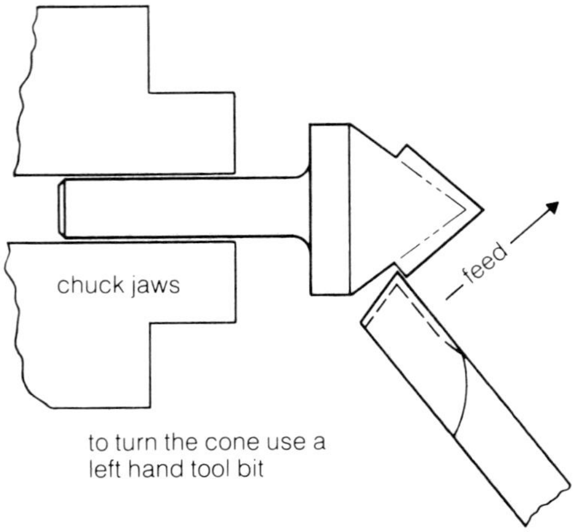

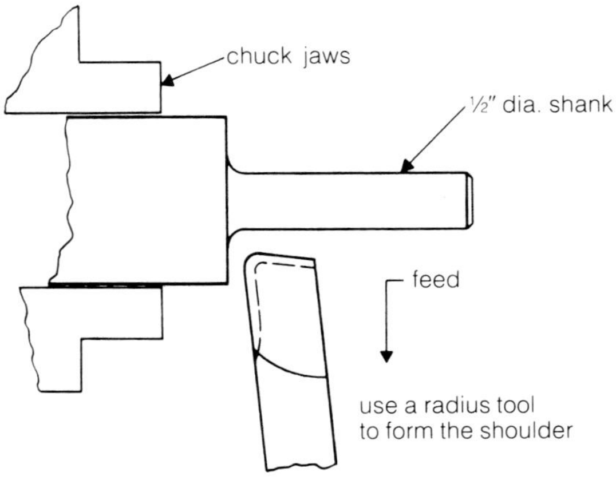

TURNING THE SHANK

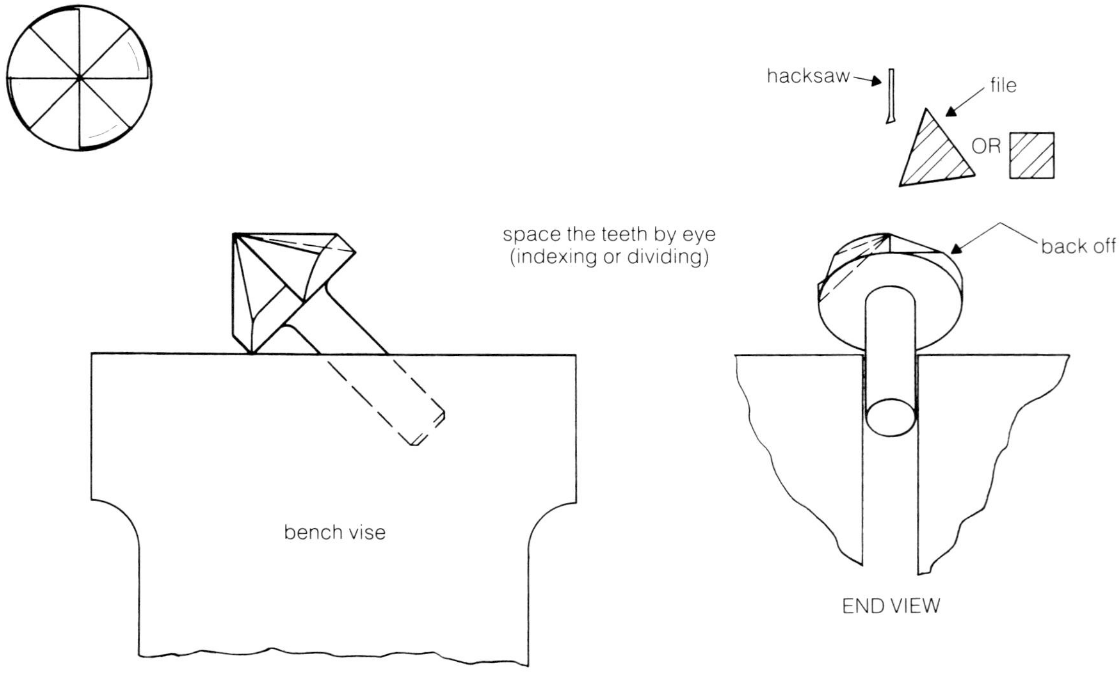

FRONT VIEW

END VIEW

When buying end mills, try to buy the double-end cutter or the single-end type with two flutes.

COUNTERSINKS

Most countersinks are made with an included angle of 82° (41° with the center line) to suit flathead screws.

The photograph shows a large 60° countersink I made.

Begin by turning a 1/2" diameter shank. Reverse the work in the chuck, and hold by the shank to turn the angle. When turning the angle, use a left-hand turning tool, and turn from the circumference toward the center. This will avoid the problem of enough side clearance near the point of the countersink. To make the flutes, place the countersink in the vise at an angle.

The flutes are made by filing the gullet and then backing off by file. Large flutes can be started by using a hacksaw to form the front of the teeth. Space the teeth by eye. The irregular spacing will reduce the risk of chattering.

To harden, heat the countersink until it becomes non-magnetic and then quench in oil or water to suit the steel. Temper to a light straw color.

Sharpen on the front face of the teeth with a narrow grinding wheel or an oilstone.

Keep the speed low when using tools made from drill rod and similar carbon tool steels. A speed of about 30 feet per minute will be suitable.

COUNTERBORES

Counterbores have two principal uses. They are used to spot face a casting so the surface under a screw head or a nut is smooth and square with the hole. Spot facing enables a screw head to seat properly. The other use, and the name, comes from the counterboring operation. The clearing hole for a screw is

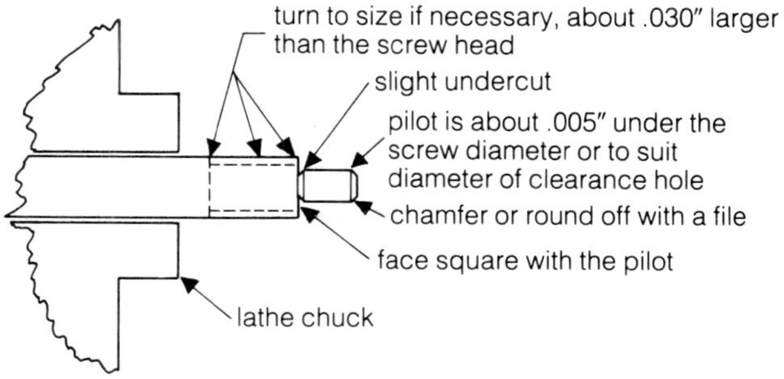

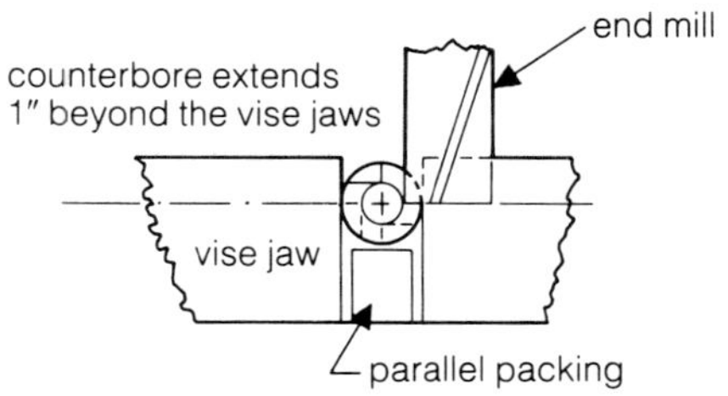

Hold the counterbore horizontal in the vertical milling machine vise. Bring the end mill to the center line and mill to the pilot. Rotate the counterbore in the vise to index "by eye". Use 3 or 4 flutes. I would suggest, however, that you at least lay out to an approximately equal spacing.

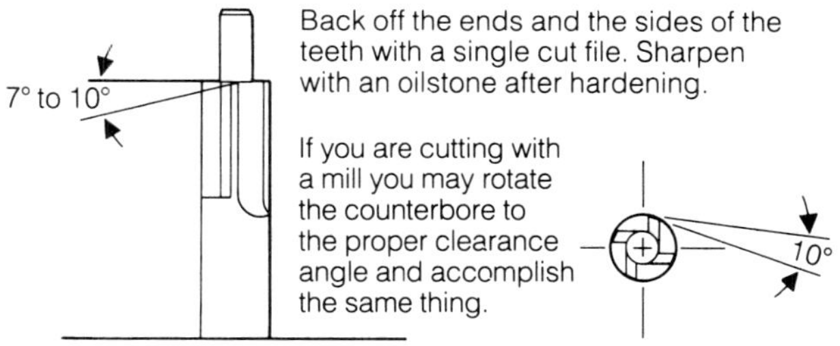

Harden the same way as the toolmakers reamer.

CUTTING THE FLUTES ON A VERTICAL MILLING MACHINE OR WITH A LATHE MILLING ATTACHMENT

Counterbores with flutes cut by end milling.

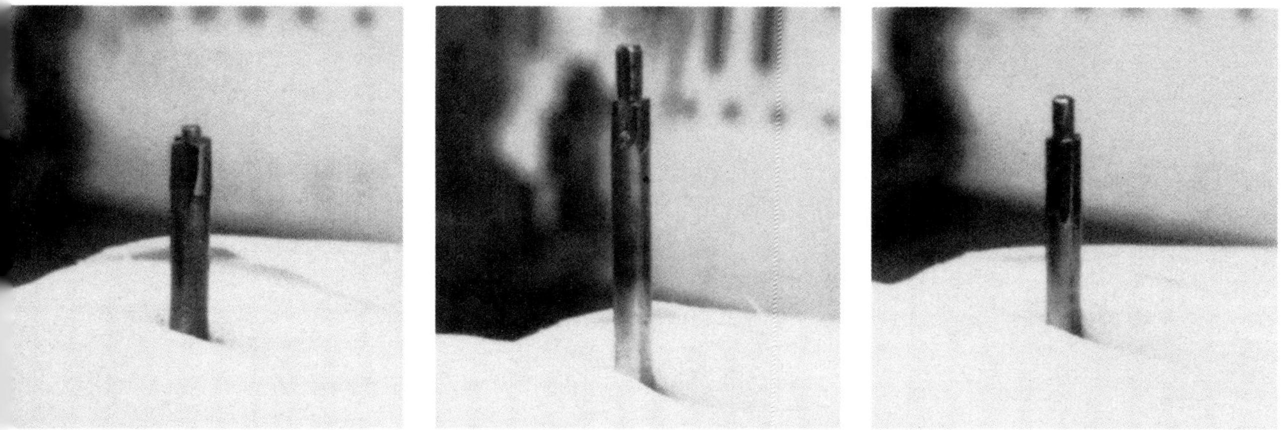

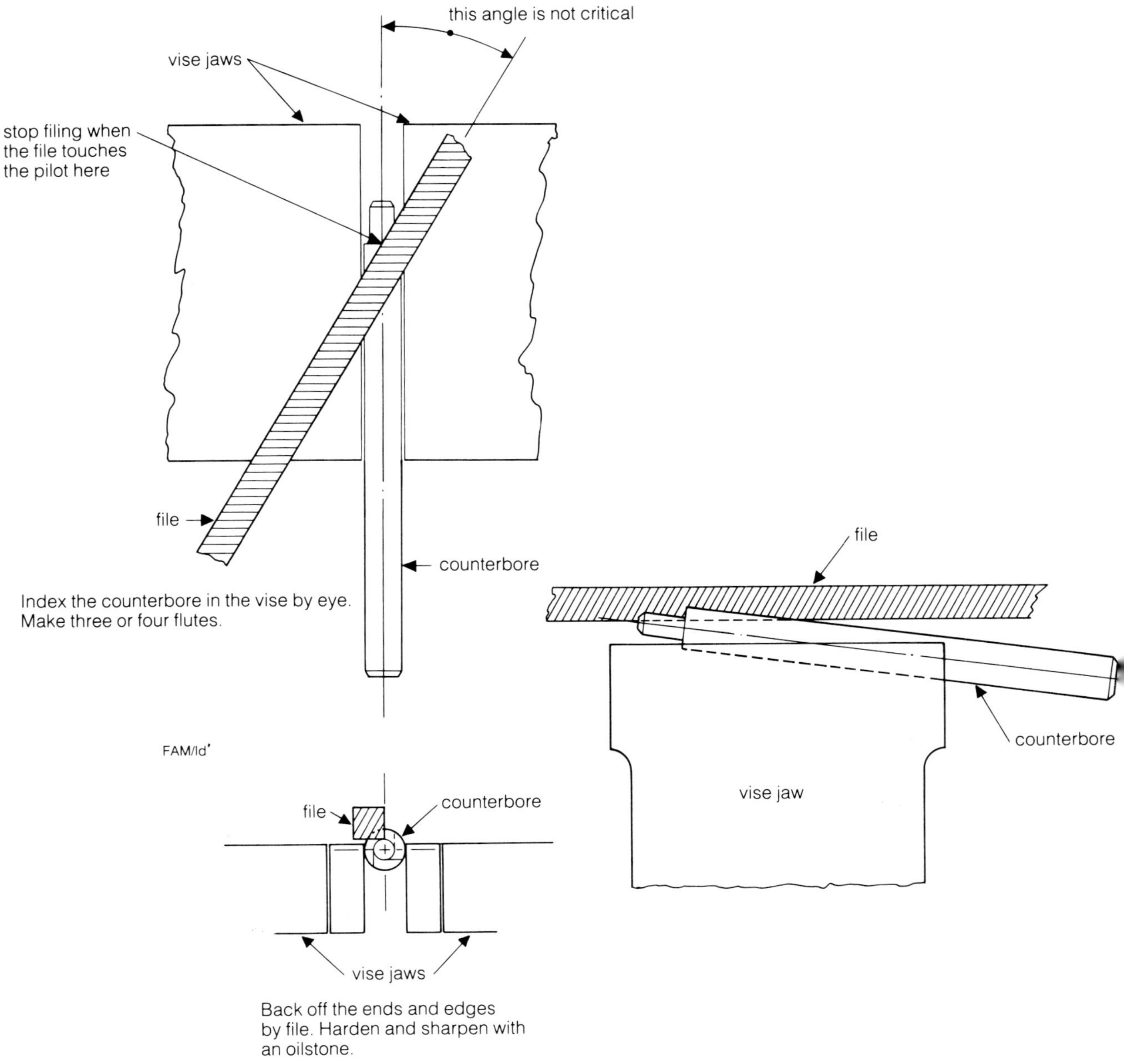

MAKING THE FLUTES BY FILING

Counterbores with flutes formed by hand filing.

enlarged (counterbored) to permit a filister head screw (cheese head) or a hexagon socket head (Allen) screw to sit flush with or slightly below the surface of the work.

Commercial counterbores with interchangeable pilots are available from the usual suppliers and undoubtedly these are best.

For occasional use, and for the small screw sizes used by model makers, very serviceable counterbores can be made from drill rod. These are fun to make and several can easily be made during an evening.

The flutes are easily machined on a vertical milling machine or on the lathe with a milling slide. The spacing or indexing is not important.

Bring the end of the milling cutter level with the center line of the pilot and move the cross slide over until the cutter just touches the pilot. Be sure to feed the work against the direction of the cutter rotation.

The flutes can be made by file alone. Place the counterbore in the bench vise at a slight incline (10°). Use a square file and continue filing until the file touches the pilot. Rotate the counterbore in the vise and repeat until three or four flutes are made. Back off the ends and the edges with a single cut file. Harden as usual and finish sharpening with an oilstone.

A Hollow Mill

The photographs show a hollow mill or "running down cutter" I made to repair broken welding tips. The tips had the threaded portion broken off. They had to be reduced in diameter to 5/16" and then threaded 26 threads to the inch. The bend in the tip makes them very difficult to hold in the lathe chuck.

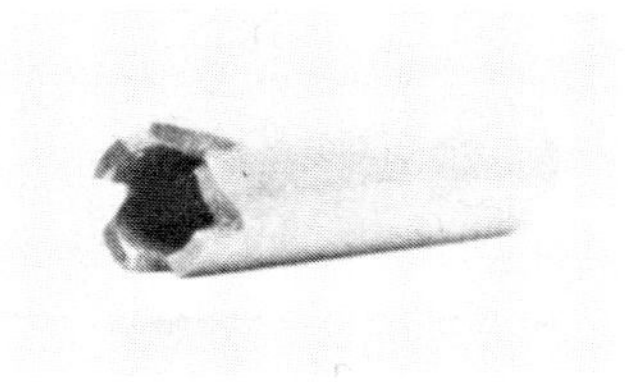

1

The hollow mill was made from a 2" length of 1/2" diameter drill rod.

The drill rod was held in the lathe chuck and both ends were faced. The bore was drilled with small drills and then reamed 5/16" with a chucking reamer. Next, I drilled up to within 1/4" of the far end with a 21/64" drill for clearance. If the hollow mill is to be used for more accurate work, such as turning a batch of small screws on the lathe, then the bore should be slightly tapered inside with a taper reamer.

2

To make small screws, hold the hollow mill in the tailstock drill chuck. Turn, thread and part off.

The teeth are filed by hand with the cutting edges backed off about 5°. Harden in oil and temper to a medium straw color. Stone the ends of the teeth and on the front face and the tool is ready to use. I made the tool in the photos in about half an hour.

These hollow mills can be purchased from the Union Twist Drill Company, Butterfield, and other manufacturers but, for our needs, these homemade cutters will serve.

The hollow mill can also be used to cut out a broken tap in a blind hole. Check first to see if there is a company in your area with an EDM (Electric Discharge Machine) or with a "tap disintegrator." There is usually a flat rate charge for removing broken taps and finishing tapping the hole. The charge will be less than replacing a casting.

If a hollow mill is used, you will have to tap a larger thread and make up a threaded plug. The plug will be drilled and tapped after it is in place.

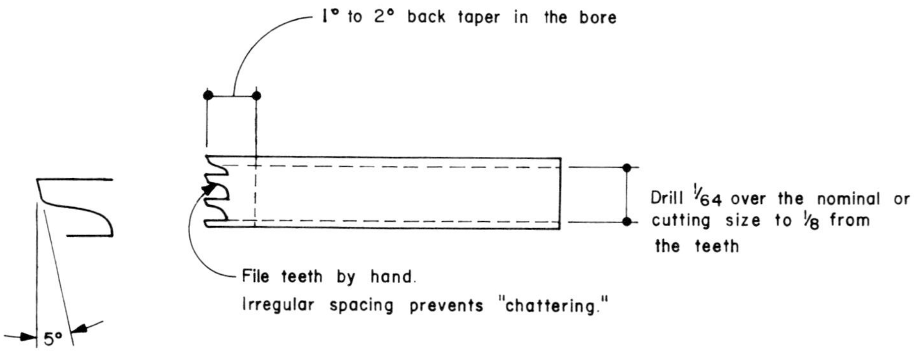

HOLLOW MILL

3

4

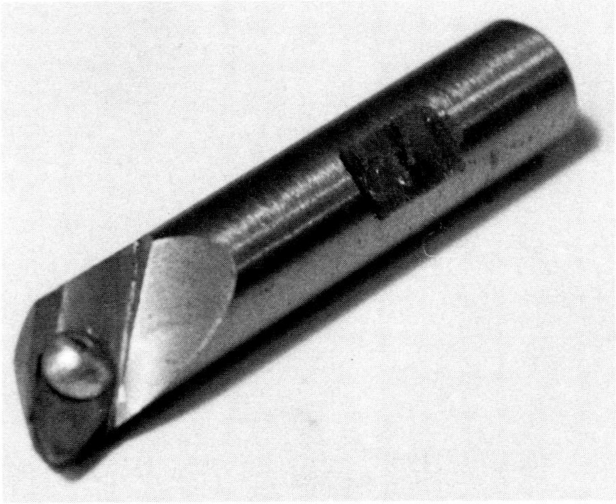

The carbide-tipped end mill.

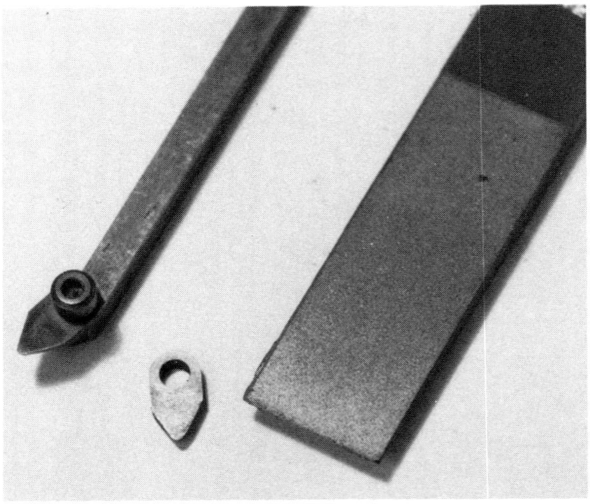

Shown here are the diamond hone, a spare tip and a 3/16" square bar used to hold the tip for grinding.

A Carbide End Mill

Most end mills with replaceable carbide tips have two or more tips and are quite large in diameter. The end mill shown on the drawing is suitable for the small mill/ drill machines. I needed a small carbide end mill to machine some iron castings with a few hard spots in them. The regular carbide end mills were too large, so a little thought and a pleasant evening's work produced this miniature end mill with a replaceable carbide tip.

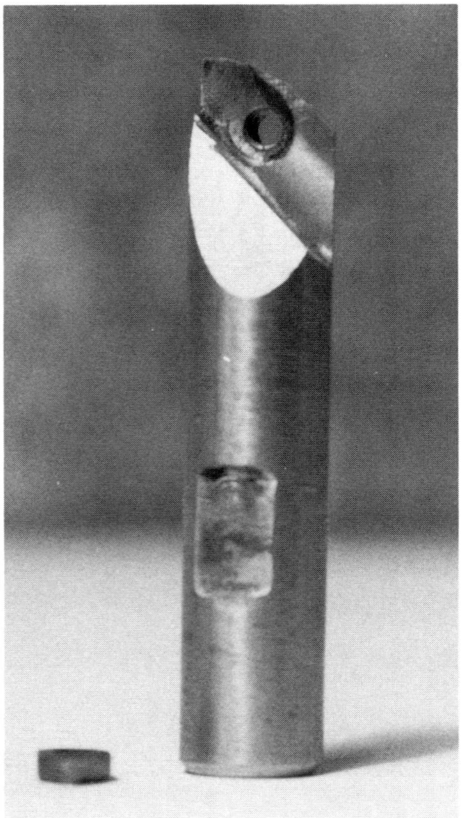

End mill or fly cutter.

Milling a cast iron block with the carbide end mill.

The tips came from a brake drum lathe. If the tips available are a different shape, then alter the design to suit. These can be obtained from your local garage. When they are dull, they are discarded.

Mill the 1/2" rod down to the center line, and then mill the recess for the tip with a 1/4" end mill. Locate the clamping screw hole by drilling through the carbide tip with a drill that just fits in the hole in the tip to form a "dimple" in the recess. Then drill and tap in the usual way.

The tip will be ground when in place in the bar. Use a green wheel for grinding, and then finish with a diamond wheel or use one of the inexpensive diamond hand hones (I use one) to finish the cutting edge.

A nice clean cut has been made on the cast iron.

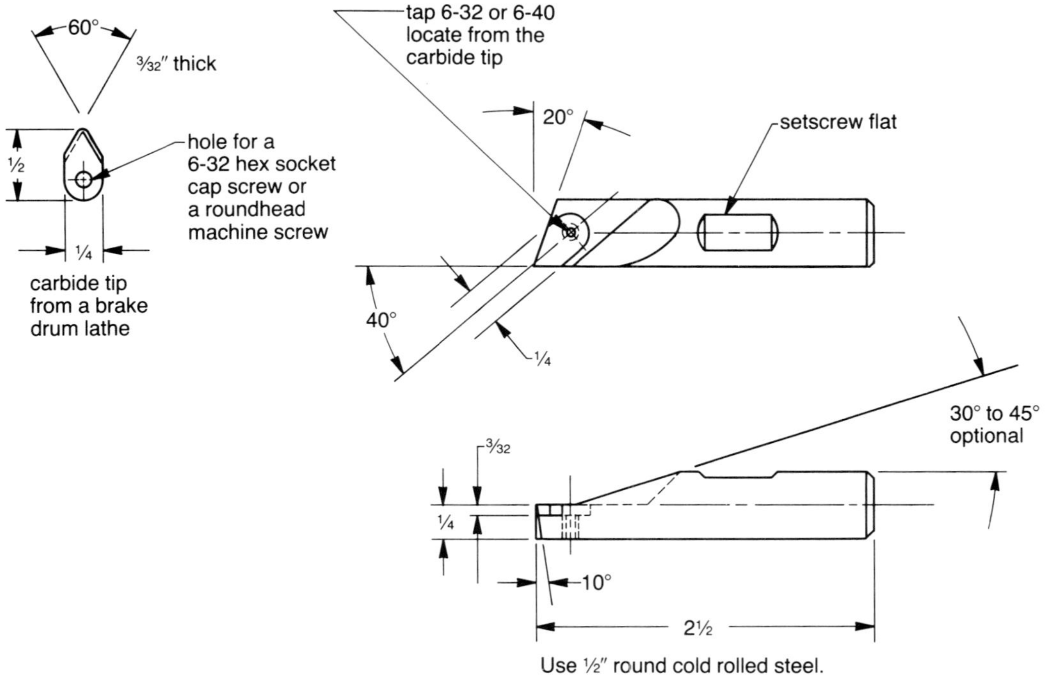

CARBIDE TIP FLY CUTTER (END MILL)

Fly Cutters and Arbors

Making your own fly cutters is a good way to overcome the high cost of commercially made cutters.

These tools are available from the advertisers in *The Home Shop Machinist*, but for those readers who prefer to make their own accessories, these three fly cutters may be of interest.

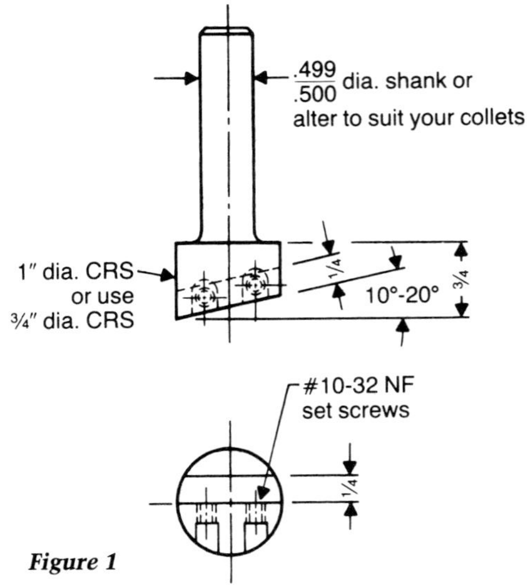

Figure 1

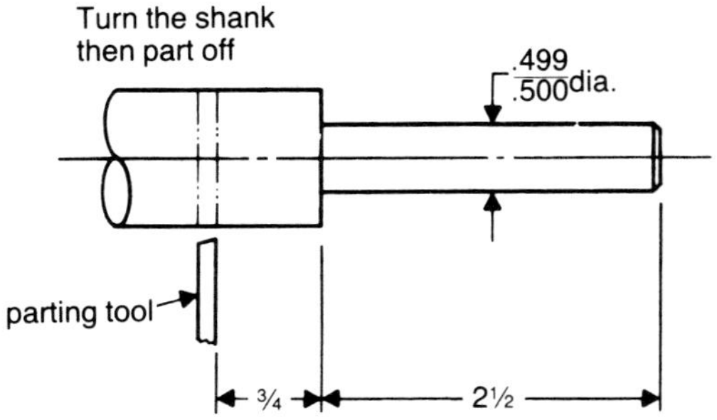

OPERATION NO. 1
Figure 2

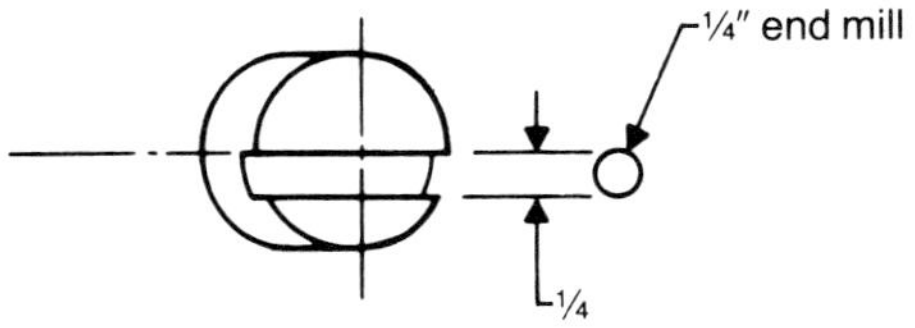

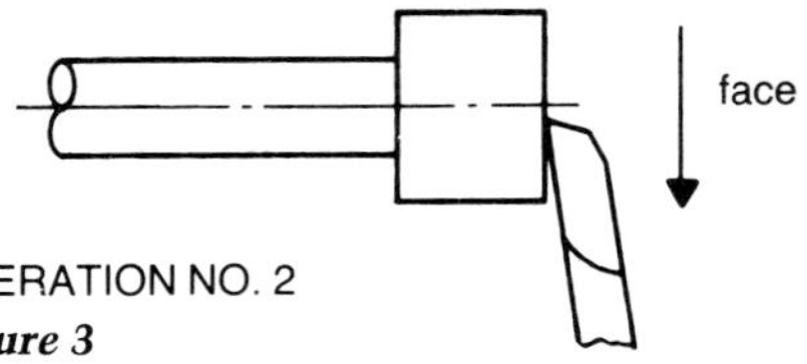

OPERATION NO. 2
Figure 3

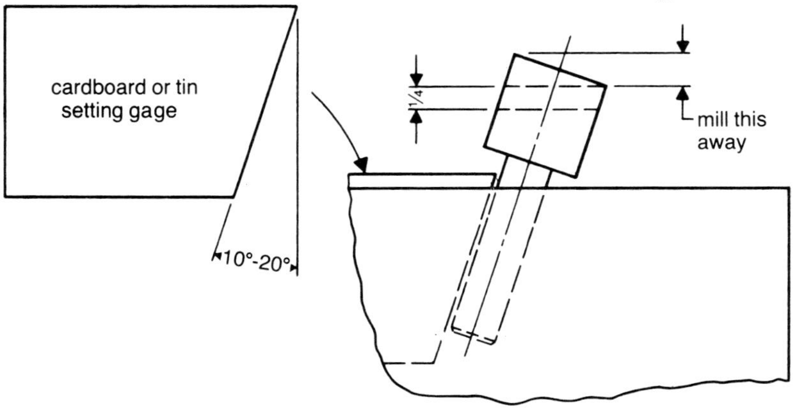

Set the blank to the angle with the gage. Locate the cutter from the center mark left from facing. Mill the slot ¼″ deep. Use the same cutter or a larger cutter to machine the face level.

OPERATION NO. 3
Figure 4

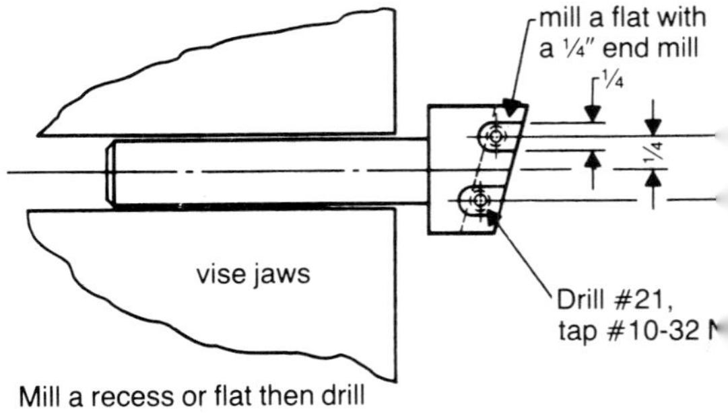

Mill a recess or flat then drill and tap for set screws.

Figure 5 TOP VIEW

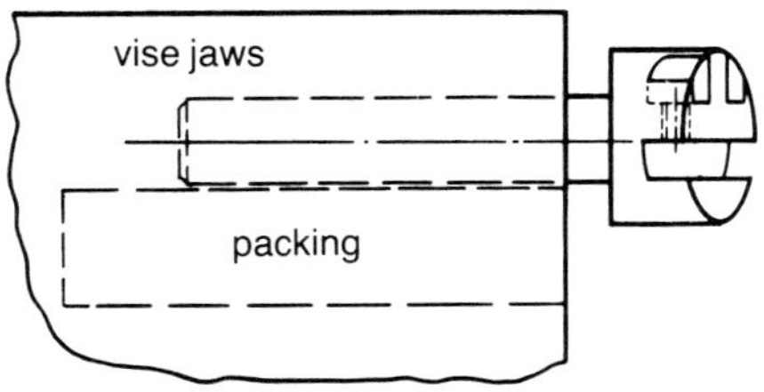

Figure 6 FRONT VIEW

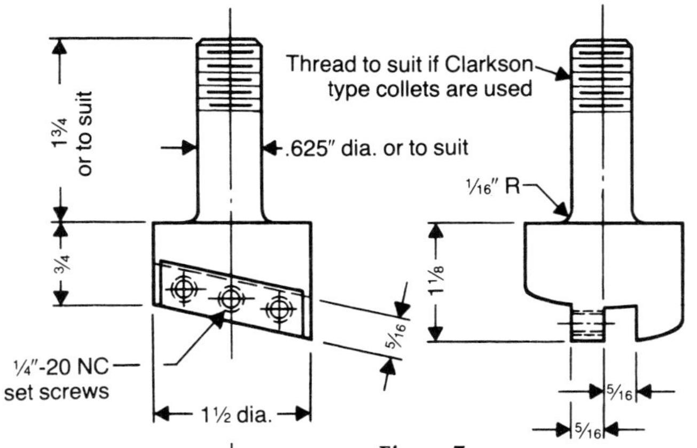

Figure 7

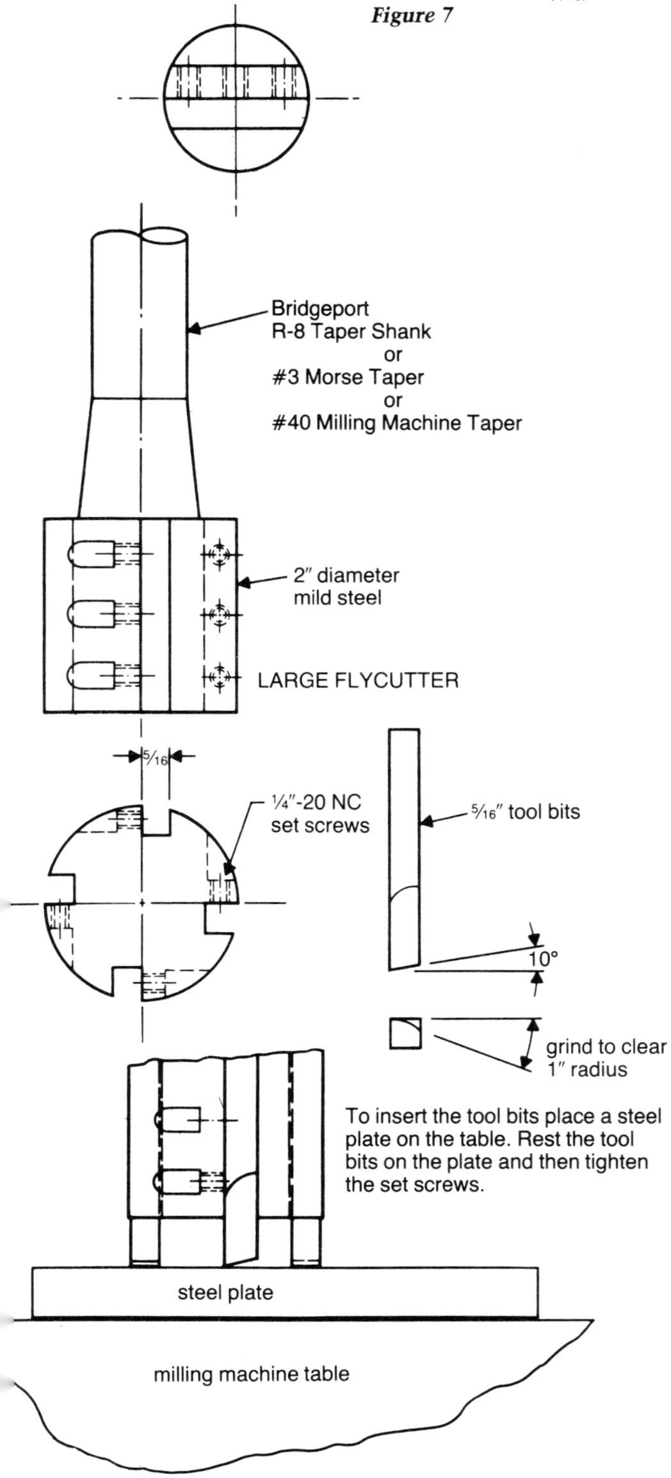

Figure 8

1

2

SHORTCUTS TO MAKING AN ARBOR

These photographs show a few tricks that will facilitate making an arbor used to hold form cutters for making gears, radii, splines and other odd jobs.

I set up a drill in the three-jaw chuck so I could set the taper attachment quickly and accurately without having to make a series of trial cuts (Photo 1).

3

4

5

6

7

Photo 2 shows the taper being tested with chalk lines and a drill sleeve.

Photo 3 and Photo 4 demonstrate how the tapered shank can be held in the milling machine vise with two pieces of wood. They should both be the same type of wood. The wood will compress and will hold the tapered shank very well.

The tool slot and the flat for the setscrews are shown after milling (Photo 5).

The sixth photograph shows how to align the arbor in the vise so the arbor will be correctly aligned for drilling and tapping the setscrew holes shown in Photo 7.

HOMEMADE ARBORS

Readers who have recently purchased mill/drill-type vertical milling machines will soon realize that a full complement of arbors, fly cutters, etc., will be necessary to utilize the full potential of the machine.

Most small vertical milling machines have a No. 3 Morse taper spindle socket with a 3⁄8-16NC drawbar. You will probably need arbors to hold the following sizes of end mill shanks: 5⁄8", 1⁄2", 3⁄8", and 3⁄16" (plus 1⁄4" for the throwaway end mills). The small sizes can be held in the 5⁄8" or 1⁄2" arbor by means of a sleeve until you make or buy suitable arbors.

An arbor to hold slitting saws is desirable, and also several types of fly cutters. I suggest you buy the No. 3 Morse taper blanks available from Busy Bee, Grizzly Imports, or other suppliers advertising in *HSM*. They are inexpensive and well made.

Most home workshop lathes have a No. 3 Morse taper in the spindle so the arbor blank can be set in the lathe spindle socket with a soft hammer. Use a speed of about 1,000 rpm for drilling a center hole and the pilot hole, about ¼" diameter.

Continue drilling in ⅛" steps until the hole is about 1/32" undersize. Now take a light cut with a boring bar to make sure the hole is running true, and then bore out to 1/64" undersize for reaming with a machine reamer, or bore to size. This boring step is very important if the hole is to run perfectly true, so don't omit this step to save time or trouble.

To make the arbors from scratch, cut a 5" length from 1" diameter bar stock. Hold the bar in the three-jaw chuck, face the end, and drill a center hole. Those readers who have a taper attachment can support the work with the tailstock center. The setting for the taper attachment will be about ⅝" per foot taper. The taper is about 3° taper from one side to the other or 1½° with the axis of the work.

8 ***Slitting saw arbor with collars for cutters. Holes are 1" diameter.***

The taper can also be turned by means of the compound slide rest set 1½° back of the lathe axis.

The drill chuck supplied with your milling machine has a No. 3 Morse taper shank, so make your arbors the same size at the small end. To drill out the arbor for tapping the ⅜-16NC thread for the drawbar, use a 5/16" drill as recommended on the tap drill size tables. For easier tapping, a 21/64" drill may be used without any great loss in the size of the threads.

If you want to turn the taper by offsetting the tailstock, follow this procedure:

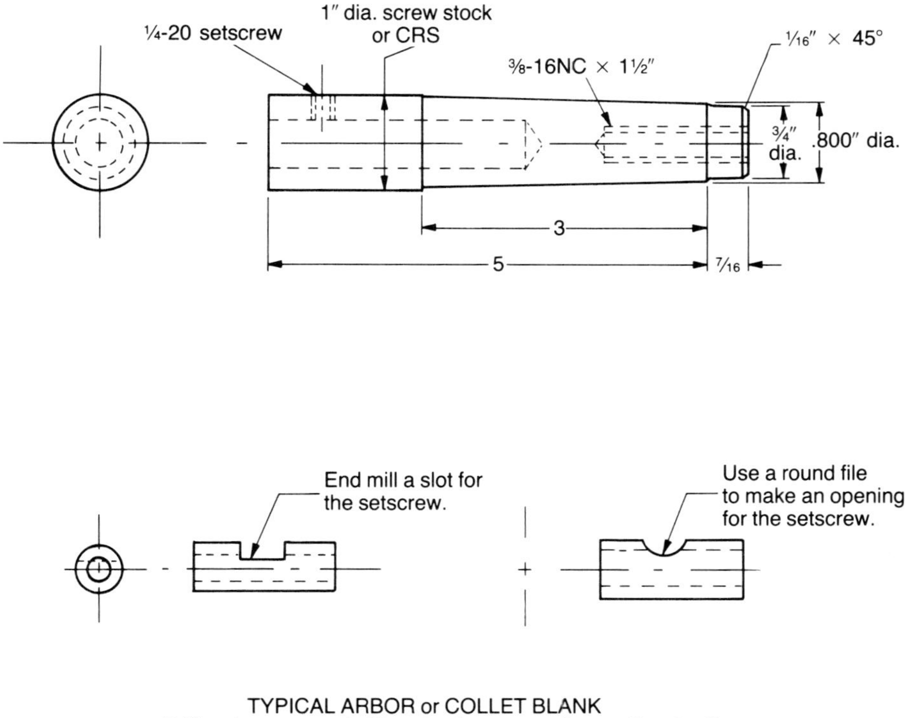

TYPICAL ARBOR or COLLET BLANK
Drill and ream to suit. Chamfer the corners for small end mills.

9 ***Collets 3/8" and 1/2".***

10 ***A Collet or arbor blank (right) and a finished 1/2" collet (left).***

11 ***A 5/8" collet and end mill adapters of 3/16" and 1/4".***

First, calculate the amount the tailstock will be set over to turn a No. 3 Morse taper on a bar with an overall length of 5".

$$\text{Tailstock Offset} = \frac{\text{Taper per foot} \times \text{Overall length}}{24}$$

$$= \frac{.60235 \times 5''}{24}$$

$$= \frac{3.01175}{24}$$

$$= .1254895'' \text{ or } .125'' \text{ or } 1/8''$$

The taper per foot for No. 3 Morse taper can be obtained from the *Machinery's Handbook*, or any machine shop textbook.

Loosen the nut that clamps the tailstock to the lathe bed. Bring the back of the tool holder to touch the tailstock spindle with a piece of paper between. The paper should be firmly held but still able to be pulled out. Set the cross slide dial to zero. Now back out the cross slide a bit more than .125" to allow for the play in the nut and screw, and then return to the .125" reading.

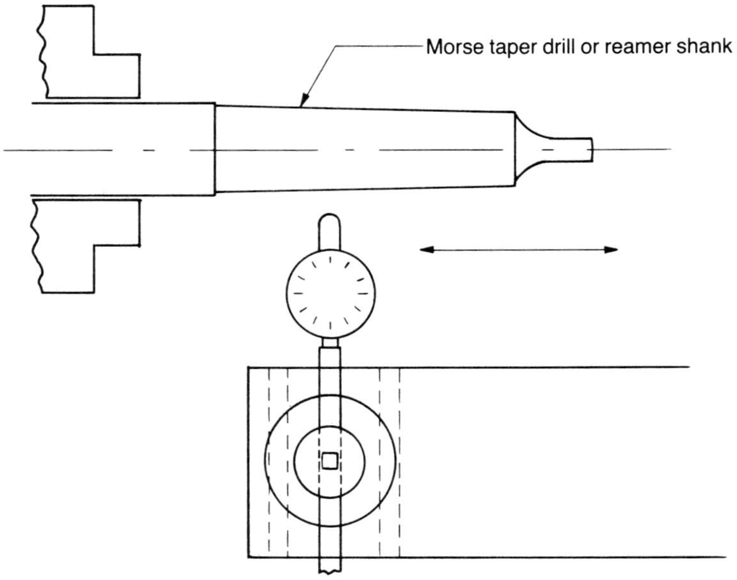

A dial indicator may be used to align the compound slide rest to the correct angle for turning a Morse taper.

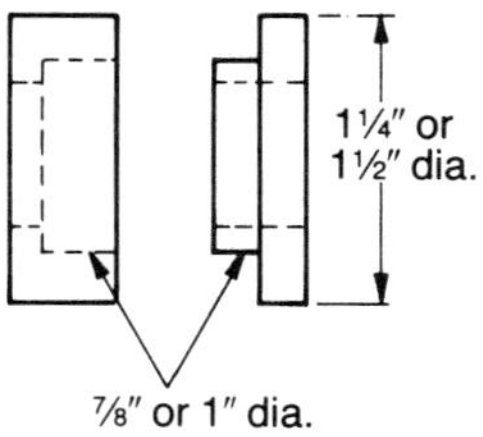

COLLARS to hold cutters with 7/8" or 1" diameter holes

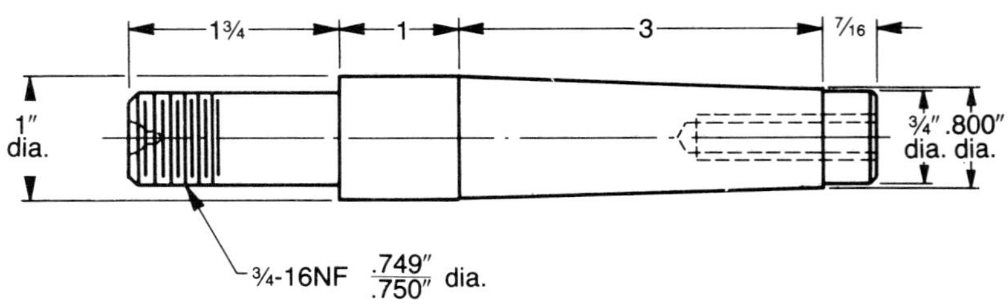

SLITTING SAW ARBOR

Loosen the setscrew at the back of the tailstock and then tighten the setscrew in the front of the tailstock to move the tailstock to the front of the lathe. Bring the tailstock over to touch the back of the tool holder and adjust until you have the same drag on the piece of paper you had when the tailstock was on center. Now tighten the back setscrew until it locks the tailstock up against the front setscrew.

The same basic procedure may be followed using a dial indicator. However, I thought you might like to know how the old-time machinists did this when they did not have dial indicators.

The arbors that have the taper turned between centers will have to be placed in the three-jaw chuck for drilling and tapping the 3/8-16 draw bolt thread. The photos show several of the tools whose arbors have been given the proper taper.

12 ***Three sizes of fly cutters.***

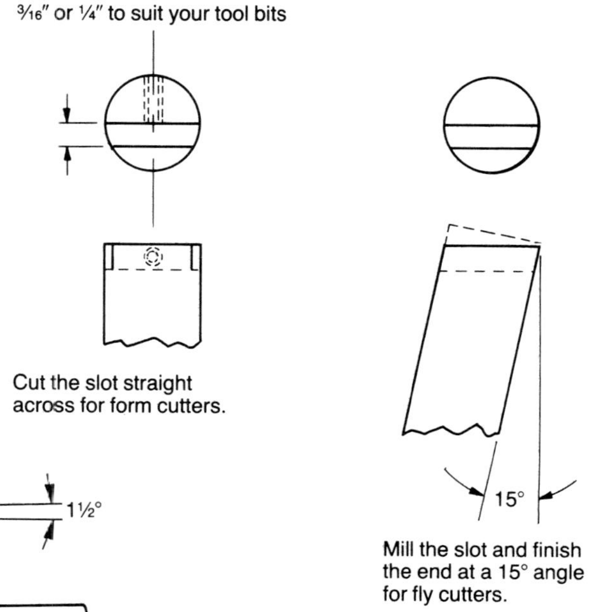

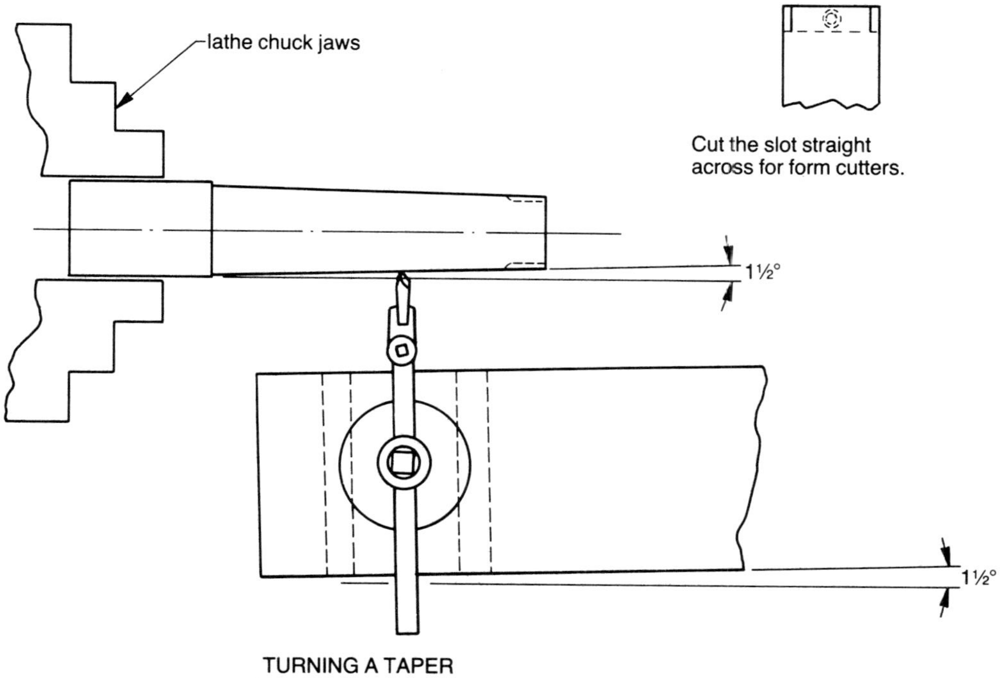

TURNING A TAPER
with the compound slide rest

Expansion Link Milling Fixture

This milling fixture is intended to be used on a vertical milling machine. Now that the small "mill-drill" or "post"-type milling machines are available for a reasonable price, the owner of a home workshop should have one instead of trying to do all of the milling operations on a lathe with a vertical milling attachment.

Most of the small expansion link milling fixtures that are used on the vertical milling attachment rely on a lever to swing the work in an arc to machine the curved slot in the expansion link. The screw feed on this fixture will give a more uniform feed and a smoother finish as the cutter passes through the scalloped sections left from the drilling operation to remove most of the metal before the milling operation.

Mounting the milling fixture on the milling machine table is much more rigid and safer than using a fixture on a vertical slide.

Start by laying out (marking) both expansion links. Make a complete layout – the center line and both edges of the curved slot, the two mounting holes where the arms from the eccentric straps are attached to the link, and the complete outline. Saw, mill and file the outside of the links to the correct shape. Drill and ream the mounting holes for the arms.

To prepare the fixture to hold the expansion links, we will have to locate, drill and tap two holes to clamp the expansion link to the top or moving plate. We could use two holes located

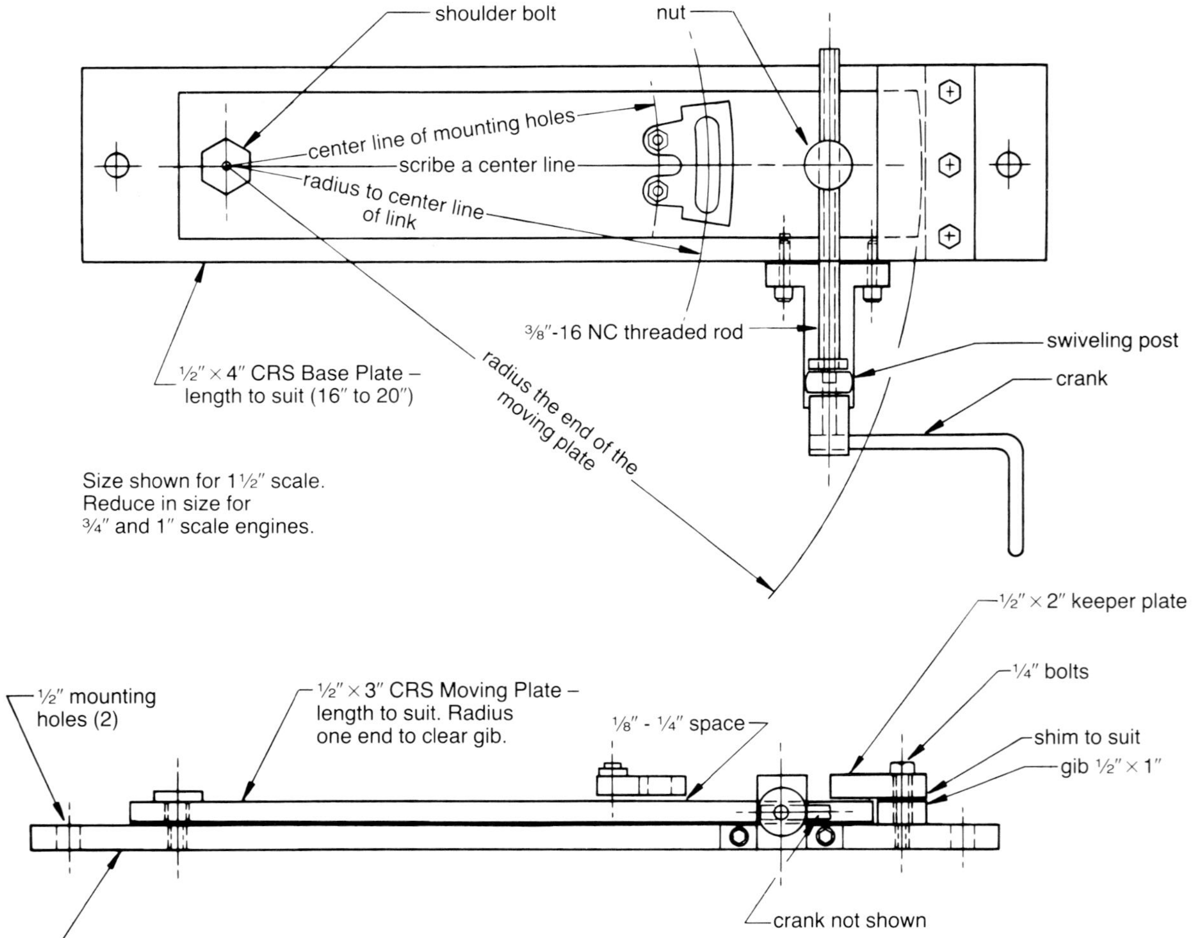

EXPANSION LINK MILLING FIXTURE

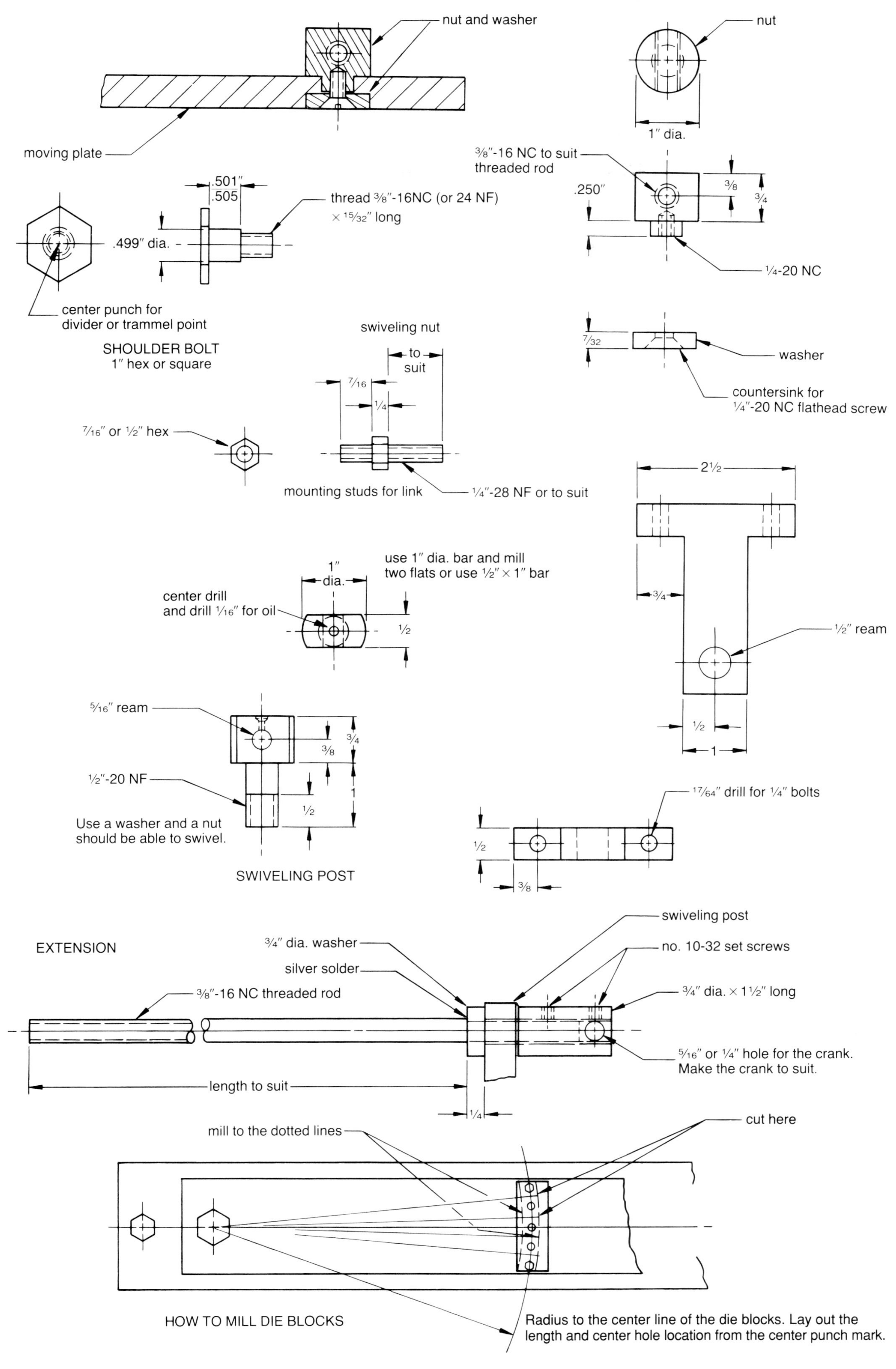
nut and washer
moving plate
nut
1" dia.
3/8"-16 NC to suit threaded rod
.250"
3/8
3/4
1/4-20 NC
.501"
.505
thread 3/8"-16NC (or 24 NF) × 15/32" long
.499" dia.
center punch for divider or trammel point
SHOULDER BOLT
1" hex or square
swiveling nut
to suit
7/16
1/4
7/32
washer
countersink for 1/4"-20 NC flathead screw
7/16" or 1/2" hex
mounting studs for link
1/4"-28 NF or to suit
2 1/2
3/4
1/2" ream
1/2
1
1" dia.
use 1" dia. bar and mill two flats or use 1/2" × 1" bar
center drill and drill 1/16" for oil
1/2
5/16" ream
3/4
3/8
1
1/2
1/2"-20 NF
Use a washer and a nut should be able to swivel.
SWIVELING POST
17/64" drill for 1/4" bolts
1/2
3/8
swiveling post
EXTENSION
3/4" dia. washer
no. 10-32 set screws
silver solder
3/4" dia. × 1 1/2" long
3/8"-16 NC threaded rod
5/16" or 1/4" hole for the crank. Make the crank to suit.
length to suit
1/4
cut here
mill to the dotted lines
HOW TO MILL DIE BLOCKS
Radius to the center line of the die blocks. Lay out the length and center hole location from the center punch mark.

in waste material outside the ends of the link. However, this involves more work, a waste of material and may require a wider top plate, so we shall use the two mounting holes for the eccentric strap arms.

Use the center punch mark in the shoulder bolt as the center to scribe an arc on the top plate to locate the two mounting holes. Use the scribed center line on the top plate to locate the distance between the holes. Remove the top plate from the fixture to drill and tap these holes.

Make up the special studs to hold the link or use Allen cap screws and washers between the link and the top plate. Remove the vise from the milling machine, and clamp the fixture to the milling machine table. Be sure the crank clears the table.

Use a wiggler in the milling machine drill chuck to locate the center line of the slot under the axis of the spindle. Lock the table so it cannot move. Use a center drill and then a drill nearly the size of the slot to drill a chain of holes to remove as much material as possible before milling.

Start the milling operation with an end mill 1/32" to 1/16" smaller in diameter than the width of the slot. Use the correct spindle speed and take very shallow cuts (say, .025"). Remove the shavings with a jet of compressed air or with a small paintbrush. Continue to take shallow cuts until the cutter breaks through the bottom of the link.

To machine the slot to the correct width, there are two options. One is to use a cutter the correct size and mill both sides of the slot at the same time. Use shallow cuts again instead of trying to mill the full depth of the link in one pass. The other choice is to use the cutter first used to open the slot and machine one side at a time. Unclamp the table and move the table over until the cutter just touches the layout line. Use shallow cuts as before. When the cutter is at full depth, move the table .001" so the cutter can take a light finishing cut. When one side is finished, move the table to machine the other side of the slot. Remember to traverse the work against the direction of the cutter rotation (plain milling).

The die blocks or link blocks can be machined on this fixture. Attach the material by two studs or machine screws located at the ends of the material. Use the center punch mark on the shoulder bolt as the reference point to scribe the arcs to lay out the inner and outer radii and the center line of the blocks. Mark out the ends of the die blocks and locate the center holes while the work is still on the fixture. Use the center punch mark on the shoulder bolt to locate one end of the rule so the ends of the die blocks will be radial (pie shaped). Mill the blocks to the layout lines, as usual. It might be a good idea to make two extra blocks at this time.

Use the die blocks as a gage when filing out the ends of the links square as usually called for on the drawings.

The fixture should be made from cold rolled steel plates. However, hot rolled mild steel can be used (I did) if cold rolled steel is not available. Alter the dimensions to suit your milling machine and the material available. There is no need to be very careful or fussy when making this fixture. Just be sure the radial distance from the center punch mark on the shoulder bolt is long enough to suit your links.

Convention calls for a left-hand thread on the screw so that when the crank turns to the right (clockwise) the moving plate will move away from you (advance into the cut). However, for a fixture that will be used only a few times, a left-hand thread requires too much effort and expense. Use a threaded rod available from a hardware store and tap the nut to suit.

The keeper plate should exert a slight drag on the moving plate to reduce chattering, so adjust this by using a cigarette paper shim between the keeper plate and the gib strip (or spacer).

The nut and the post have to be able to swivel slightly as the moving plate moves through an arc. The shoulder bolt should have a slight clearance between the head and the moving plate. Place some grease or oil between the plates to reduce friction.

Use this fixture with common sense, and you will be pleased to find how easy it is to mill the curved slots. Why file the curved slot to size as advocated by some older writers when, with a little effort to make a fixture, the links can be made more accurately on a milling machine and the fixture can be used to machine any other curved slots later on?

Machinist's Screw Jacks

These little jacks are used to support the outer ends of long work held in the vise on a drill press or a milling machine. When used as a set of three, they may be used to support an irregular casting on a surface plate for making the layout. They are also useful for raising the frames of a locomotive slightly so the wheels may be removed by hand.

The size of the parts is not important, so use whatever material you have that suits. The base may be square, hexagon or round. I shall give two methods of making the screw: one simple and one more difficult.

To make the *Base:* face, counterbore, drill and tap the thread without removing the work from the chuck so the threaded hole will be perpendicular with the base. Rechuck to finish the top end.

The *Screw* can be a hex- or square-head bolt turned back to leave a boss about 3/16" diameter. Center drill the boss to locate a ball bearing and solder the ball bearing in place.

The *Cap* is turned as shown on the drawing. The hole should be an easy fit over the ball bearing. Turn down the outside diameter until the lip is about 1/32" thick, or slightly less.

Hold the threaded part of the screw loosely in the tailstock drill chuck while the ball bearing is in the hole. Set the roller forming tool against the lip and "swage" the lip around the ball bearing using a slow spindle speed. Stop this operation when the swaging has progressed enough to prevent the pad from falling off the ball. Now, still supporting the thread portion loosely in the drill chuck, part off the pad from the bar. Reverse the pad in the lathe chuck with the screws inside the chuck, to face and form grooves in the top of the pad or cap. Note: Locate the pad perpendicular (or flat) in the universal chuck with the jaws of the tailstock drill chuck, by pressing on the face left after parting. Tighten the universal chuck just enough to hold the pad. Force the pad flat with the tailstock chuck and then finish tightening the universal chuck.

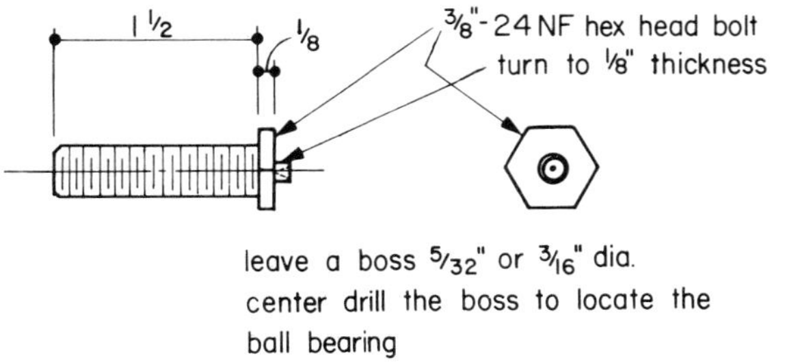

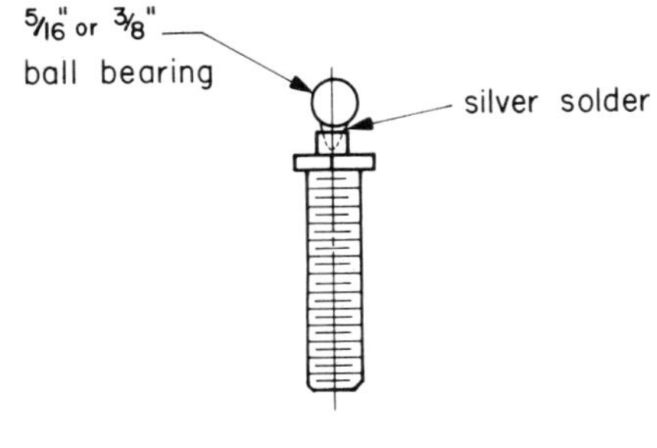

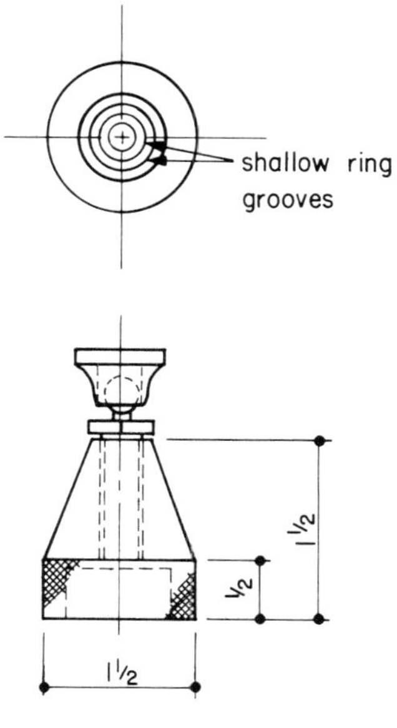

BASE

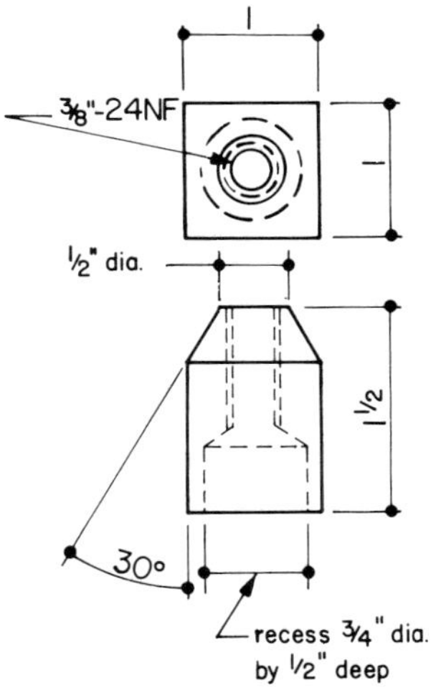

ALTERNATE BASE DESIGN

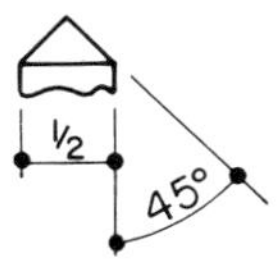

Cone top

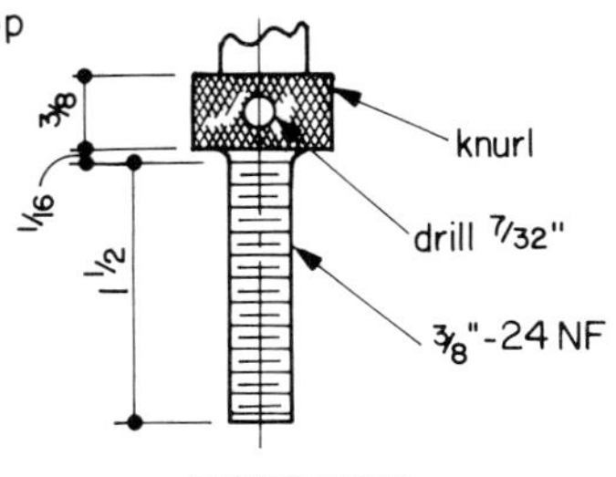

ALTERNATE SCREW DESIGN

TURNING THE CAP

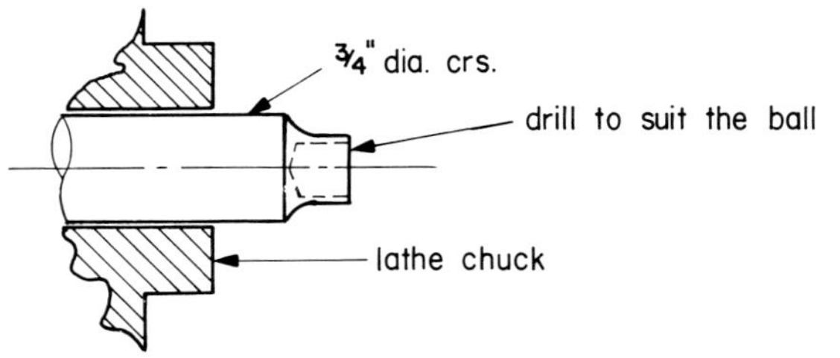

Step 1

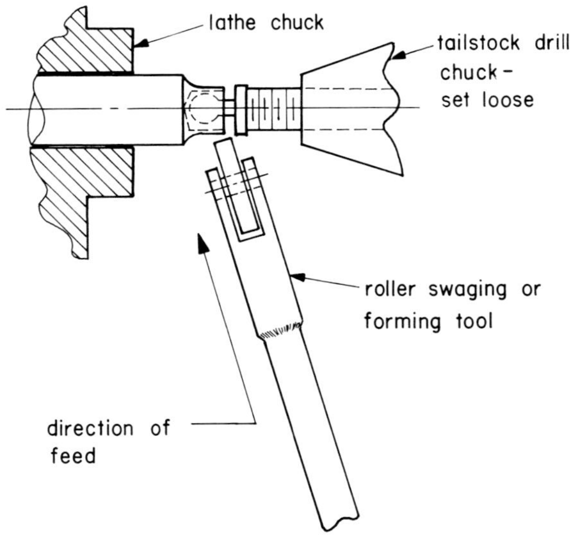

Step 2

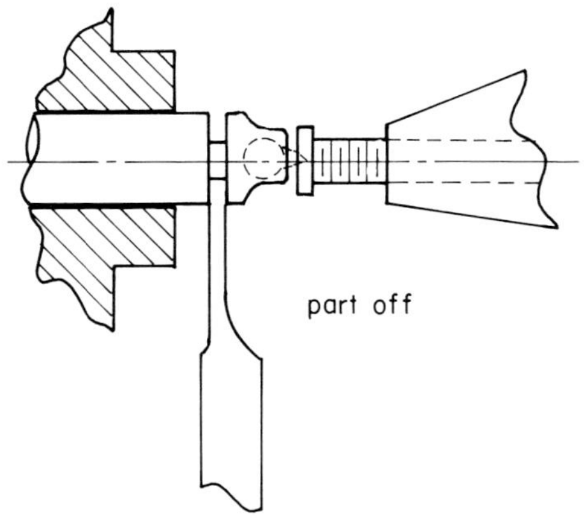

Step 3

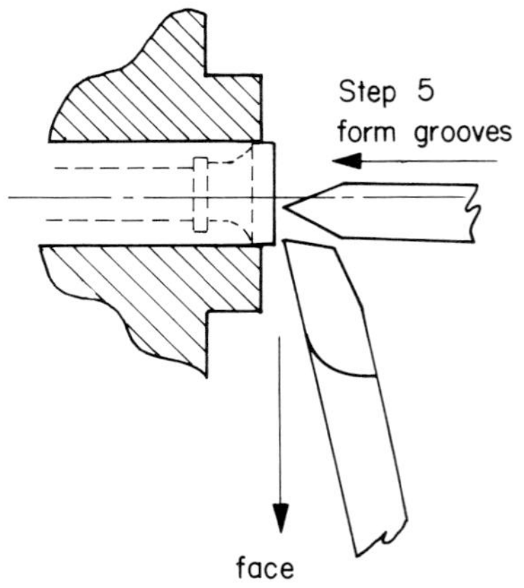

Step 4

To make the screw from hexagon bar stock, you will first need to make a forming tool (see drawing). The best material to use is oil hardening tool steel, either as hot rolled or ground (gage plate) flat stock. Use a piece about 3/4" wide, 3/8" or 1/4" thick and 3" to 4" long.

Lay out the shape of the ball and the neck. Drill and ream a hole the correct size for the ball. Turn the work over and ream the hole with a 10° taper reamer to give a front clearance of 5°. Now cut the unwanted portion away and file or mill a 5° clearance as shown on the drawing.

Harden as usual and temper to a light straw color. Use a round oil stone to sharpen the tool when required.

Now to make the screw. Rough from the ball and neck with a narrow round-nose tool bit by manipulating the carriage handwheel and the cross slide handle at the same time. This is easy to do, but it requires a little practice. Next, place the forming tool in an English-style flat toolholder. The top of the forming tool must be at the lathe center height. Use a slow spindle speed and feed in the cross slide slowly. Use sulfurized cutting oil for a fine finish. Cut off the bar stock to give a length of 1-1/2" for the thread. Reverse the screw in the universal chuck with about 1/4" protruding beyond the jaws. Face and center drill with a small center drill. Draw the screw blank out of the chuck until 1/8" remains in the chuck jaws. Support the outer end on the tailstock center and retighten the chuck jaws. Note: Many 3-jaw universal chucks have a mark beside one of the openings for the chuck key. Use this place for the final tightening to have the best accuracy from the chuck.

Turn the screw to about .005" under size (.375" − .005" = .370") for threading. Thread with a single-point tool or use a die held in a tailstock die holder.

Riser blocks can be made up, if there is a need for a greater height. For rough castings, a conical point is sometimes better than a swivel pad.

For best results, make up a set of four of these jacks.

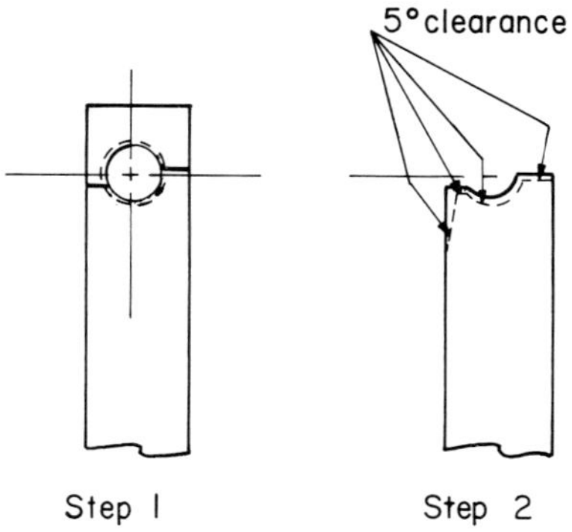

MAKING THE FORM TOOL
Lay out and drill.
Turn over and taper ream.
Cut and file clearances.
Harden and oil stone.

Hold in a flat, English-style tool holder (*Live Steam*, December '80)

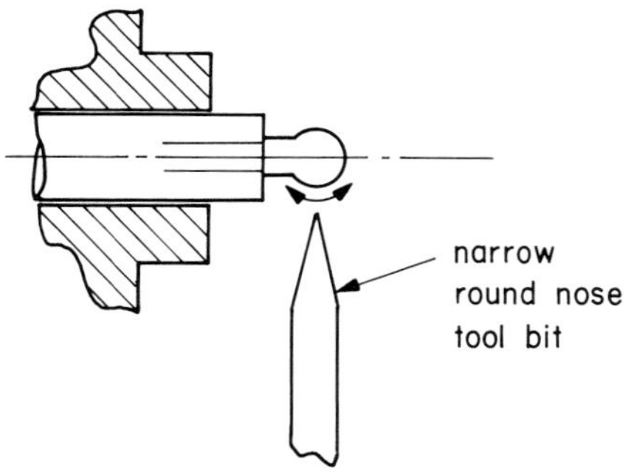

Step 1 - Rough form

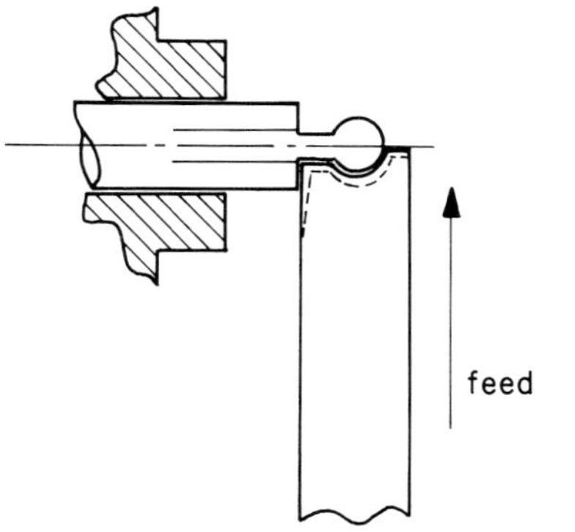

Step 2 - Use a form tool

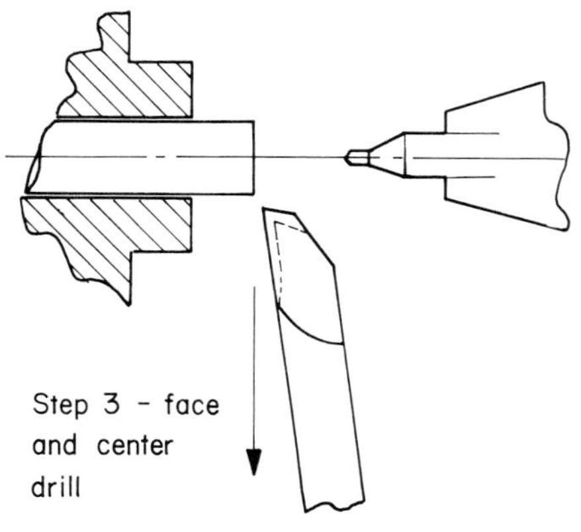

Step 3 - face and center drill

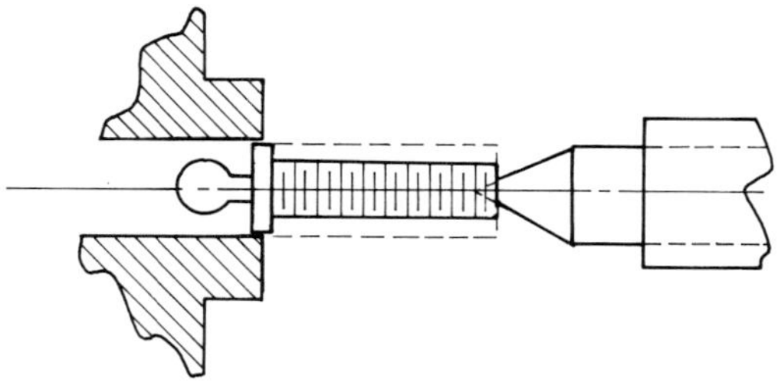

Step 4 - turn to size 8 thread

A Shell End Mill Arbor

A shell end mill is a very useful cutter for machining flat surfaces, and for squaring the ends of plates and large bar stock; it cuts smoothly, unlike the carbide cutter usually supplied with the milling machine, and the shell cutter leaves a fine finish on the work.

Many of the small vertical mill-drill machines have a No. 3 Morse taper spindle socket so they can hold taper shank drills. This is the same taper used in the headstock spindle of most bench lathes, such as Atlas, South Bend, Logan, or Boxford. This means that good arbors and all sorts of tooling for the milling machine can be made, and you will be sure the tools will be concentric when they have been finished by turning in the Morse taper socket of the headstock spindle.

The slots in the spindle nose of my vertical milling machine are made to accommodate a 10 mm wide key (.3937″). It does not matter if they are wider than the key, so I used 3/8″ square cold rolled steel for the keys. (This can be purchased in short lengths as key stock from some merchants.) The 3/8″ end mill I used to mill the slots in my arbor had been

1 *Turning the taper*

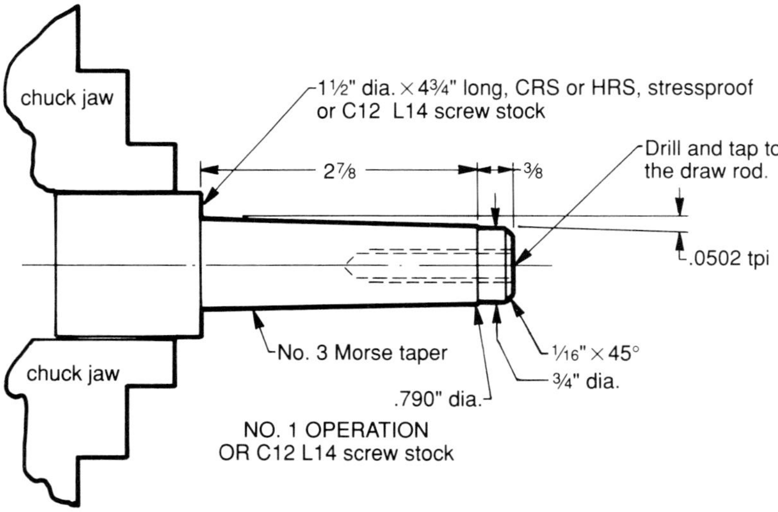

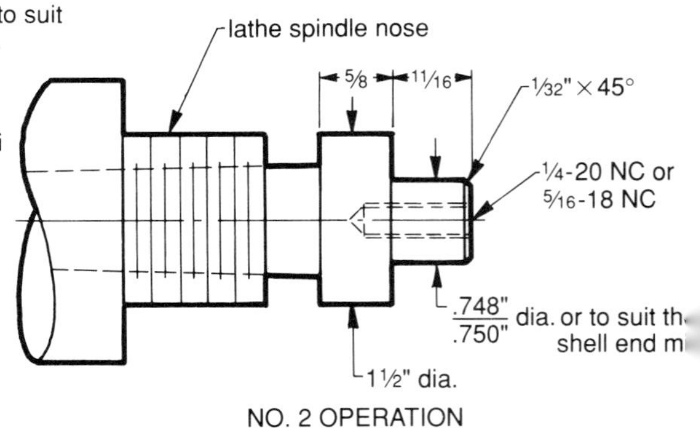

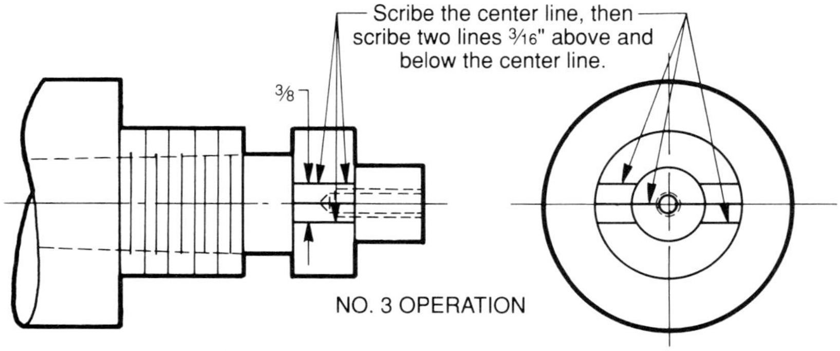

2 ***The finished taper.***

3 ***The end of the taper is reduced in diameter.***

4 ***The end is being tapped.***

sharpened once, so it was slightly undersize, and left a slot that was a press fit for the keys.

The ends of the keys had to be reduced in width to 5/16″ to suit my shell cutter, so this was done prior to pressing the keys into place on the arbor press. (Note: every shop needs a small arbor press – 1 to 3 tons.)

The screw holes were then drilled, counterbored, and tapped. Follow the directions shown in the drawings, and I'm sure you will be pleased with the finished arbor.

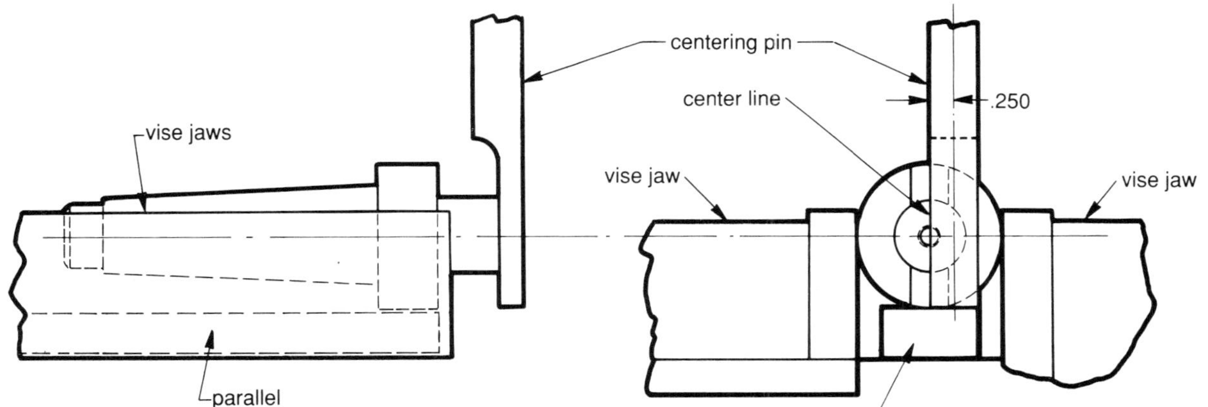

NO. 4 OPERATION

When the center line on the work has been aligned with the edge of the centering pin, move the table over .250" to bring the axis of the spindle over the center line of the work.

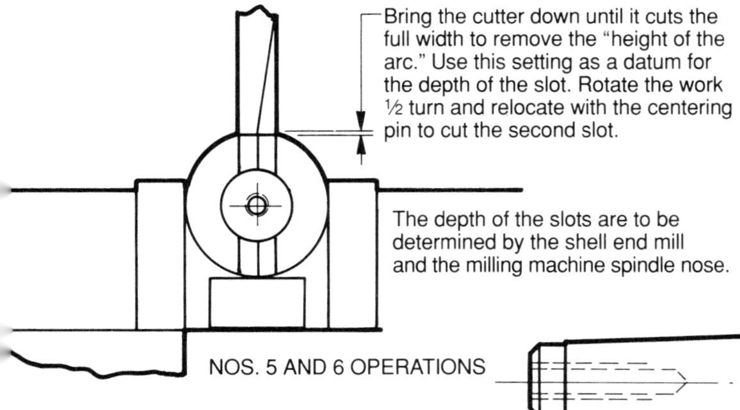

The depth of the slots are to be determined by the shell end mill and the milling machine spindle nose.

NOS. 5 AND 6 OPERATIONS

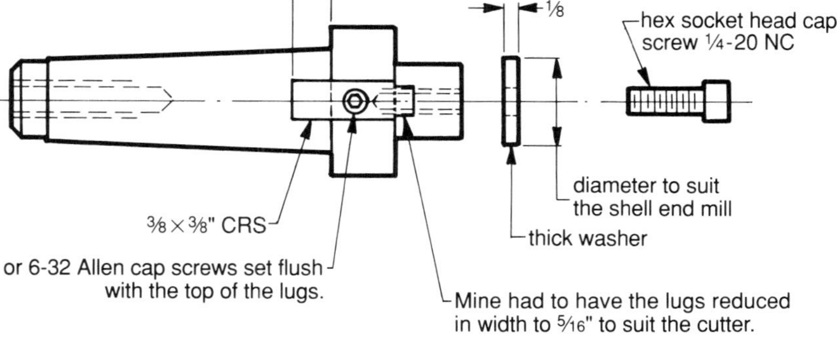

5 *and* 6 ***Measuring with a depth gage after facing to length.***

7 ***Placing the scriber in contact with the top of the arbor. The setting was transferred to the scale on the square in the background, and then the slots were marked out.***

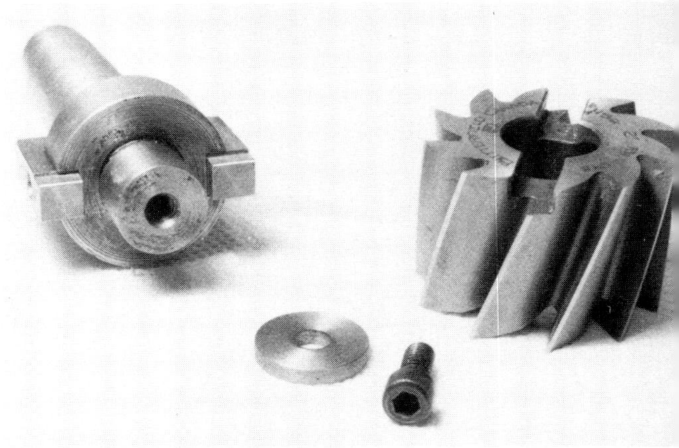

8 ***The finished arbor, washer, Allen cap screw, and cutter.***

Shop Tools & Improvements

A Sturdy Workbench

The design for this workbench was in the back of my mind for several years. The design became reality when I was given several pieces of 4 × 4" fir that had been used for skids under some new machinery being installed in a local factory.

Along with the timber came some 2 × 2 × 1/4" angle iron. The bench top could be composed of 4 × 4" fence posts or of 2 × 4" studding set on edge. The minimum for the legs would be 3/16 × 2 × 2" angle. Do not use the slotted angle unless it is very heavy.

I picked out the best timbers, removed any nails and then had them planed all over at the school carpentry shop.

Next I sat the timbers together on the floor in the order they would be when bolted together. Any holes in the timbers were plugged and the timbers placed so that the plugs were horizontal and hidden by the adjacent timbers.

The timbers are held together by long through-bolts and glue. I was able to drill 9/16" diameter holes so I used 1/2" diameter bolts. If the largest you can drill easily is 1/2" then use 7/16" diameter bolts.

Center punch the location of each bolt hole on the front timber. Now make a counterbore with a carpenter's brace and bit to clear a 1/2" washer and deep enough to clear a washer and a nut. Set this timber on your drill press, and using the center holes left from the point of the wood bit as a guide, drill right through. Now clamp the second timber to the first and place them together on the drill press. Pass the drill right through each of the holes in the first timber and make a shallow hole or "dimple" in the second timber. Remove the first timber and drill each of the holes using the dimples to start the drill correctly. Repeat for each of the other timbers using the first timber as a drill jig.

Threaded rod can be used for the bolts. I used 1/2" cold rolled steel. Hot rolled steel is also suitable. Thread one end just enough to thread on the nut. The other end needs about 2" of

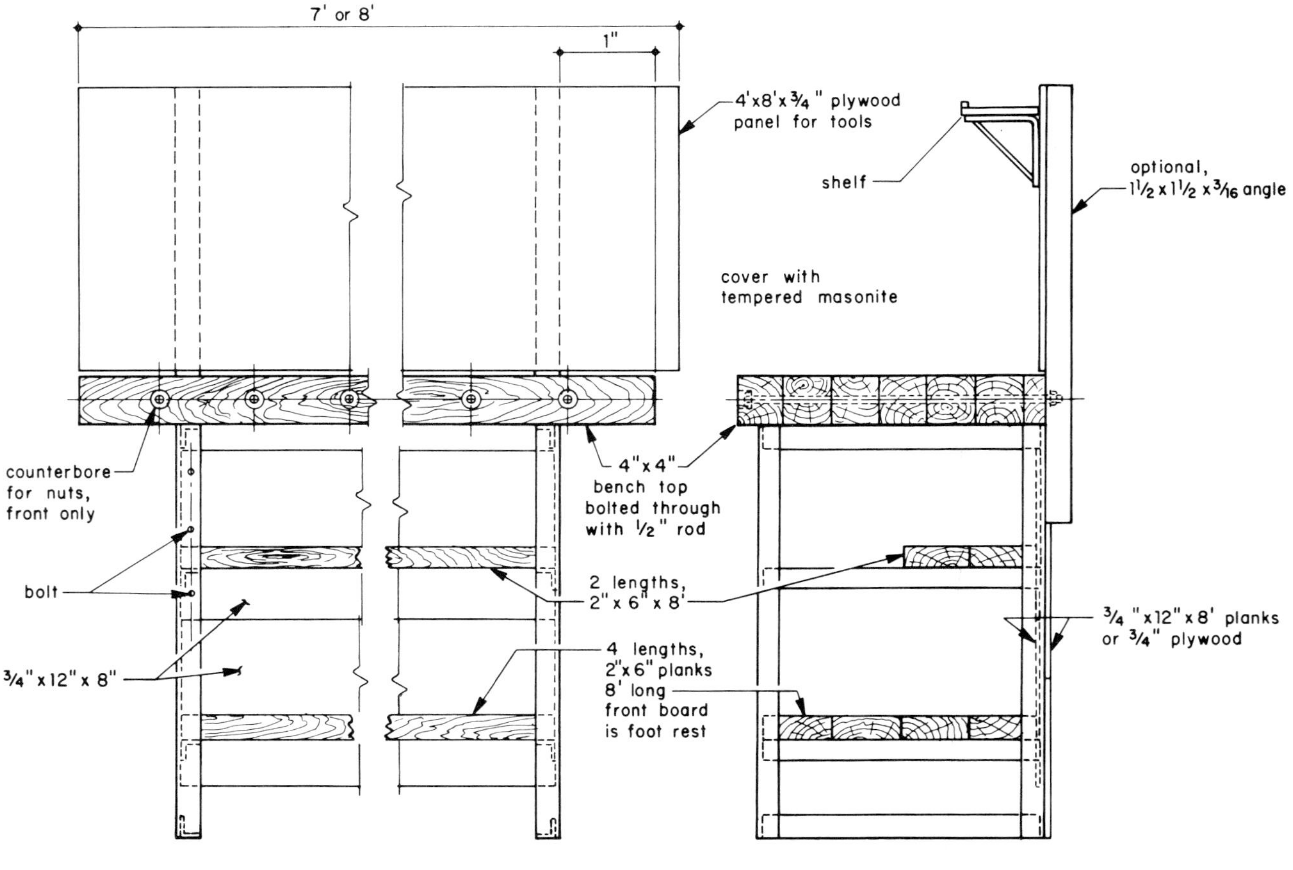

Figure 1

thread to take up any shrinkage in the timbers and to clamp a 3/4" plywood tool rack in place.

Coat the adjacent surfaces with a good wood glue and place the timbers in position on the floor. Place a washer under the nut and drive each bolt through the timbers from the front. Place a washer over the end of each bolt and tighten the nuts until the timbers are well clamped. Allow at least 24 hours for the glue to dry.

I intended to have the bench top finish planed on a 24" thickness planer, but the top came out slightly too wide. I hand planed it with a jack plane and tested for flatness and "wind" with a piece of 3/8 × 1-1/2 × 8' cold rolled steel for a straightedge and feeler gages (Figure 3).

I finished the top with several coats of varnish (top and bottom, edges and ends.) Normally I keep a piece of masonite on the top at the bench for all rough work. Then if I have to use the bench as a desk I have a clean, varnished surface to work on.

The legs are made with mitred or lapped joints and welded as shown in Figures 2 and 4.
The cross rails for the shelves can be smaller material such as 3/16 × 1-1/2 × 1-1/2".

The leg sections are made like a frame, with the top and the bottom made from the same material as the legs. There are holes drilled in the top section so that the top can be fastened to the frames with 3/8" lag bolts from underneath. The bottom section is necessary to distribute the weight of the bench over a larger area than resting on a concrete floor.
1/4 × 4 × 4" pads could be welded to the legs but a bottom section of the frame is necessary for rigidity. The plain angle resting on a concrete or wooden floor will damage the floor very quickly. The flat section will permit the bench to slide easily over the floor for positioning the bench and the leveling can be

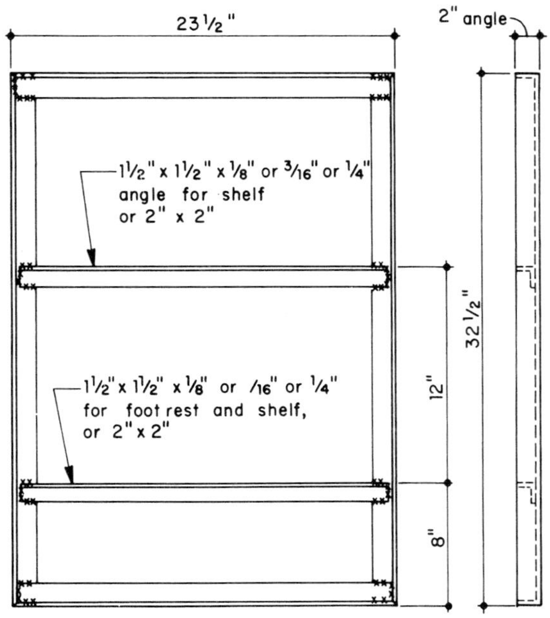

Figure 2

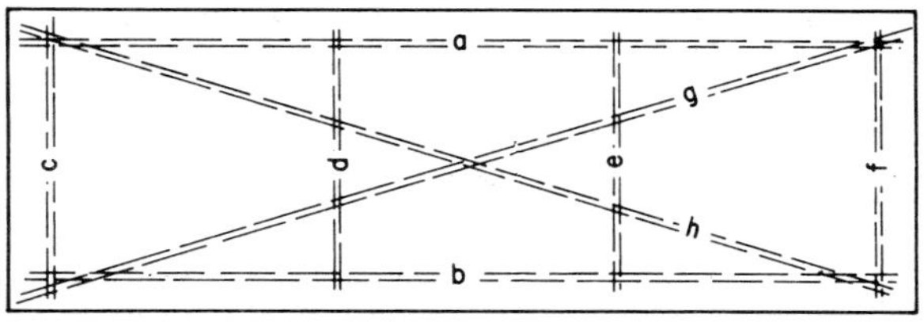

test for flatness and twisting "wind" in all the positions shown with a long straight edge and feeler gauges. for a straight edge try a piece of 3/8" or 1/2" x 1 1/2" or 2" cold rolled steel. set on edge for testing after removing any nicks or burrs by file.

HAND PLANING AND TESTING THE BENCH TOP

Figure 3

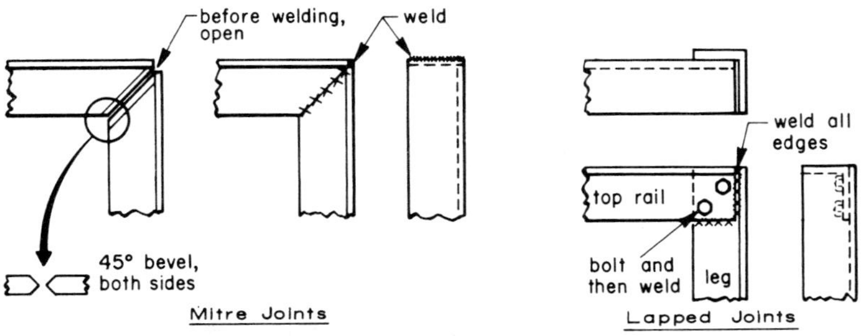

BENCH LEGS

Figure 4

done with cedar shingles or wooden wedges under the bottom sections.

The legs are braced by the 2 × 6" or 2 × 12" shelves being bolted to the shelf rails. There is a 3/4" piece of plywood or 1 × 12" planks bolted to the back of the legs. This stiffening is very important. Use machine bolts with washers or carriage bolts.

The tool rack on the back is made up of 3/4" plywood panels joined in the center that I had on hand, but a single piece would be better, of course. I was going to bolt two pieces of 1-1/2 × 1-1/2 × 3/16" angle to the backs of the legs to support the tool panel but so far this has not been necessary. The tool panel is clamped to the bench top by the long clamping bolts. The shelf on top of the tool panel is optional but I find it very handy to hold the drill stands, and other tools. My taps, dies, counterbores and reamers are set in wooden blocks and stored on the shelf.

The vise may be mounted at either end of the bench close to a leg for maximum rigidity, with the fixed jaw extending about 1/4" from the edge of the bench so that long bars may be held vertically. Please use a decent vise, that is 4" to 6" jaw width, not a toy from a local store. My vise had a swivel base that broke and, frankly, I find the vise better without the swivel base. The swivel base is less rigid and the vise jaws are too high. If the bolt holes in the vise base are too large for the mounting bolts the vise will soon loosen. The solution is to make up special bushings as shown in the drawing (Figure 5).

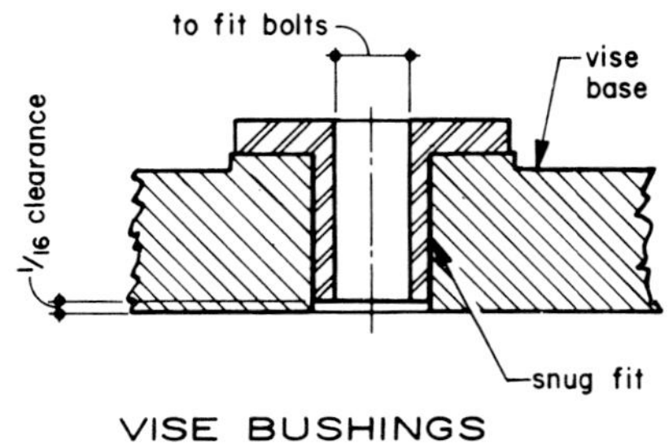

Figure 5

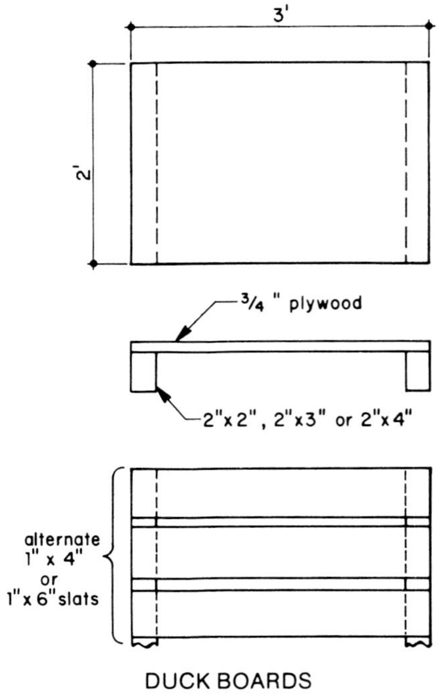

Figure 6

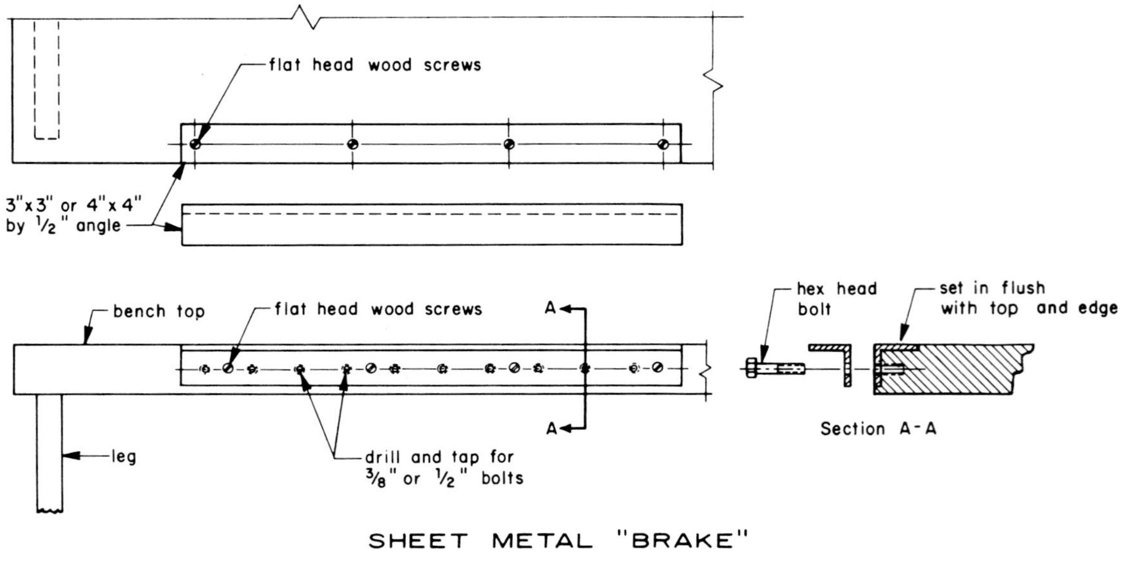

Figure 7

Usually when the bench top is a convenient height the vise is too high to saw or file comfortably. A small wooden platform called a duck board is required (Figure 6). This is a piece of plywood or wooden slats supported by 2 × 2" or 2 × 4" strips underneath.

The height of the duck board has to be determined by the person who is to use the vise, for the top of the vise jaws should be about the same height as the user's elbow. Don't make the duck board too large (say 2' × 3') for when the vise is not being used, the duck board can be stored at one end or underneath the bench.

This bench can be dismantled in about two hours into convenient sized sections if you have to move the workshop. The 2" thick shelves will permit you to store boxes of short ends of steel, brass, etc., and any heavy tools under the bench. This weight will help to hold the bench down if you have to do any heavy bending at the vise.

A piece of 2 × 2" or 3 × 3" angle iron may be "let" into the front edge of the bench to bend sheet metal (Figure 7).

If you're new to the hobby, the construction of your workbench will be a strong, positive step in the right direction, so get at it! And best of luck!

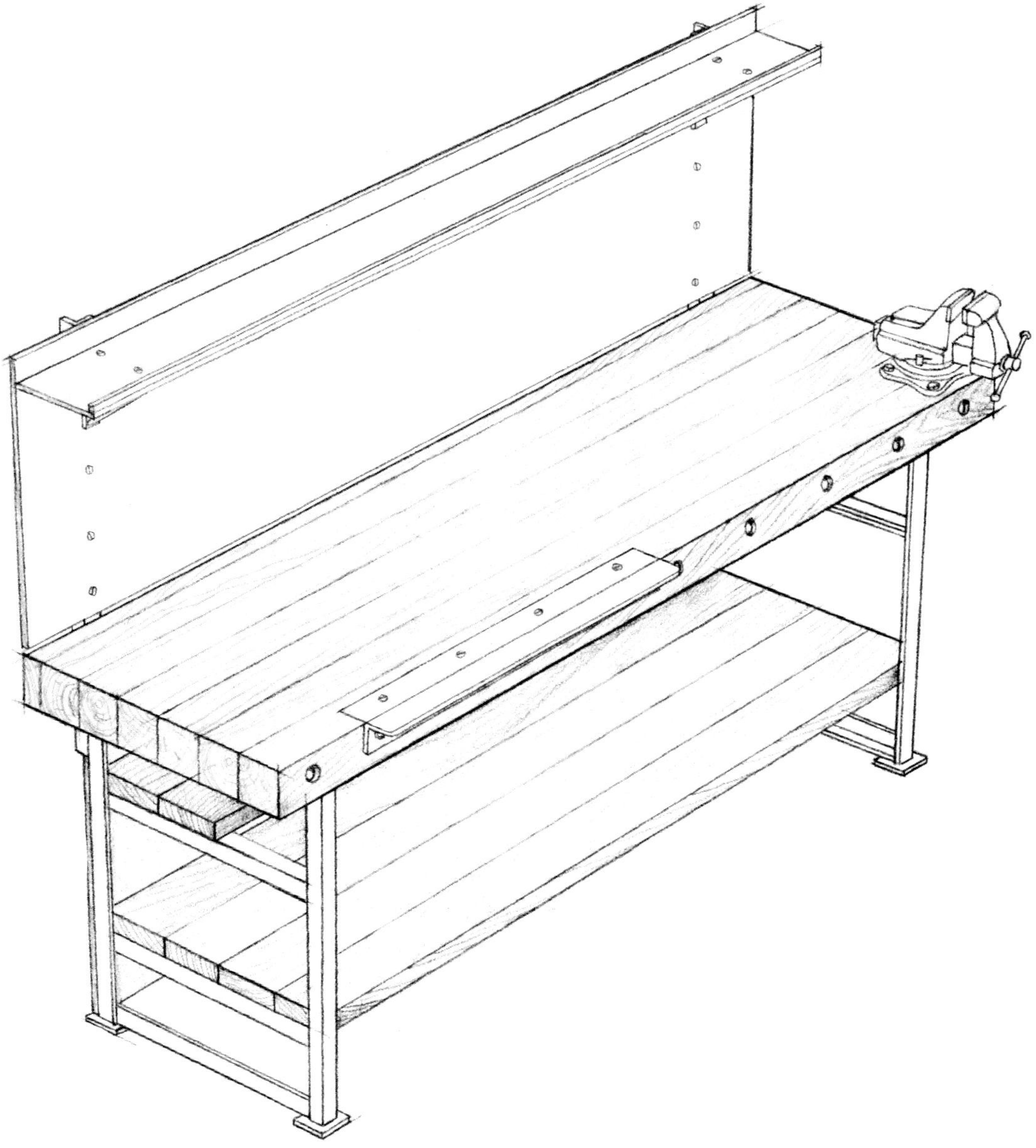

How to Make a Special Tap

Eureka! Nineteen years of teaching high school students is enough!

I started a new job in July developing new products for Lee Valley Tools Ltd. in Ottawa, Canada. This company sells fine woodworking tools by catalogue and from their stores in Toronto, Ottawa and Vancouver. My time is very limited for I have many design projects ahead of me. I have been writing the "Beginners' Column" in *Live Steam* Magazine for nine years this September; however, my interest is mainly in machine shop work rather than live steam so I shall now write only the "Scrap Box" column.

My basement workshop is no longer just a hobby shop; I now use my shop for development work, therefore the area has now expanded into another room. I've kept the vertical mill shown in the September/October issue of *The Home Shop Machinist* and the company bought me a larger vertical milling machine to be described in a later issue.

I no longer have the use of a school machine shop so I bought a combination vertical/horizontal band saw from Busy Bee. I can use this saw for cutoff work and then raise the saw to the vertical position and fasten the table in place for contour sawing. This machine takes a ½″ wide blade and has three speeds to suit the material being cut. One of the first tasks was to cut slices off a 4″ × 4″ bar of aluminum. This was done very easily and accurately with a little cutting oil applied by brush from time to time to prevent the blade from binding in the cut. I find that steel cuts more easily than aluminum because the latter seems to make the blade bind in the saw kerf unless lubricated. Recently at a machine tool show I saw the same model of saw cutting 1⁄16″ thick slices off a 3½″ diameter mild steel bar. I find that the blade lasts through several months of occasional use. I always release the tension on the blade after use to avoid metal fatigue and I open the guard to clean out the cuttings around the wheels. There are two wheels on the back legs so the saw can be kept in an out-of-the-way corner and then rolled into the central floor area for use.

My saw is model no. B-013 bought from Busy Bee in Toronto; however, they are also available from Grizzly Imports, Westchester Tool Supply, and other advertisers in *HSM.* I should have bought one of these saws years ago; they are not expensive and the ability to do contour sawing is a great asset when space is limited and one machine can do the work of two machines. Please note: I have no connection with Busy Bee or any other supplier except as a satisfied customer.

Recently we needed a 1¼-8 NS tap to retap some chucks for wood lathes that had been tapped slightly undersize. Now, $120 for a special tap to retap four chucks seems a bit expensive, so I decided to make a tap. The

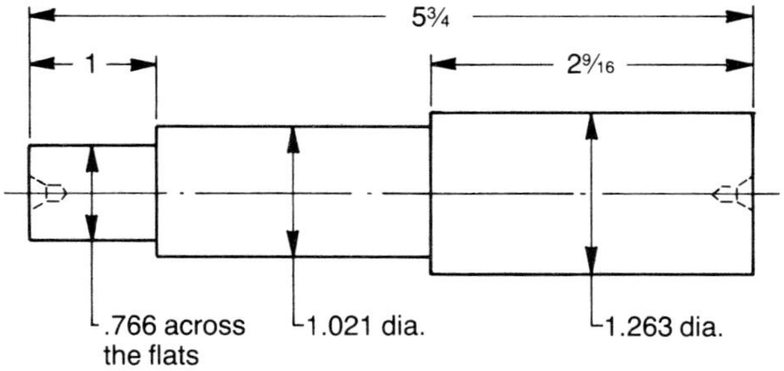

Operation 1
Cut to length. Use oil or water hardening, tool steel or drill rod.
Face and center drill.
Turn to size between centers.
Mill the square on the shank. Use a dividing head if possible.

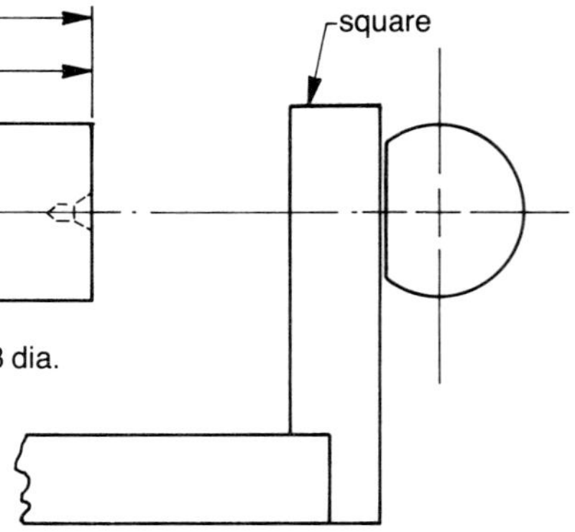

Alternate method of indexing: Mill one flat. Depth of cut = 1.263″ – .766″ = .497″ ÷ 2 = .248″. Turn 90° and align with a square. Repeat for the other side.

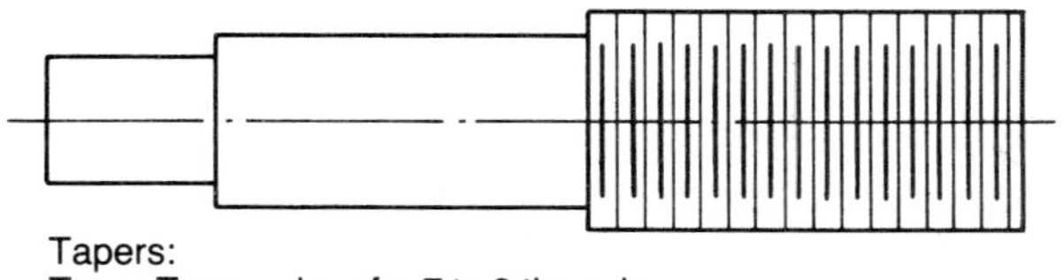

Tapers:
Taper Taps – chamfer 7 to 9 threads
Plug Taps – chamfer 2½ to 5 threads (preferred taper for general use)
Bottoming Taps – chamfer 1 to 1½ threads

Operation 2
Set up between centers. Wedge the tail of the dog in the slot of the faceplate with a piece of wood. Cut the thread to size. Measure by the three wire method.
Turn the taper with the compound slide rest.

cost was $4 for a bit of oil hardening tool steel and two pleasant evenings spent on an interesting job.

The 1¼ is the nominal diameter of the tap; 8 represents the number of threads per inch; N means that the thread is the National form (60° with a flat at the crest and the root); and the S stands for a special number of threads per inch, because the standard number of threads per inch for 1¼" diameter is 7.

Although the nominal diameter of the tap is stated at 1¼", the diameter is actually 1.263". This size and the general dimensions of the tap are taken from the *Machinery's Handbook*. You may wish to own a copy of this book for I shall be referring to it in future articles and it is an excellent reference when making small tools. The *Handbook* is available from some of our advertisers; however, an older copy bought from a secondhand bookstore will do very well since the information for our purposes will not be out of date. The sizes of this tap have been taken from the *Handbook*.

Follow this procedure and you will make a tap that is nearly as good as a commercial tap and good enough for occasional use:

1) Cut a piece of tool steel to length, face and center drill both ends.

2) Turn the tap blank to size between centers.

3) Mill the square on the shank using a dividing head if it is available, or use the "square method" as shown in Operation 1.

4) Put the tap back in the lathe between centers. Be sure to wedge the tail of the dog in the faceplate slot with a piece of tapered wood.

5) Cut the thread as usual; however, you should measure the thread size by the three-wire method as shown in the *Machinery's Handbook*.

6) Turn the taper or chamfer with the compound slide rest or use a taper attachment. The taper or chamfer for plug taps will be the most suitable for general use if you want to make only one tap.

7) Now the flutes will have to be milled. Most people have vertical milling machines, so they will have to use a ball mill (half-round end mill).

To choose the size of cutter most suitable, draw center lines on squared paper and then draw a circle the same diameter as the tap. From *Machinery's Handbook* you will find the number of flutes required for the size at the tap – in this case, four.

Use a hole template and try several size holes until you find one that seems to be a suitable size. This is not too critical. If you don't have a ball end mill this size, one slightly larger or smaller can be used. An old end mill with chipped corners can be reground to the half-round form quite easily by hand on a grinder. Use a radius gage to check the shape as you proceed and then back off the cutter by grinding. This is more difficult to describe than do.

The depth of the flutes is usually a bit deeper than half the diameter of the cutter to provide a bit of positive rake to the cutting face of the

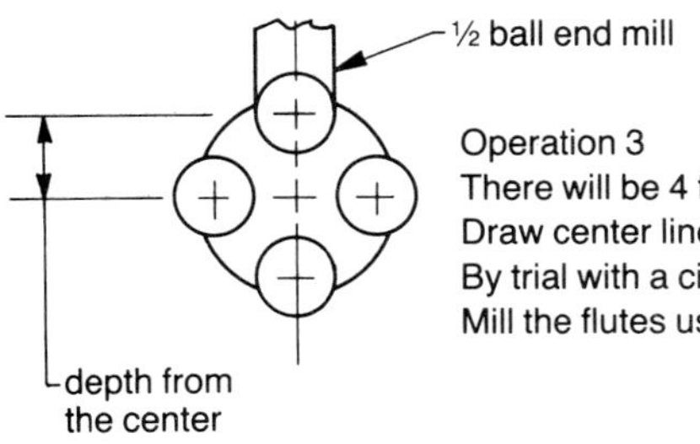

Operation 3
There will be 4 flutes on a 1¼" tap *(Machinery's Handbook)*.
Draw center lines and a 1¼" diameter circle.
By trial with a circle template choose a suitable size ball end mill.
Mill the flutes using a dividing head or other means for indexing.

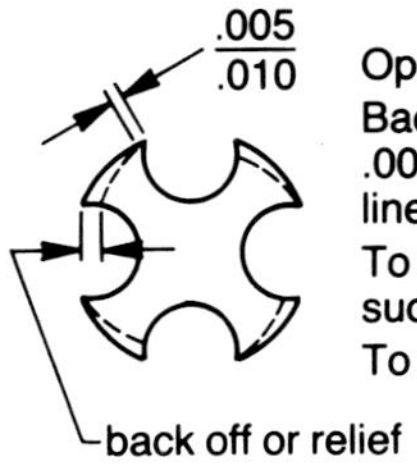

Operation 4
Back off each thread with an extra slim three-corner file. Leave a land about .005″ to .010″ wide. Coat the tap with layout blue before filing so a thin blue line remains at the land. Stamp the size on the shank. Harden and temper. To sharpen the tap after heat treating, grind the flutes with a hand grinder such as a Dremel. Choose a mounted point the same size as the flutes. To remove the burrs, tap a thread in a piece of scrap material.

teeth. Try to use an index head, if possible; otherwise, hold the tap shank in the milling machine vise and locate for indexing from the square end at the shank.

8) We will have to perform the backing off or relieving operation by filing as we do not have a relieving lathe. Most taps are made commercially and ground to size on a special grinder that grinds a relief as the tap rotates. The cut taps are relieved the same way with a relieving lathe such as those made by Hendy, Pratt & Whitney, etc.

Coat the tap threads with layout blue. Now file an eccentric relief at each thread with an extra slim three-corner file and leave a narrow land at .005″ to .010″ to maintain the correct size.

9) Stamp the size of the tap on the shank.

10) Harden and temper as usual.

11) Sharpen the tap with a hand grinder such as a Dremel. Choose a mounted point to fit the flute and run the stone along the flute until all the black has been removed.

12) To remove the burrs on the edges of the teeth, drill a tapping size hole in a piece of mild steel scrap and the burrs will be removed as the tap passes through the metal. Don't forget to use cutting oil or one of the tapping fluids.

The next issue will continue to describe modifications and improvements for vertical milling machines and I shall also have some photographs of my shop.

Transfer Screws

Transfer screws provide a simple solution to the problem of locating the position of the screw holes when you have to make a replacement part for a water pump, a cover for a steam chest, or any other job requiring bolt holes to match blind, tapped holes in the main casting. Toolmakers use them when they have to make a replacement part in a die.

Transfer screws are available in all the common machine screw and bolt sizes from commercial suppliers. Mine were made by the Heimann Manufacturing Company, Orbana, OH. They come in a little hexagon case that holds six screws. One end of the case is a hexagon socket wrench to drive the screws.

The procedure is to place a screw in each of the tapped holes with the hex end of the screws flush with the work so the center punches on the screw all project the same distance. Now lay the cover or replacement part in place and strike once with a lead or brass hammer. The location of the holes is exact. Use a wiggler to locate for the drilling of the center punch. Center drill, and then drill to the clearing size of the bolt.

The drawings show how to make your own transfer screw.

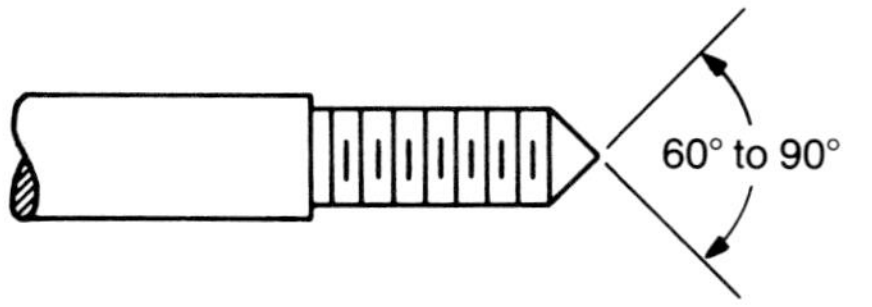

Step 1 Turn a piece of drill rod to the major diameter of the thread. (For machine screw)

Step 2 Thread with a tailstock die holder.

Step 3 Turn the point.

Step 4 Hold the shank in the bench vise to file a flat on each side of the point for a wrench.
Alternately leave the work in the lathe chuck for filing, but be sure to put a piece of plywood on the lathe bed to protect the lathe bed in case the file slips and to prevent the filings from landing on the lathe bed and then being trapped under the carriage and tailstock.

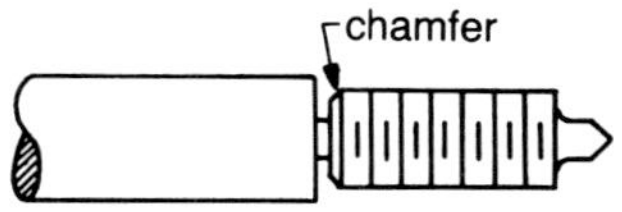

Step 5 Part off part way, chamfer the thread with a tool bit or a file.

Step 6 To harden, hold by the shank.
Heat the point with a torch until bright red; quench in oil.

Step 7 Cut the neck with a hacksaw.

This procedure will prevent damaging the threads.

HOW TO MAKE YOUR OWN TRANSFER SCREWS

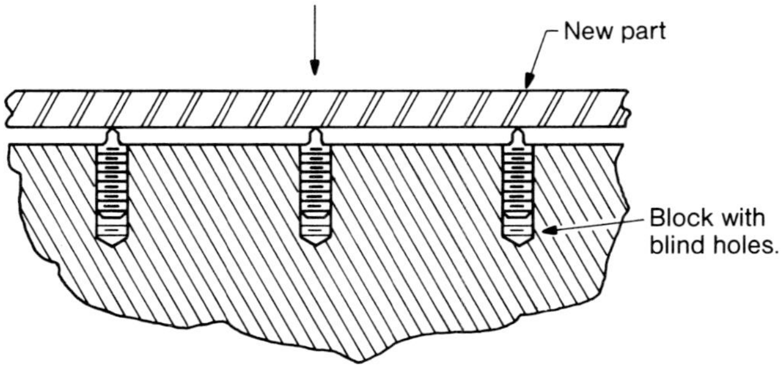

The points of all the transfer screws should be the same height above the surface.

A Power Hacksaw

This saw can be made by a novice. The experience gained will be invaluable when a more complex project such as a locomotive or steam launch is undertaken. The saw will be a useful addition to your shop equipment, for the saw uses ordinary 12" hand hacksaw blades for economy. If a micro switch is fitted to be operated by the stop arm, the saw will shut itself off when the cut is finished so you will be able to do something else while the saw is working. I find that the blades last from three to six months with intermittent use.

During the planning stage, I had the rotation of the crank clockwise when viewed from the crank side so that the friction from the aluminum sliding blocks on the square guide bar would lift the arm slightly on the back stroke. However, when I reversed the blade in the saw frame when one end of the blade was dull, I found that the saw cut just as well. So much for theory.

My saw was made from materials I had on hand. Short ends of brass rod were used for the bushings, the 1/3 hp motor came from a discarded clothes dryer, the pulleys from the junk box, and I made the gears in the school machine shop. The total out-of-pocket cost was slightly over three dollars for a vee belt and a switch.

A good place to start is with the *Gears* (Details 1 and 2). I have given all the information necessary to make the gears on the drawings for those who have the opportunity to make them during an adult education machine shop course at their local high school or college. They are easily made and are an important milestone on the road to mastering the arts of a machinist.

For those who choose to buy the gears, try to buy solid gears without spokes. If you have to use spoked gears, then one of the spokes will have to be drilled and tapped for the crankpin. A better method would be to bolt a plate to the gear for the crankpin if the spoke is too small for the thread. If you have gears with a slightly different ratio, use them and alter the pulley ratio to suit. (See the note on the drawing.) Sprockets and a chain drive could also be used.

The next part to make is the *Gear Box* (Detail 10) using 1/4" and 1/2" mild steel plates. Either cold- or hot-rolled will do. I used hot-rolled and then took a light cut all over on the shaper after welding. Those who have electric or oxy-acetylene welding equipment, bevel the plates 45° for 1/4" and bronze- or fusion-weld.

Those who do not have suitable welding equipment may assemble the plates with 1/4" flathead screws passing through the 1/4" plates and threaded into the 1/2" plates. Use three screws for each joint and locate the screws so they will clear the bushings.

1

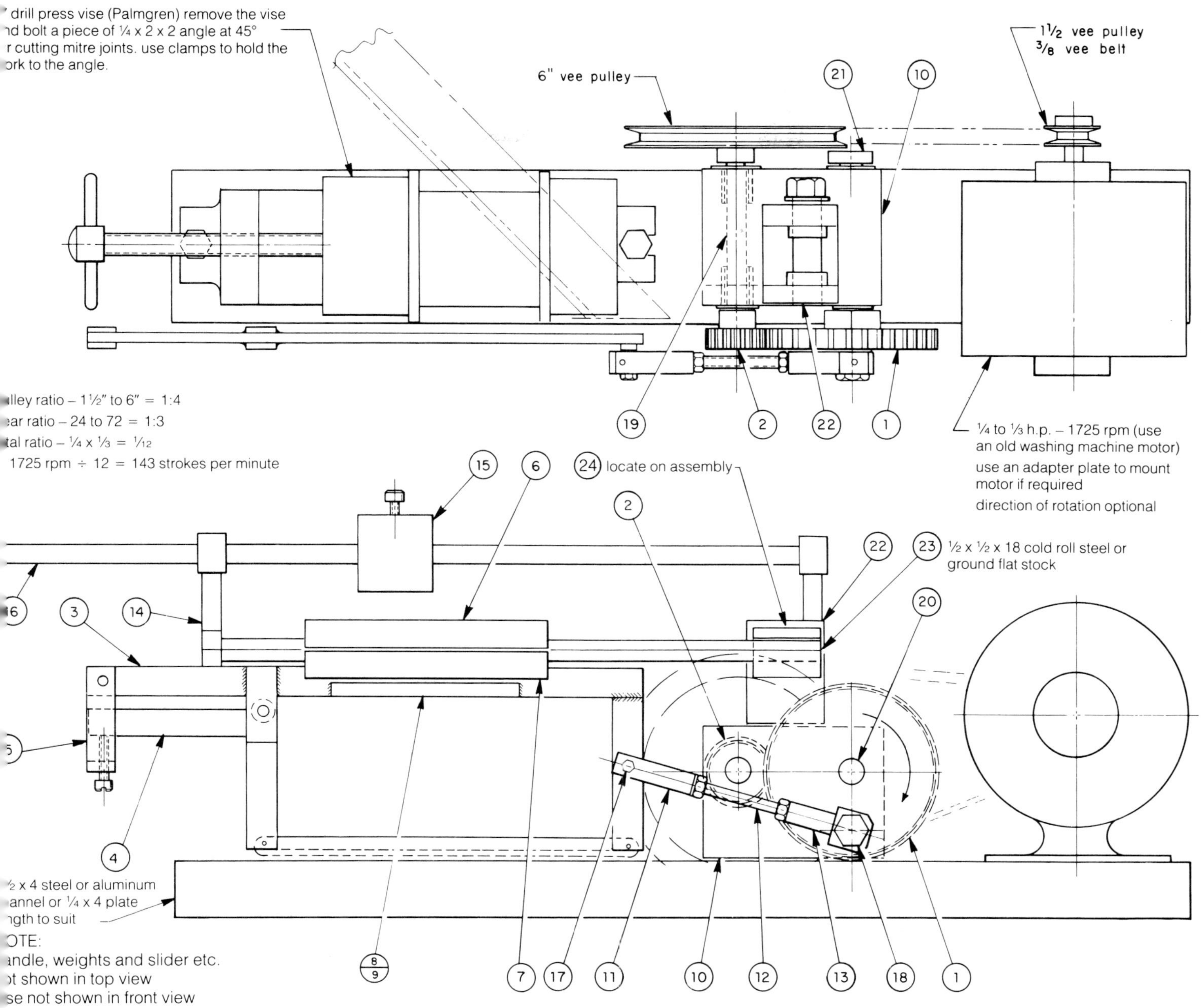

12" POWER HACK SAW ASSEMBLY

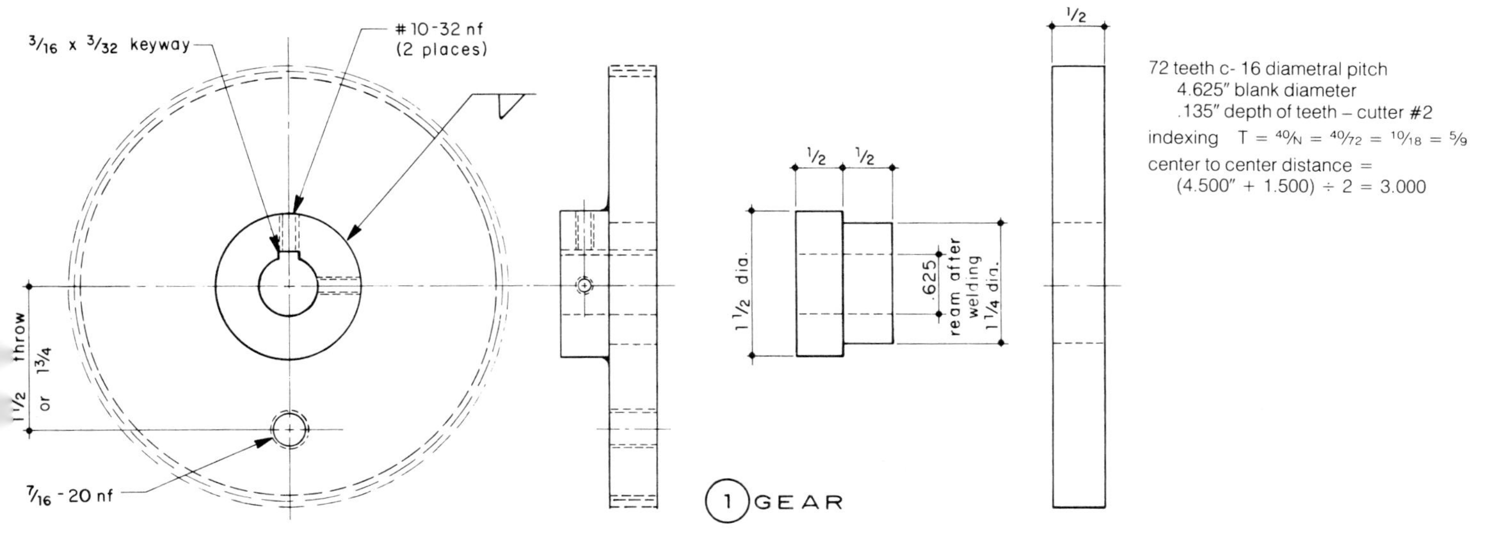

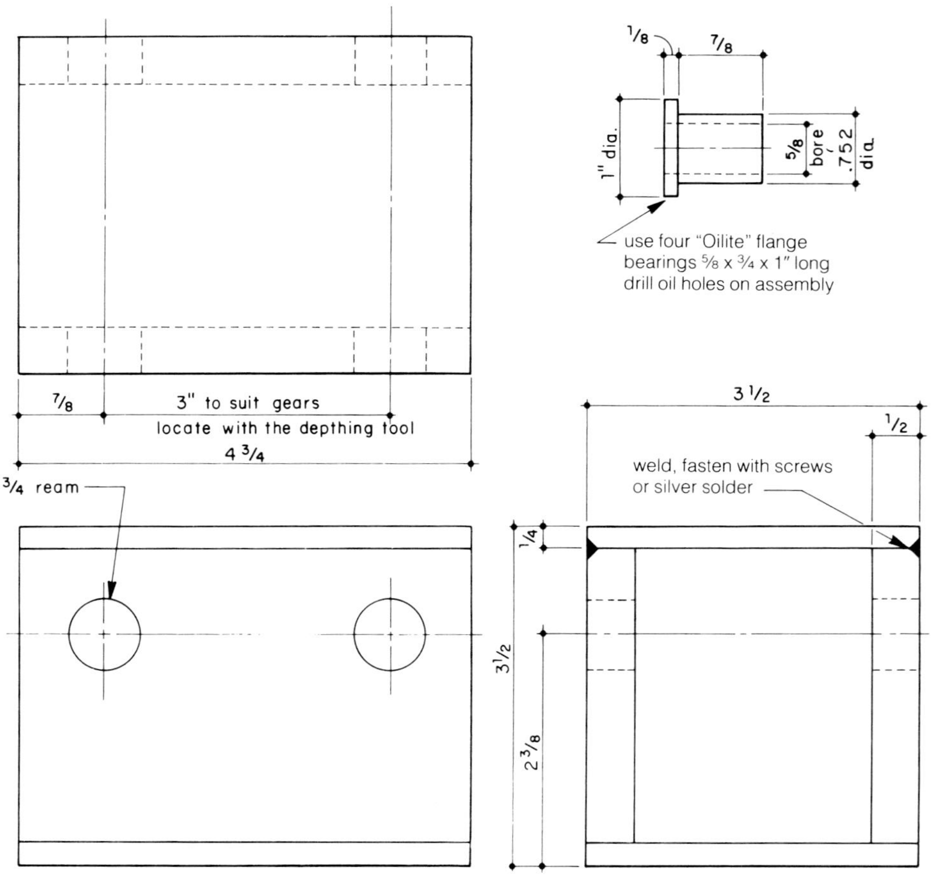

(10) GEAR BOX

On the first saw I made, I used cold rolled steel plates and carefully clamped them in position on an angle plate for drilling and tapping. I used hexagon socket flathead screws (Allen type), tightened until they "squeaked." They are visible in Photo 2. The box was square and did not need to be machined on the surfaces. Be sure to drill the clearing size, tapping size, countersink and then tap the thread without moving the work on the drill press so that concentricity will be maintained. Tap the thread by holding the tap in the drill chuck and pull the belt by hand while maintaining a slight pressure on the feed handle. Be sure to use cutting oil or one of the special lubricants for threading. If a spiral flute tap is used, the shavings will feed up the flutes the same way as when drilling.

If ordinary slotted flathead screws are used, then the plates should be brazed or silver soldered on the inside corners after the screws have been tightened.

If the sides of the box are not quite square, take a light cut on the shaper or milling machine. For those who have to use the lathe, hold the box in the four-jaw independent chuck. After the first side has been faced, use this side as a reference face. Place the box in the four-jaw chuck and test the alignment with a small square held against the chuck face and the reference side of the box. Repeat for each face.

Lay out the holes for the shafts with a surface gage or a combination square. A good method of locating the centers of the gears is to use a clockmakers' *Depthing Tool* (see drawing of the tool). Place the gears on the *Spindles* and adjust the clearance between the teeth until there is a slight amount of "play" between the teeth.

Clamp the thumb nut and place the points of the spindles on the centers laid out on the box. Strike the tops of the spindles with a soft hammer.

The bushing holes may be bored on the vertical milling machine or the lathe. Use a wiggler or center finder to locate the work accurately. Use several drills and then bore or ream to size (Photo 3). Be sure to do both sides at the same setting so the bearings will be in line.

The *Bearings* (bushings) may be purchased or made according to the drawings. The outside diameter should be about .002" oversize for a press fit.

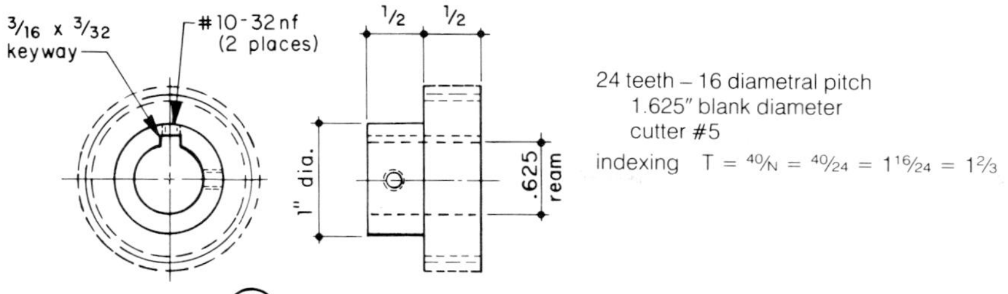

(2) GEAR

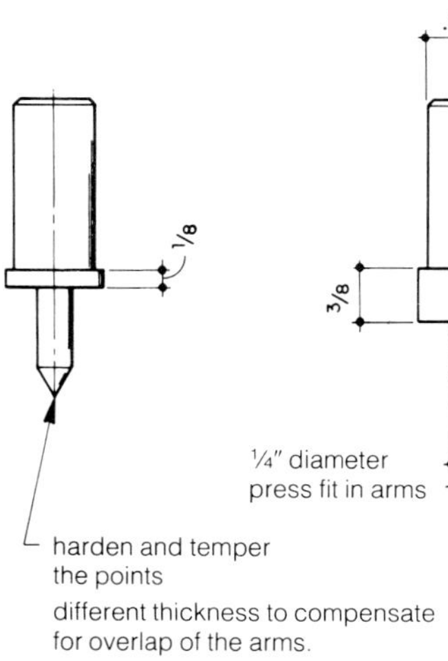

Spindles

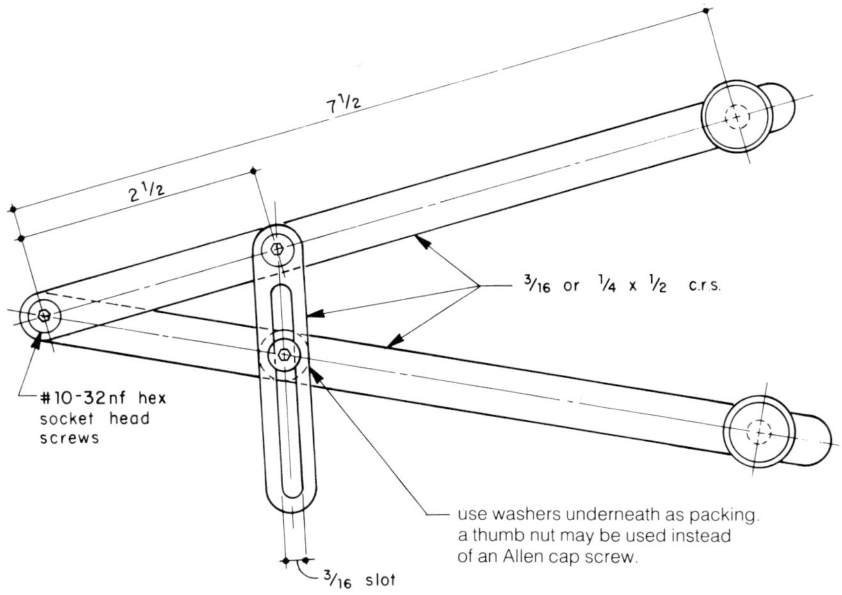

DEPTHING TOOL

Oil holes should be drilled on top of the bearings at the junction of the flanges and the sides of the box.

Tilt the box to an angle of 10°-15° and drill with a 3/32" drill into the bore. (Note: tilt the drill press table or use a tilting vise.)

Now pass a 5/8" hand reamer from one bearing into the other to make sure the bores are in line and on size. This will also remove the burrs from the drilling operation.

Cold rolled steel is usually about .001" undersize, so the clearance will be about right for long life and good lubrication. (I used LaSalle stress-proof steel.)

The shafts are cut and faced to length in the usual way. The key seats are machined by milling.

The gears will have to be keyed to the *Shafts*, or taper pins or spring pins may be used. The keyways in the gears will have to be broached, or they can be cut on a shaper or on the lathe. Those fortunate enough to have a 3/16" broach will need to buy or make a suitable slotted bushing. Most broaches need to be passed twice through the gear with a shim placed at the back of the broach for the second pass.

2

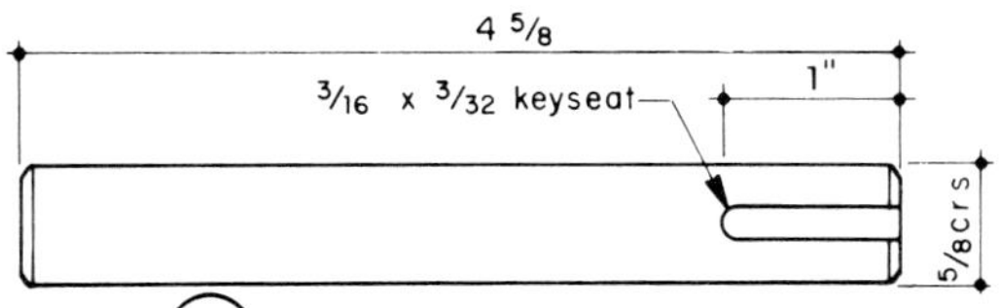

(19) DRIVEN SHAFT

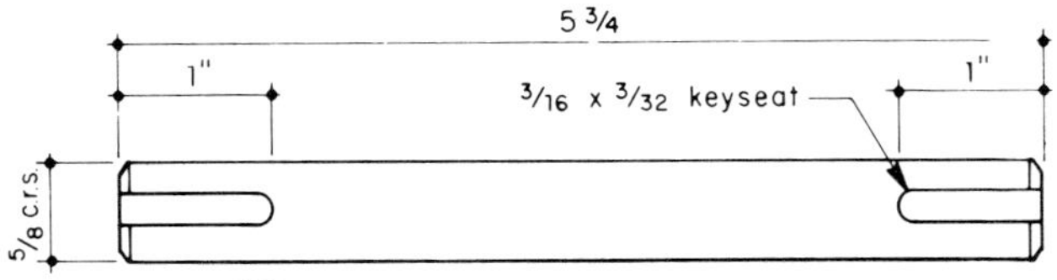

(20) DRIVER SHAFT

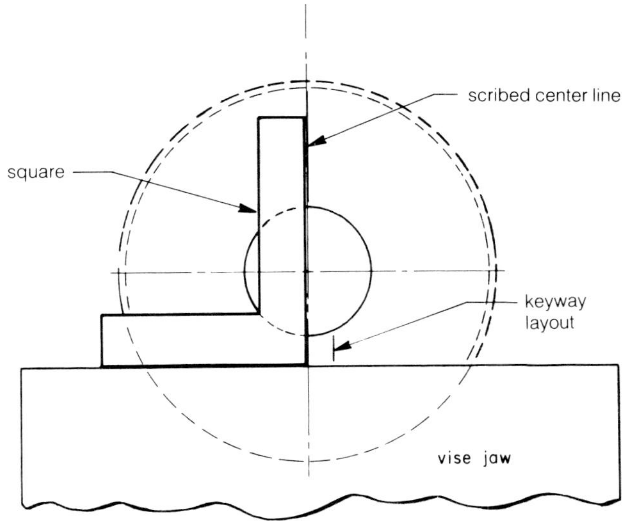

SETTING UP ON A SHAPER

Those with a shaper should make up an *Extension Tool Holder* as shown in the drawing. The bolt is set up tight against the tool slide to prevent the clapper from lifting. The tool bit is placed in the bar so the cut will take place on the back stroke.

The time to lay out the keyways is when the gears are still in the lathe. I use a piece of steel plate set on two parallels to clear the vee strips on the lathe bed. Coat the face of the gear with layout fluid. Use a surface gage or a vernier height gage. Bring the scriber to just touch the bottom of the bore. Raise the scriber to one-half of the bore diameter and scribe a center line across the face of the gear hub. Now scribe lines 3/32" above and below the center line.

3

When the gear is being set up in the shaper vise, use a square to align the center line vertically. The bottom edge of the bore should clear the vise jaws by about 3/16" to 1/4" (see the drawing). Bring the corners of the tool bit to touch the bore at the layout lines and reset the dial of the tool slide to zero. Use a slow speed and feed about one thousandth (.001") per cut. The depth of the keyway should be 3/32" (.093") plus a bit of clearance, say .100" total (see Photos 4 and 5).

This extension tool is meant to cut on the back stroke so the bar will spring away from the cut instead of "digging in" as usually happens with an extension tool holder in a small shaper. Photo 6 shows the laying out of the keyways prior to cutting.

The keyway can also be cut while the gear is in the lathe chuck. Use a 1/2" boring bar with a 3/16" square tool bit ground like the tool bit in the shaper extension tool (see drawings). This bar will be set in a vee block raised with packing strips (shims) until the center of the tool bit is at the lathe center height. Check with the layout on the gear. When the tips of the tool bit touch the bore, reset the cross slide dial to zero. Traverse the tool bit with the carriage handwheel. The tool bit is set to make a "draw cut" so that the bar will spring away from the cut like the shaper extension tool. Feed about one-thousandth each cut. The total depth of cut will be about .100".

The square hole in the boring bar or the shaper extension tool can be made in a few minutes. Lay out the center line and the 3/16" square hole on the bar with a surface gage or a vernier

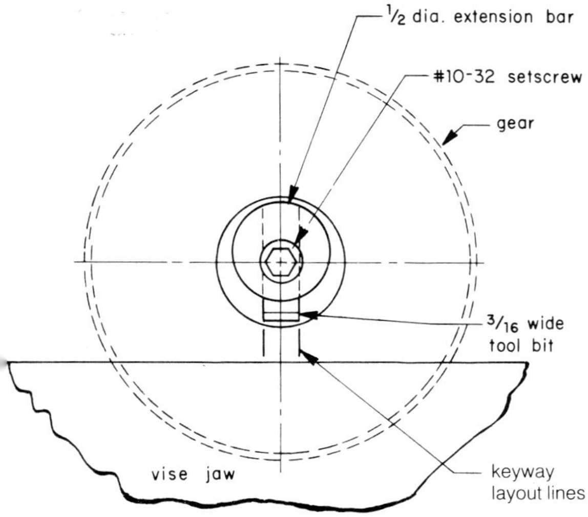

SET UP ON A SHAPER

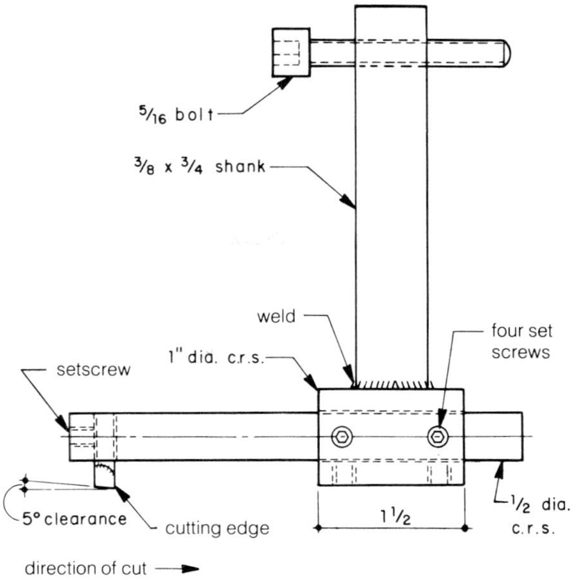

SHAPER EXTENSION TOOL HOLDER

HACKSAW BUSHINGS
Brass or Bronze

1. Chuck with 1½" projecting beyond the jaws.
2. Face the end.
3. Turn to size and ⅞" long.
4. Center drill and drill nearly to size.
5. Bore to be sure the hole runs true.
6. Ream to size.
7. Part off with 1/32" extra for facing.
8. Reverse in the chuck and face the flange.

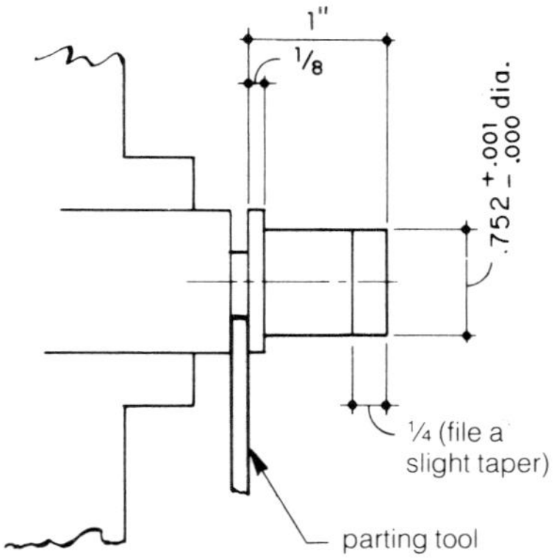

OPERATIONS
1. turn the outside diameter
2. part off
3. reverse in the three jaw chuck with a .010" to .020 shim placed between the bushing and one of the jaws (#1)
4. drill, bore and ream .625" dia. mark the bushing with a center punch on the flange at the jaw with the shim (#1)

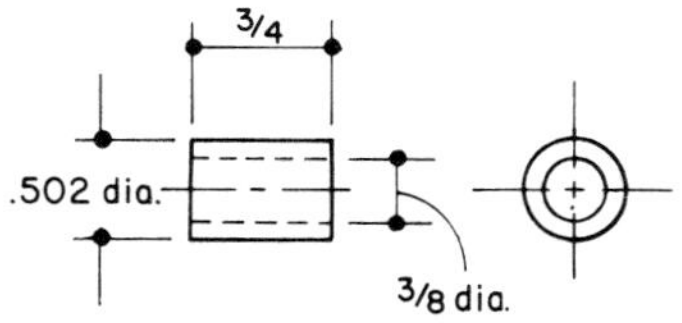

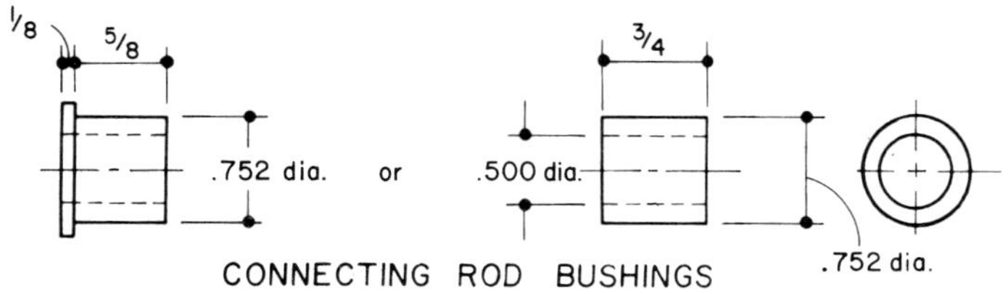

CONNECTING ROD BUSHINGS
Bronze or Brass

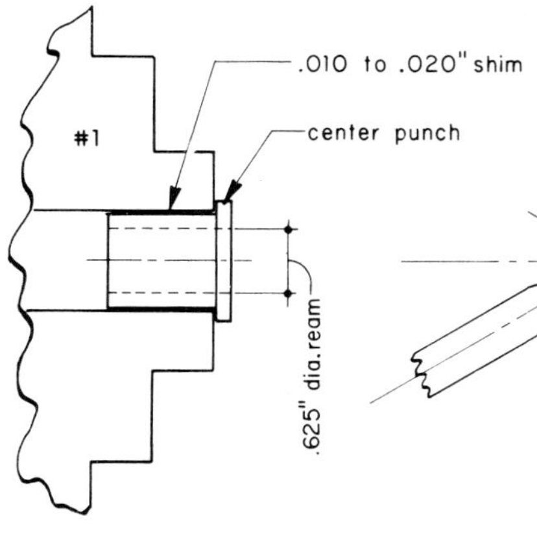

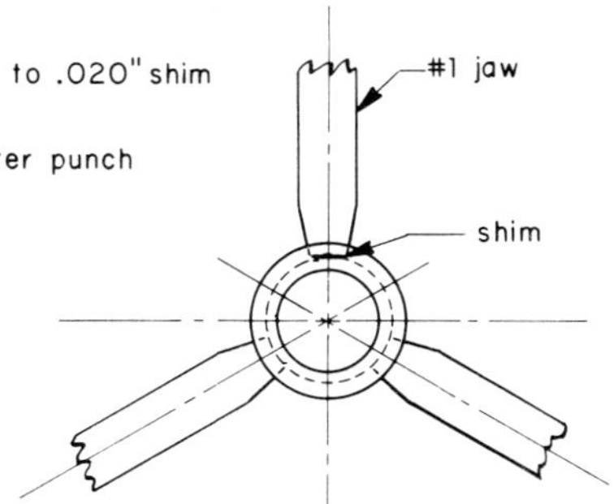

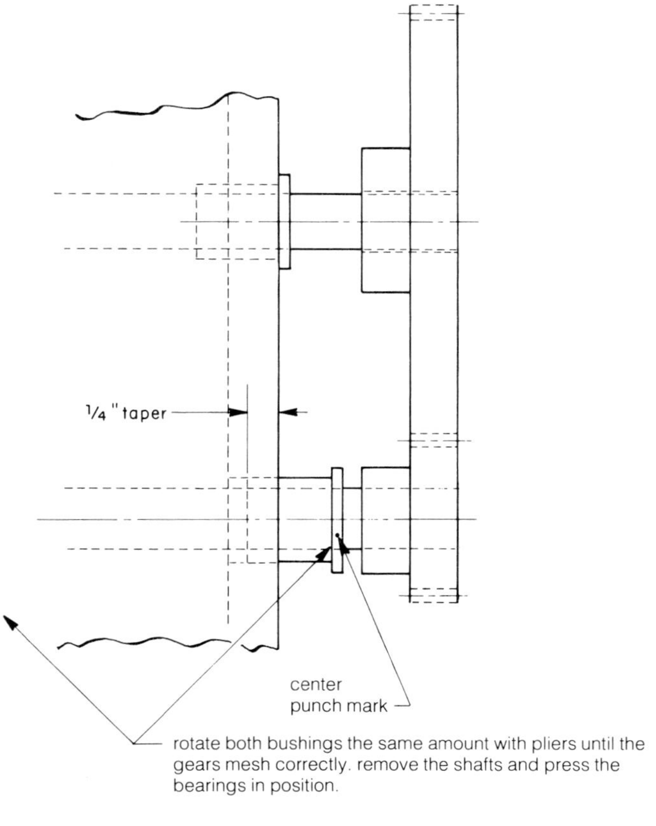

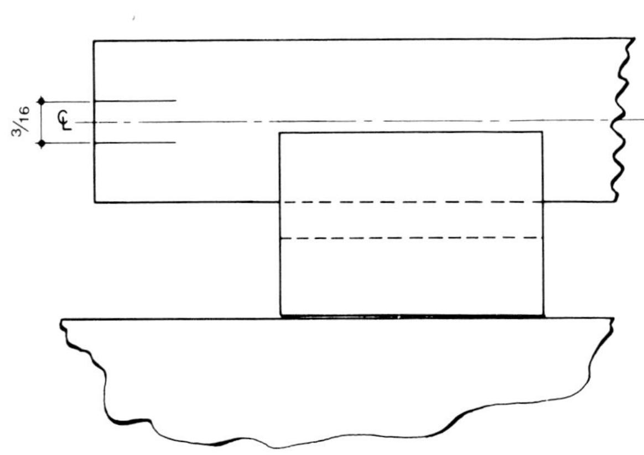

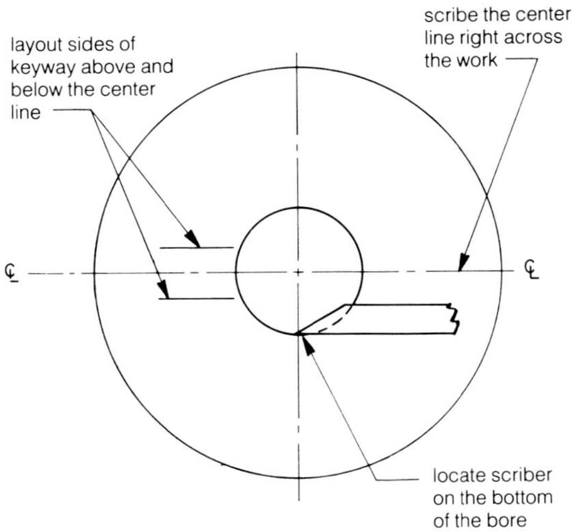

LAYOUT OF KEYWAY ON THE LATHE

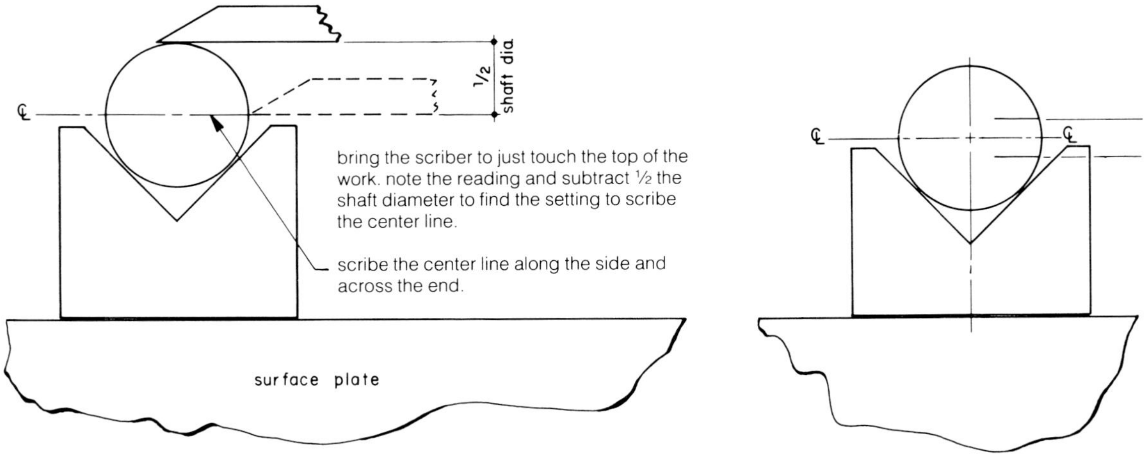

use a Vernier height gauge or a surface gauge and transfer the measurements to a steel rule clamped to a small angle plate.

HOW TO LAYOUT A SQUARE HOLE ON A BORING BAR

height gage (Photo 7) with the bar placed in a vee block (see drawing). Now set the bar in the drill press vise and locate the center of the square with a wiggler. Drill with a center drill, 1/8" or 5/32" drill and finish with a 3/16" drill.

Now prepare a 3/16" square tool bit by grinding the ends square with the sides. File out the hole with a small square file until you reach the layout lines. Set the bar in a vee block or a vise and place the 3/16" square tool bit on the layout. Force the tool bit (now called a broach) into the bar about 1/32". Remove the broach with a small brass rod and a hammer. File out the burrs and repeat the broaching and filing operations until the hole is complete. Then drill and tap the setscrew hole in the end of the bar. The broach should be pushed into the bar with an arbor press, not the drill press. If this is not available, use a strong bench vise or a brass hammer. (Do not use a steel ball peen hammer on the hardened broach.)

The *Saw Frame* is made from 1/4 × 3/4" or 1/4 × 1" cold- or hot-rolled mild steel (Detail 3). The joints can be bronze or fusion welded. Silver soldering could also be used, but a small gusset 1/8" thick should be added to both sides of the joint at the back end of the saw frame.

The *Bell Crank* is used to tighten the blade without twisting as sometimes happens when the tensioning is done with a nut.

The *Aluminum Blocks* (Details 6 and 7) that slide along the guide bar will have to be placed in a shaper or a milling machine to machine out the 90° vee slots. Whatever method is used, be sure to align the vise or the block carefully with a dial indicator so the vee groove will be parallel with the edges of the blocks.

If a shaper is used, set the tool slide to 45° and offset the apron away from the cut (see Photos 8 and 9).

Move the table sideways for each cut (about .020") and feed with the tool slide. Continue until you reach the layout line. Now, without

4

5

6

7

moving the table, reverse the block in the vise and finish the other side of the vee groove at the same setting.

The 1/4" wide slot in the bottom block is machined on a milling machine in the usual way. Be sure the slot is parallel with the sides of the block and the vee groove, or the saw will not work properly. The 1/4" slot can be machined on a shaper (Photo 10). Grind up a

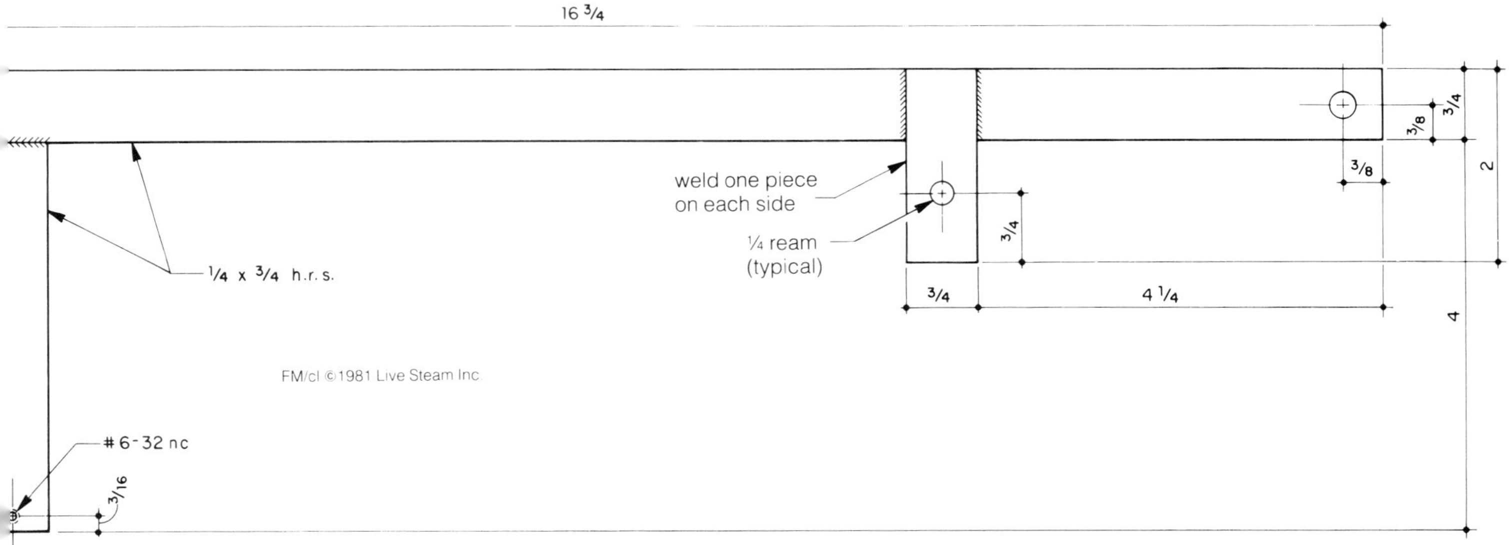

DETAIL 3

8

9

tool bit like a parting tool and take a cut down the middle of the groove. Then take a cut on each side as required to bring the groove to a snug fit on the saw frame.

The *Saw Frame* is held in the groove by two or three No. 6-32 or No. 8-32 hexagon socket head cap screws. Drill and counterbore from the vee groove in the bottom block. I have shown two pieces of 1/8 × 1/2" angle on each side of the frame (Details 8 and 9). These are for extra rigidity and are optional. Drill an oil cup with a center drill and a 3/32" drill into the vee groove on the top block.

The blocks are clamped together with No. 10-32 screws and shims are used between the blocks to adjust the clearance. If the 90° vee grooves are too deep, machine a little off the joining surfaces.

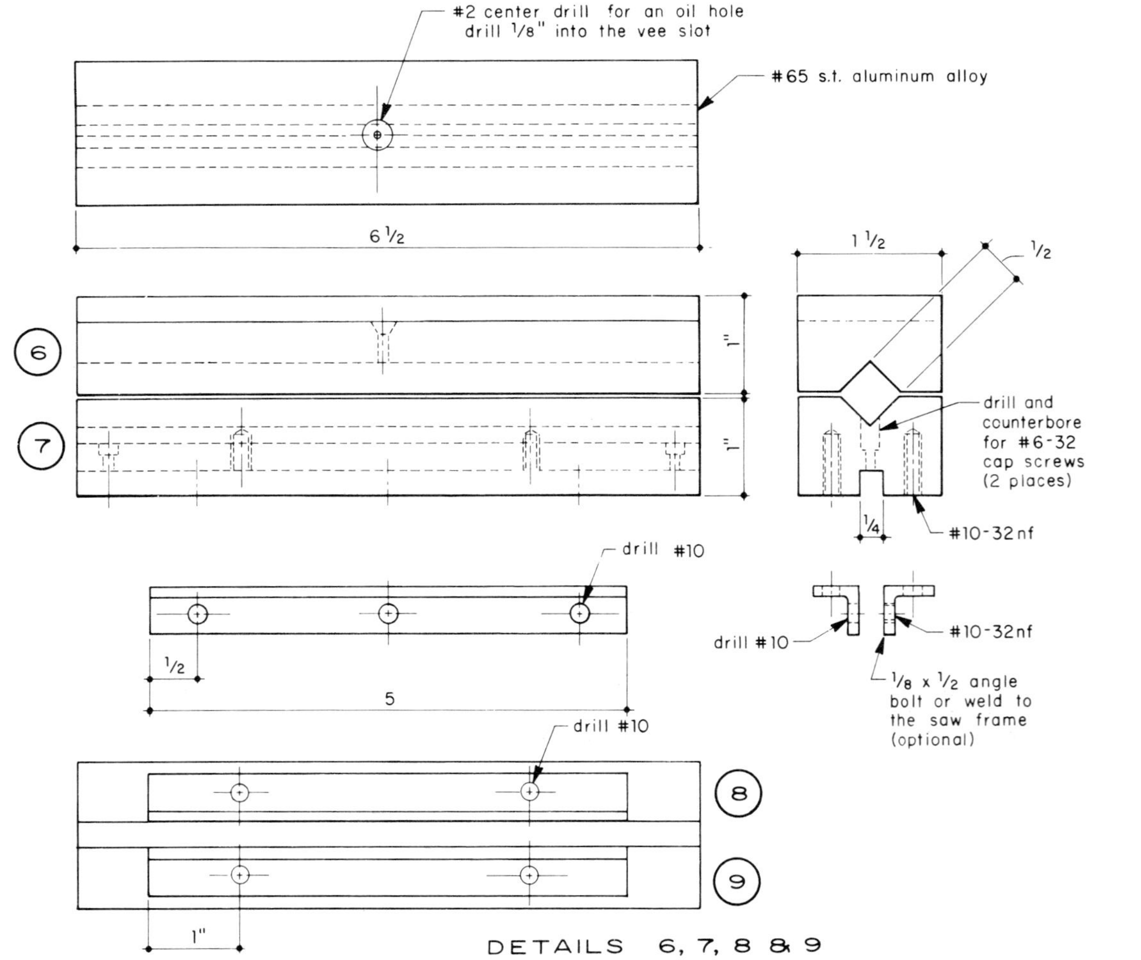

DETAILS 6, 7, 8 & 9

The *Guide Bar Shaft*, or pivot (Detail 23), is made from a piece of 1-1/2" square mild steel. Hold one end in the four-jaw independent chuck. Face the end and measure from the sides to the center mark left from the facing. Adjust the chuck jaws and reface until the centering mark is equidistant from the sides. Center drill and support with a tailstock center for turning and threading. The thread will have to be screw-cut on the lathe, if the nut is to bear evenly on the flange of the bushing.

Lay out the other end with a center line and the 90° vee groove. Mill away half the block to the center line. Tilt the shaft to 45° in the milling machine vise for milling the vee groove. See Photo 11.

If you are milling the shaft on the lathe, swivel the milling attachment to 45°. Be particular about this vee groove or the saw frame will not be vertical when the machine is assembled.

The *Cap* (Detail 24) is made from 1/4" plate and is fastened with four No. 10-32 screws. Drill the No. 9 clearing holes in the *Cap* and then clamp the cap to the block for spotting, drilling and tapping the screw holes.

Drill and tap the two No. 10-32 NF holes in the underside of the pivot. Drill and tap the nut for a setscrew as shown on the drawing. Place a piece of brass rod under the setscrew to avoid damaging the thread.

11

The U-shaped bracket (Detail 22) for the pivot or shaft is made from three pieces of 1/2" plate, or it could be made from a solid block. Mine was welded up and then machined all over so that the sides were parallel and square with the bottom. The bushing holes may be bored on the lathe or the vertical milling machine. Support the sides with a spacer block when pressing in the bushings (Photos 12 and 13), drill the oil holes and then ream. The other parts are simple to make and do not require an explanation. Use the Assembly Drawings as a reference if the part is not labeled.

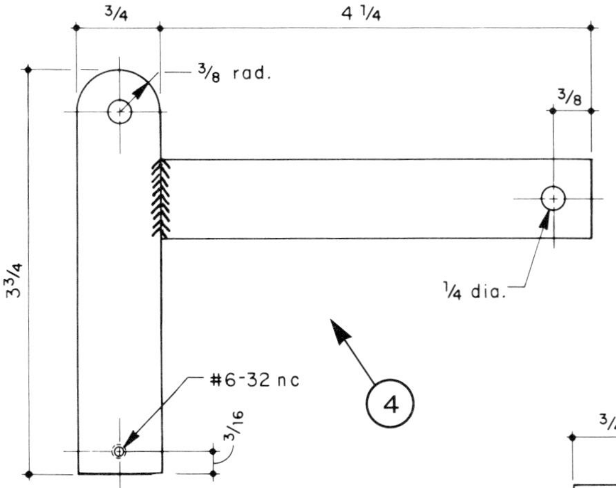

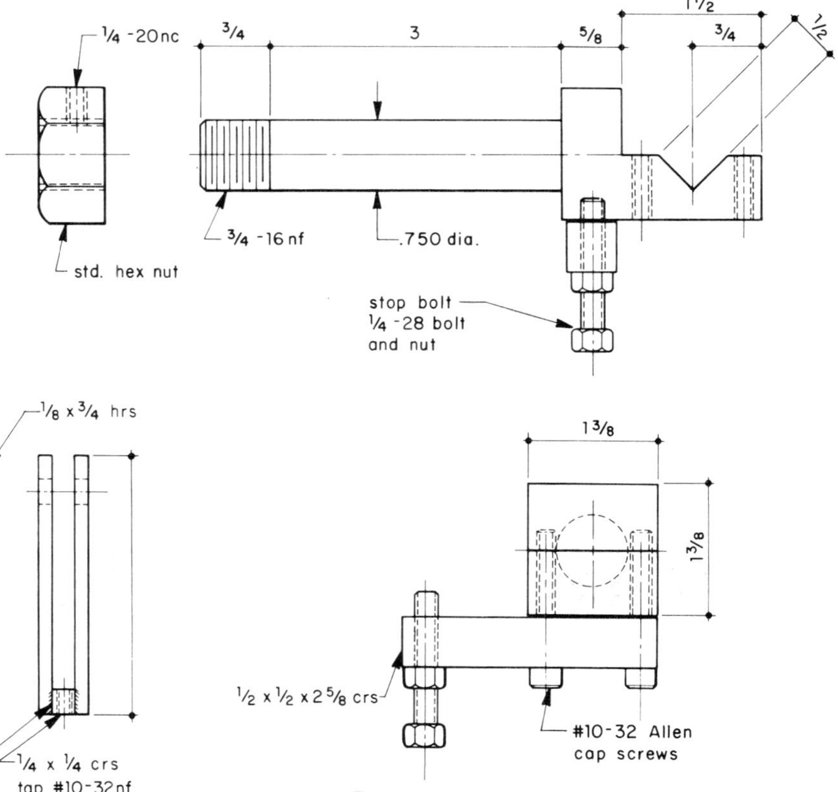

10

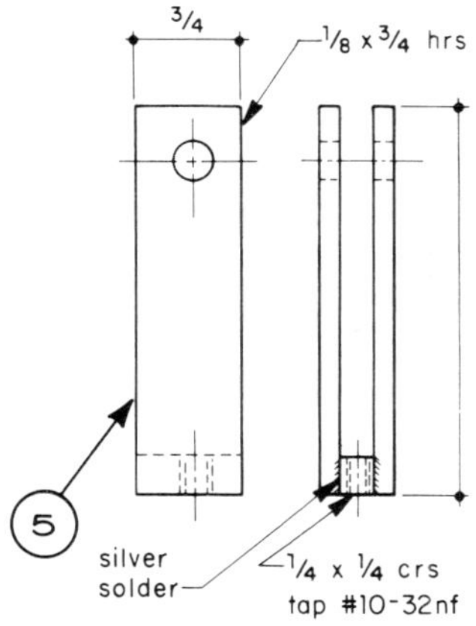

12

13

ASSEMBLY INSTRUCTIONS

Place the shafts in position in the gearbox and attach the gearbox and gears. If, through bad luck, the gears mesh too loosely or too tightly, you can make up a pair of eccentric bushings (see drawings). Make a slight taper on the end of each (outside diameter) bushing with a file for a 1/4". Place the bushings in the bores just enough so the bushings may be rotated with a pair of pliers on the flange. Assemble the gears and test the clearance with a piece of paper between the teeth. Rotate each bushing the same amount to adjust the clearance. Use the center punch mark on the flange as a guide. When the gears mesh correctly, remove the shafts and gears and press the bushings into position. Drill the oil holes as previously described. Now clamp the U bracket (Detail 22) on top of the gear box and assemble all the parts. Check the alignment of the U bracket with a square so that the pivot will be parallel with the gear shafts. Lift the saw frame to be sure the lower corners of the pivot do not foul the gears. Rotate the pulley by hand to be sure the mechanism works freely. Bolt the U bracket to the gear box.

14

Now you are ready to place the vise, saw and the motor on the baseplate. Photo 14 shows a view from above of the arrangement. I used channel because it is very rigid and lightweight and I can grasp it easily by the ends when I want to move the saw (easily transported to the metal merchant to avoid cutting costs or to the track side to cut rails to length).

Place a 4" drill press vise (Palmgren or similar) on the front of the channel close to the right-hand side. Place the saw mechanism in the forward stroke position and locate the gear box on the channel so that the back end of the saw frame is about 1" behind the fixed jaw of the vise. Make sure the saw frame is parallel with the edge of the channel. Bolt the gear box to the channel. Align the vise with a square on the fixed jaw and the edge of the channel or the saw blade and bolt the vise in position. The vise can be bolted down from the bolting lugs or Allen cap screws can be used set below the surface of the vise seat. The vise can have a plate fastened underneath with flathead screws and the four corners of the plate can be bolted to the channel.

The vise shown in the photographs is homemade from castings given to me by a friend with a backyard foundry. Photo 15 shows the machining of the vise seat on my aluminum casting, and Photo 16 is the machining of the base.

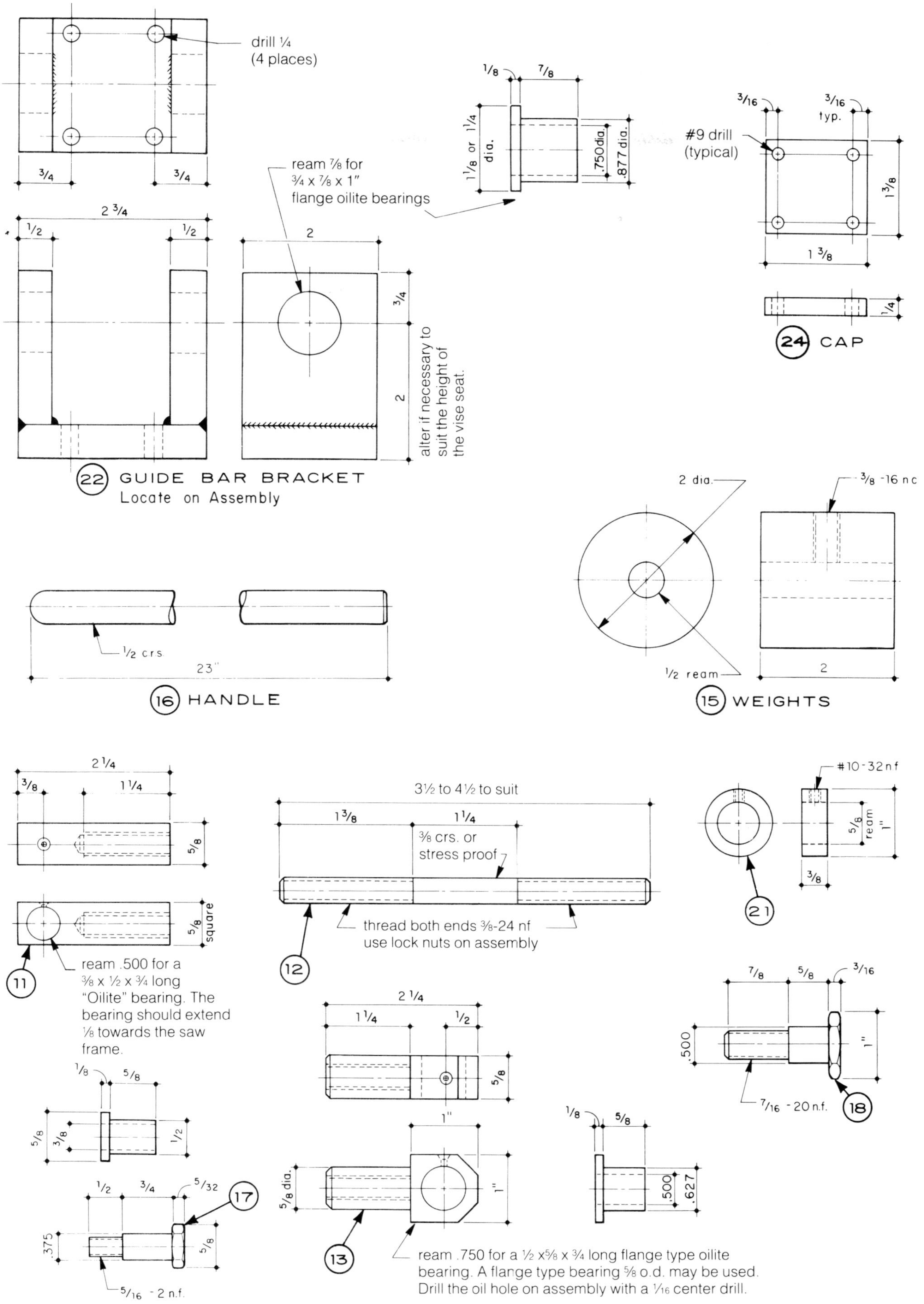

CONNECTING ROD AND ENDS

15

17

16

18

Align the motor with a straightedge across both pulleys. Place the belt in position to have the correct belt tension. Bolt the motor in place.

Now to finish up the loose ends. Make and fix the stop arm under the pivot. The arm can be extended to operate a switch. My switch is mounted on two 1/8 × 1/2" strips of steel fastened to the top of the U Bracket. Guards can be made up to suit.

I have a short length of 2 × 2 × 1/4" angle iron that I bolt to the channel in place of the vise for cutting mitres (45°). The work is clamped to the angle with two small C-clamps (Photo 17). I find this more accurate than a vise with swiveling jaws, for these seldom clamp well at 45°.

I also clamp plates directly to the channel for cutting. Recently I was making the smokebox fronts for two locomotives from 1/4" plate. The four corners needed to be cut off each plate before turning the outside diameters. I clamped the plates to the channel and made four simple cuts while I did some more interesting work on the lathe. Photo 18 shows the finished hacksaw and vise in place.

I hope you will enjoy making this little saw and will find it a work saver in your shop.

NOTE: The design of this saw can easily be altered to suit your own ideas or the material you have available.

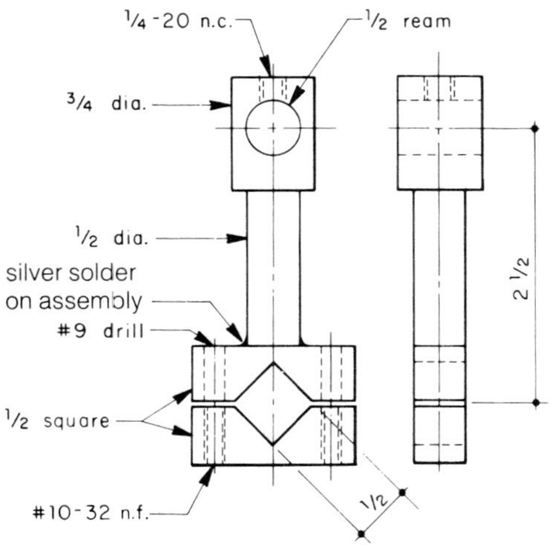

(14) POSTS

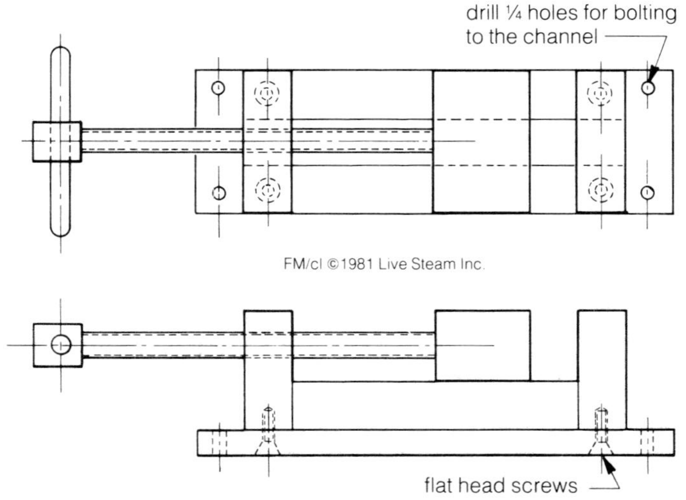

VISE CLAMPING PLATE

for use with a commercial vise or a home made vise built up from bar stock or castings.

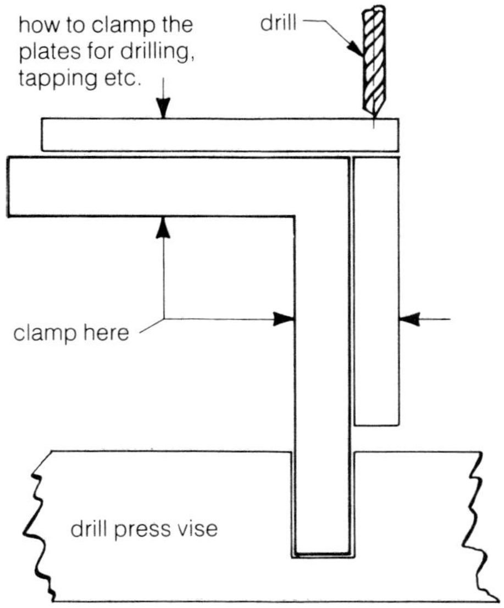

ordinary angle plate held in a drill press vise

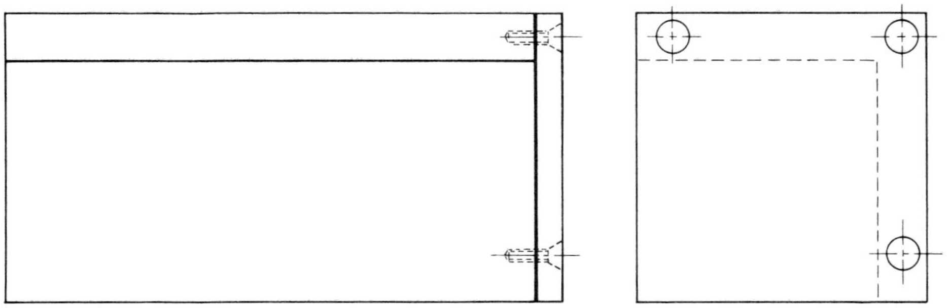

end plate added to ordinary angle plate so angle plate will stand by itself for drilling and tapping plates for box type assemblies. often used by toolmakers.

Power Hacksaw Table

The small power hacksaw was intended to be used on a workbench and stored on a shelf underneath when it wasn't required. When the saw was in use on the workbench, long bars could be supported on a wooden strip at the outer ends so the bars would be level with the bench top, and then the cuts would be square with the sides of the bars.

In 1980 we moved into a new house in the country with a 1-1/2-acre lot, so there was room to relay the "Frog Pond and Mosquito Haven Railway."

My new workshop is located in a small room in the basement that was supposed to be an extra bedroom. This room is well lighted with a large window and heated by electricity, so it is very comfortable. However, there is not enough room to use the hacksaw on the workbench, so I added some legs (Photos 1 and 2) so the saw can be placed in an out-of-the-way corner and then drawn into the main floor area when required.

The saw slides so easily on the concrete floor that the usual pair of wheels mounted on the rear legs are not necessary. The legs can be removed if I want to place the saw on a bench in the garage, where I keep most of my stock.

To make the legs easier to remove. I tapped 1/4-20 NC threads in the channel baseplate instead of drilling right through with a clearing size drill and using nuts.

If you want to make the saw more compact, saw the channel off behind the gearbox and mount the motor on a shelf underneath.

1

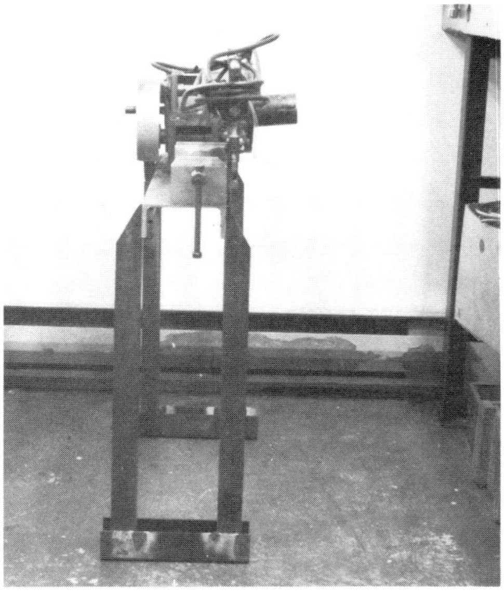

2

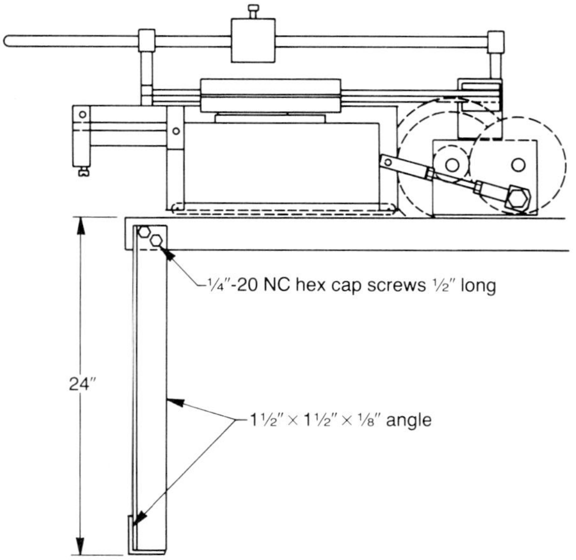
¼″-20 NC hex cap screws ½″ long
24″
1½″ × 1½″ × ⅛″ angle

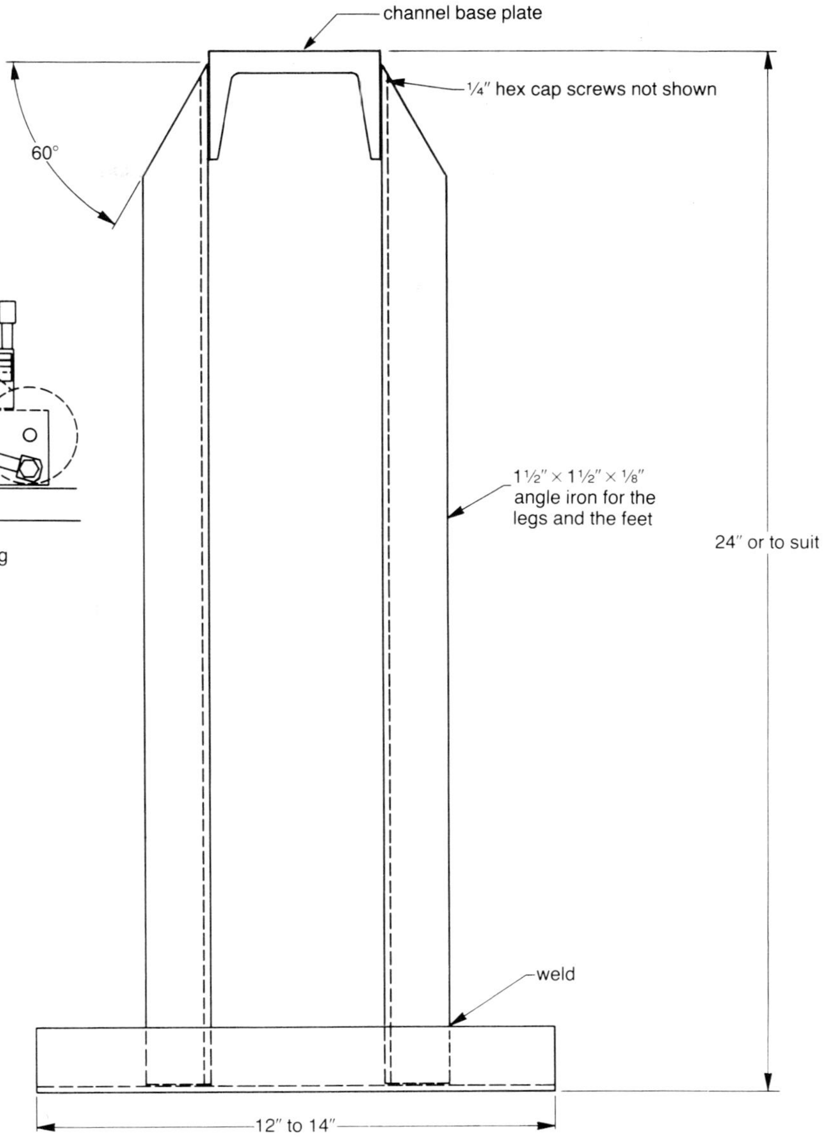
channel base plate
¼″ hex cap screws not shown
60°
1½″ × 1½″ × ⅛″
angle iron for the
legs and the feet
24″ or to suit
weld
12″ to 14″

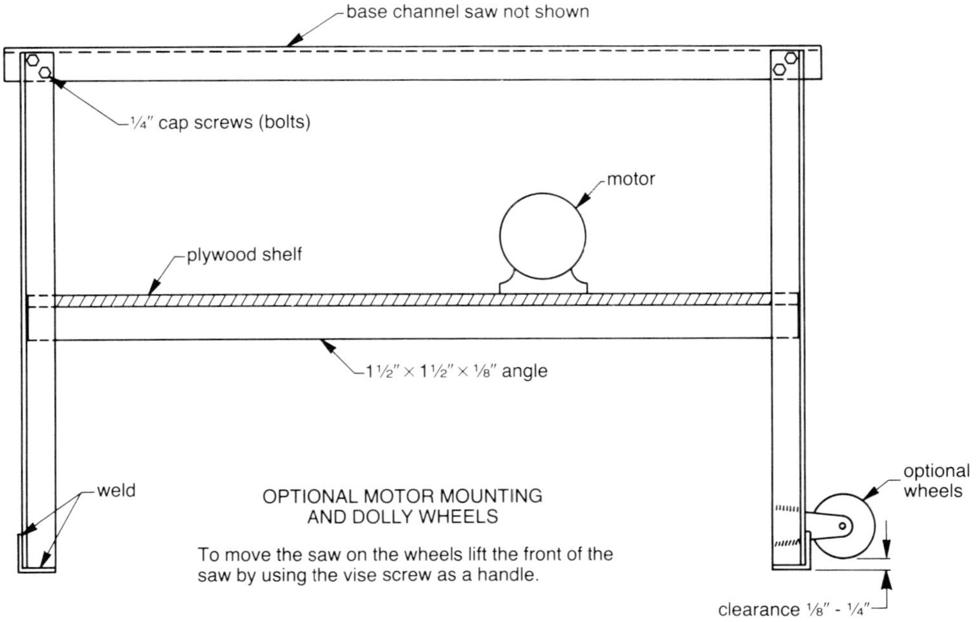
base channel saw not shown
¼″ cap screws (bolts)
motor
plywood shelf
1½″ × 1½″ × ⅛″ angle
weld
optional
wheels
OPTIONAL MOTOR MOUNTING
AND DOLLY WHEELS
To move the saw on the wheels lift the front of the
saw by using the vise screw as a handle.
clearance ⅛″ - ¼″

A Simple Rail Bender

Not long ago, I tried making a rail bender without gearing to drive the rollers. The bender worked out better than I expected, so I thought it would be suitable for many readers who are contemplating or are now building their own garden railway.

My bender was designed for the rail sold by International Model Products, St. Catharines, Ontario, Canada. This rail is intended for 3/4" and 1" scale tracks but it scales out exactly to 80 pound rail for 1-1/2" scale (7-1/4" or 7-1/2" gauge). This rail would not be suitable for a 7-1/4" or 7-1/2" gauge club track, but it works out fine for a scenic garden railway where passenger hauling is a secondary consideration. The rail sold by other suppliers would be equally suitable, but the design of the bender would have to be altered to bend the heavier rail usually supplied for 1-1/2" scale tracks.

Make the Side Plates (Detail 2) as a pair. To mill the 3/4" slot, lay out one plate, drill a 3/4" hole at both ends and then mill out the material in between.

The U-block (1) should be squared up on the mill or shaper. If these are not available to you, then use the 4-jaw lathe chuck. Lay out and then mill out the four corners so the block fits loosely between the side plates. Drill and ream the 1/2" hole before milling out the middle.

Drill and tap the 1/4-20 bolt holes in the Nut Block (3) and the Angles (6) on assembly.

To make the Screw, make sure that the lathe is turning parallel by measuring both ends of the screw with a micrometer during the first roughing cuts. The screw has a 3/8-1/2" extension which accommodates a center hole for the tailstock center. It also indicates when the screw thread is close to size. This extension is turned off after the screw is finished. The extension is turned to the minor diameter of the thread. The single depth of a National thread is .6495 divided by the number of threads per inch.

The minor, or root, diameter is equal to the major (outside) diameter of the thread minus two times the single depth of the thread.

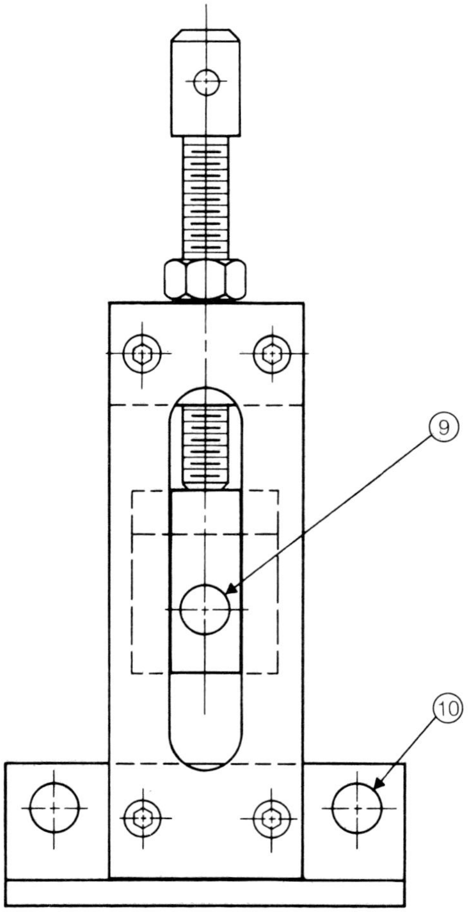

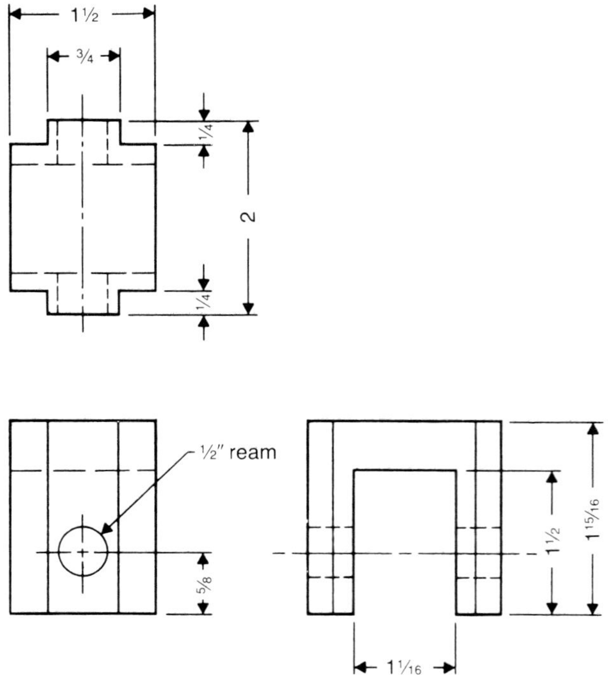

1. U BLOCK
mild steel

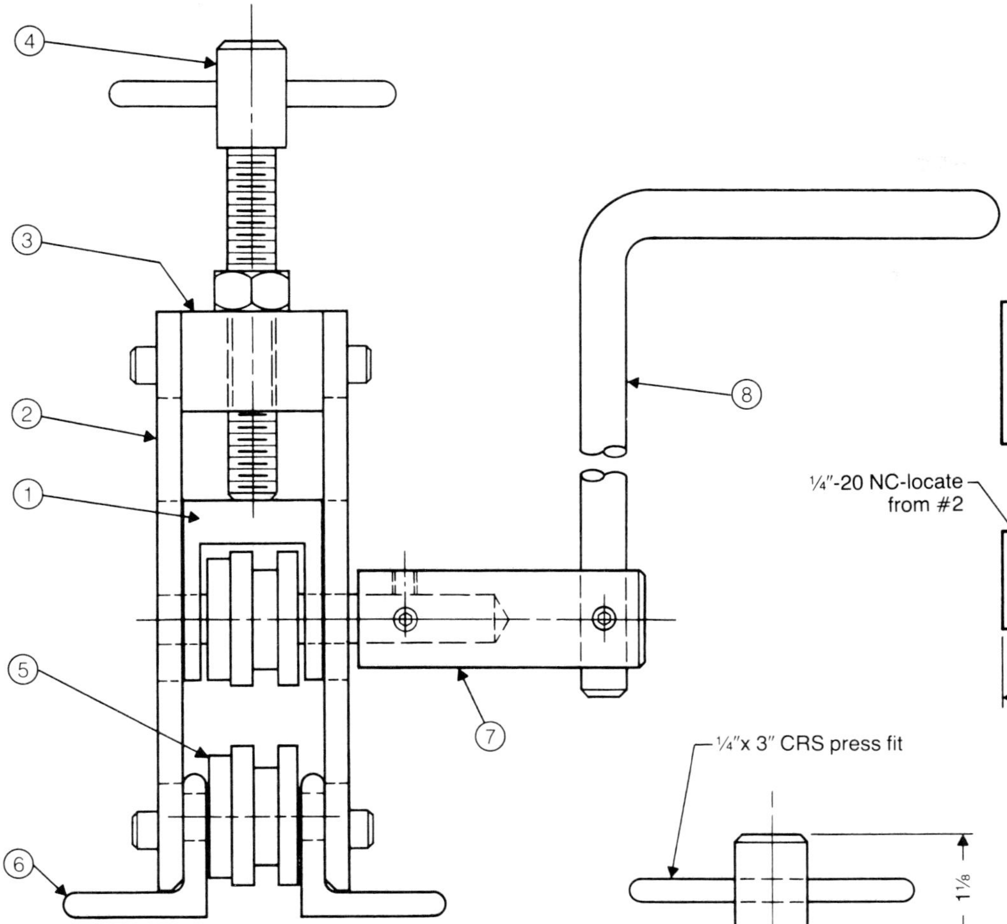

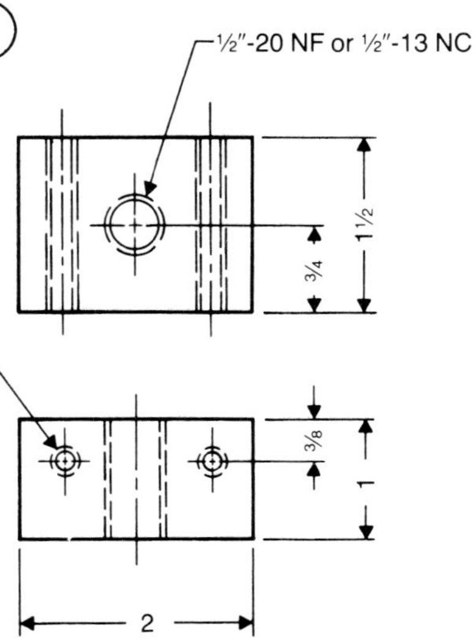

3. NUT BLOCK

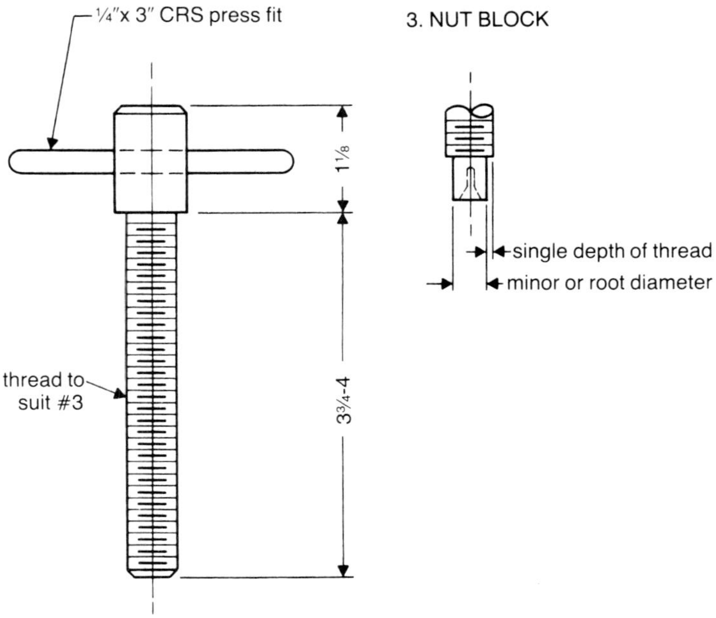

4. SCREW
3/4" CRS or hex

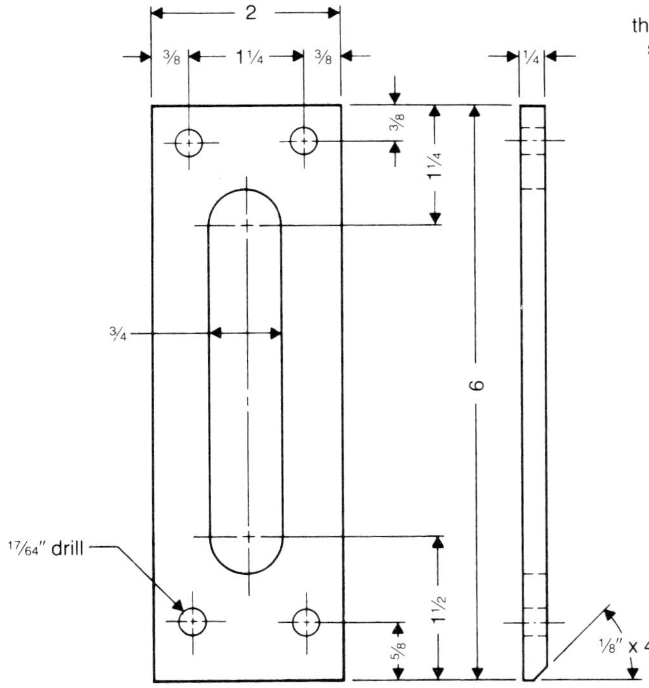

2. SIDE PLATES
mild steel, 2 required

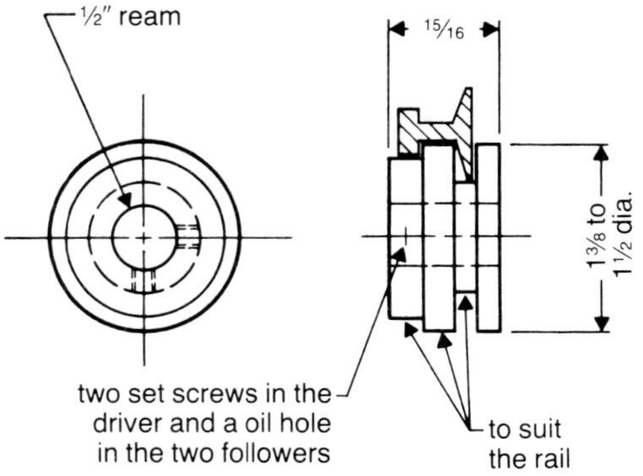

5. ROLLERS
mild steel, 3 required

Example for 1/2-20 NF:

$$\text{Single Depth} = \frac{.6495}{\text{TPI}} = \frac{.6495}{20} = .032''$$

$$\begin{aligned}\text{Minor Diameter} &= \text{Major Diameter} - (2 \times \text{Single Depth})\\ &= .500 - 2 \times .032\\ &= .500 - .064\\ &= .436''\end{aligned}$$

For easier threading, turn the screw slightly undersize (.495") so that a slight burr from the threading operation will not give a false impression when testing the thread with a nut.

$$\text{The Pitch of a thread} = \frac{1}{\text{Number of Threads per Inch}}$$

The width of the root of the thread is only .006", so this is usually ignored for fine threads ($1/8 \times 1/20 = 1/160 = .006''$).

Usually, the tip of the tool bit receives one or two strokes of an oil stone slip after grinding to remove the sharp edge. For this reason, when the tip of the threading tool touches the minor diameter, the thread will be close to size but not necessarily on size. The time required to calculate the minor diameter and to turn the extension will take less time than the usual cut-and-try method and there is no chance of making the thread undersize.

I clamped my bender to the end of a picnic table when it was in use. An assistant to support the end of the rail while you turn the crank will be helpful.

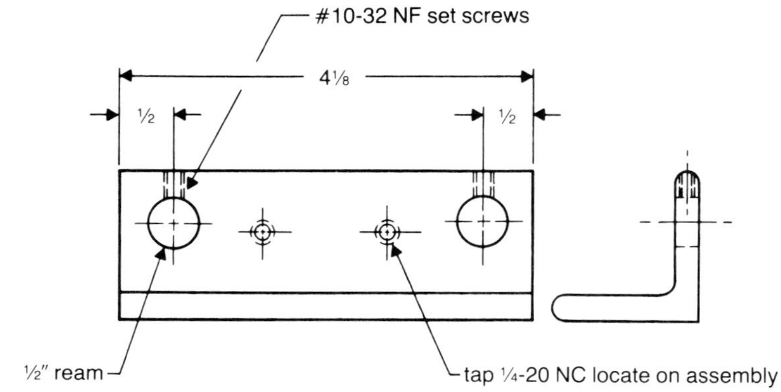

6. ANGLES
1½" x 1½" x ¼" angle, 2 required

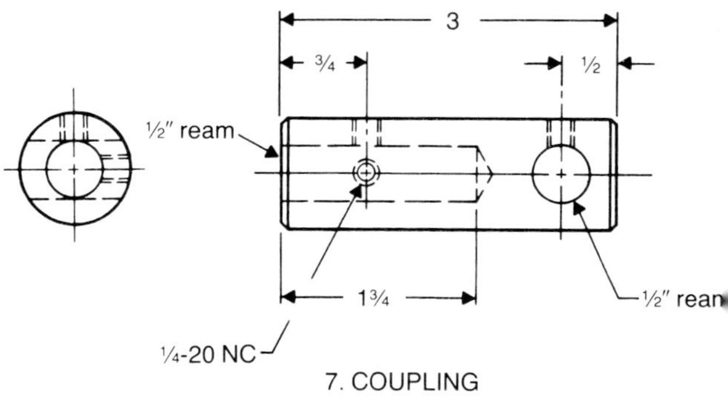

7. COUPLING
1" CRS

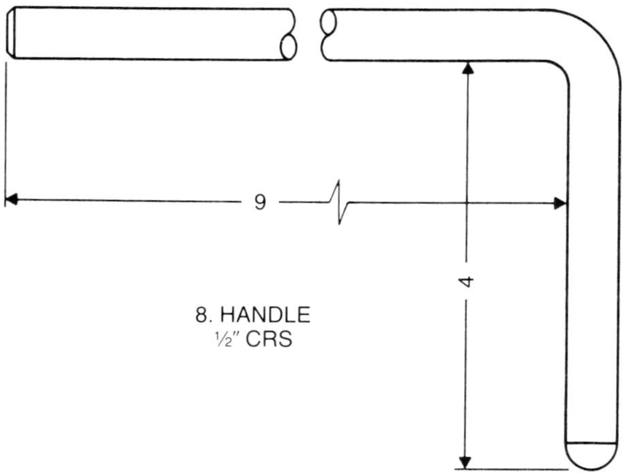

8. HANDLE
½" CRS

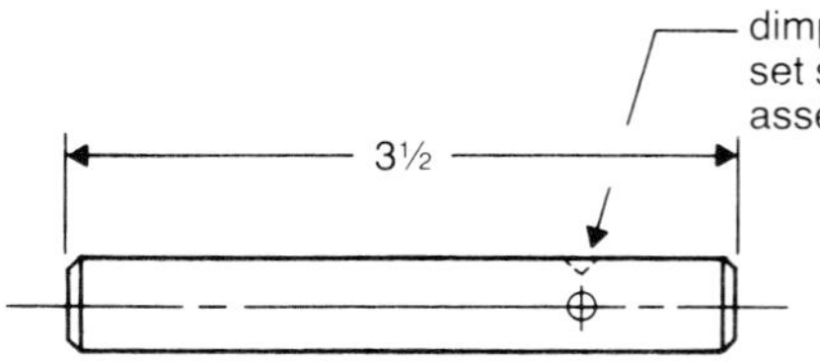

9. DRIVE SHAFT
½" CRS

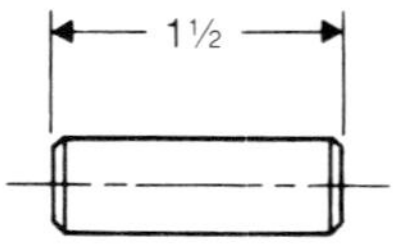

10. 2 required
½" CRS

A Grinding Bench

I prefer to keep all grinding operations out of the machine shop, so my grinding bench is set into an alcove in the basement outside the shop area. The photo shows the arrangement of the grinders.

On the right is the usual double-end grinder with a 1" wide belt sander located behind. I reach over the grinder to use the belt sander.

On the left is a double-end grinder driven by a motor located on the back of the bench. On the left is a white wheel of 80 grit used for sharpening wood chisels and plane irons. This is a very cool cutting wheel, and is not likely to burn those tools. On the right is a drill sharpening attachment.

Located between the two grinders is a wet grinder driven from the motor behind. From my own design, the wet grinder is intended for woodworking tools. It has a 10" wheel, 1-1/2" wide, that turns at 150 rpm. The grinder has a plastic hose on one side to drain the water after use.

I have an oxyacetylene welding outfit, and I have been welding for forty years, but I still cannot call myself a good welder. The stand shown in the photos and the drawing was made by a friend, Douglas Orr, who is an excellent welder.

This stand is the first thing I've seen from any welder that is absolutely square and straight. Doug uses simple layout fixtures and TIG welding to achieve this accuracy. His occupation is building theatre sets at the National Art Center in Ottawa, Ontario.

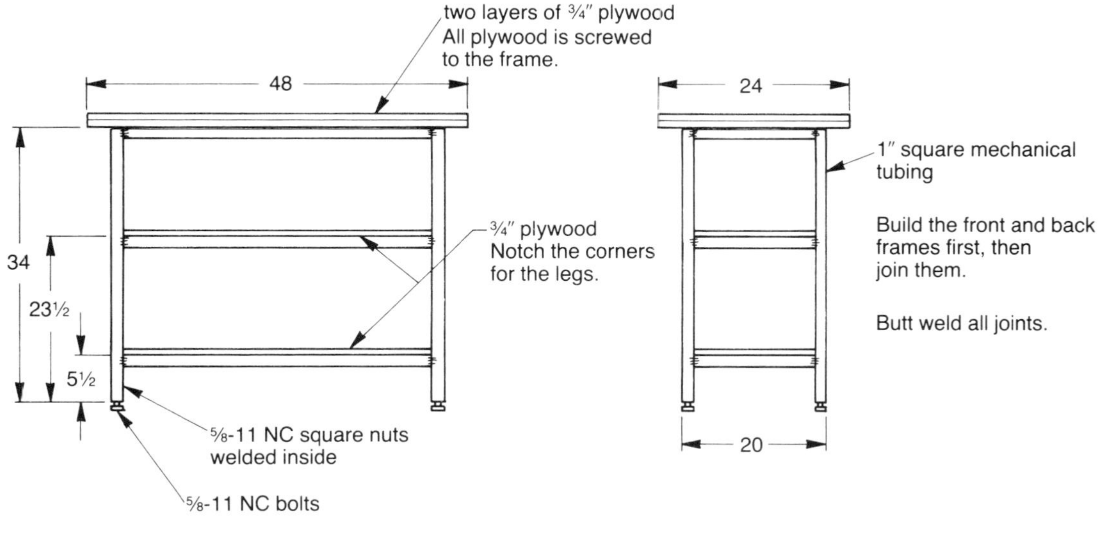

Rivet and Screw Cutters

This simple tool will cut screws off without damaging the end of the thread. I first saw these types of cutters in an aircraft factory where they were used to cut rivets or screws to length.

Make the top and bottom parts from oil hardening tool steel (ground flat stock). The handle is made from hot or cold rolled mild steel.

Locate the holes symmetrically about the pivot hole in case the top part has to be inverted to compensate for warping when the parts are hardened.

To make a rivet cutter, the holes in the top plate should be a close fit to the rivet. The holes in the bottom plate should be 1/32" larger.

Harden and temper the plates as usual. To remove the scale from the threaded holes after hardening, use a screw with fine valve grinding compound and work the screw in and out of the hole a few times. For the rivet holes, split a small rod with a hacksaw and bend the legs as shown in the sketch. This rod is a lap and should be used in a drill press at a medium speed. Work the lap up and down in the rivet holes and add fine valve grinding compound occasionally.

To use the cutter hold the bottom plate in a bench vise. Make up a collar or tube to give the correct length. (Include the thickness of the top plate to obtain the total length.)

Place the top plate so that the holes line up. Insert the screw or rivet and then give a swift pull to the handle. The sheared portion will drop through the hole in the bottom plate.

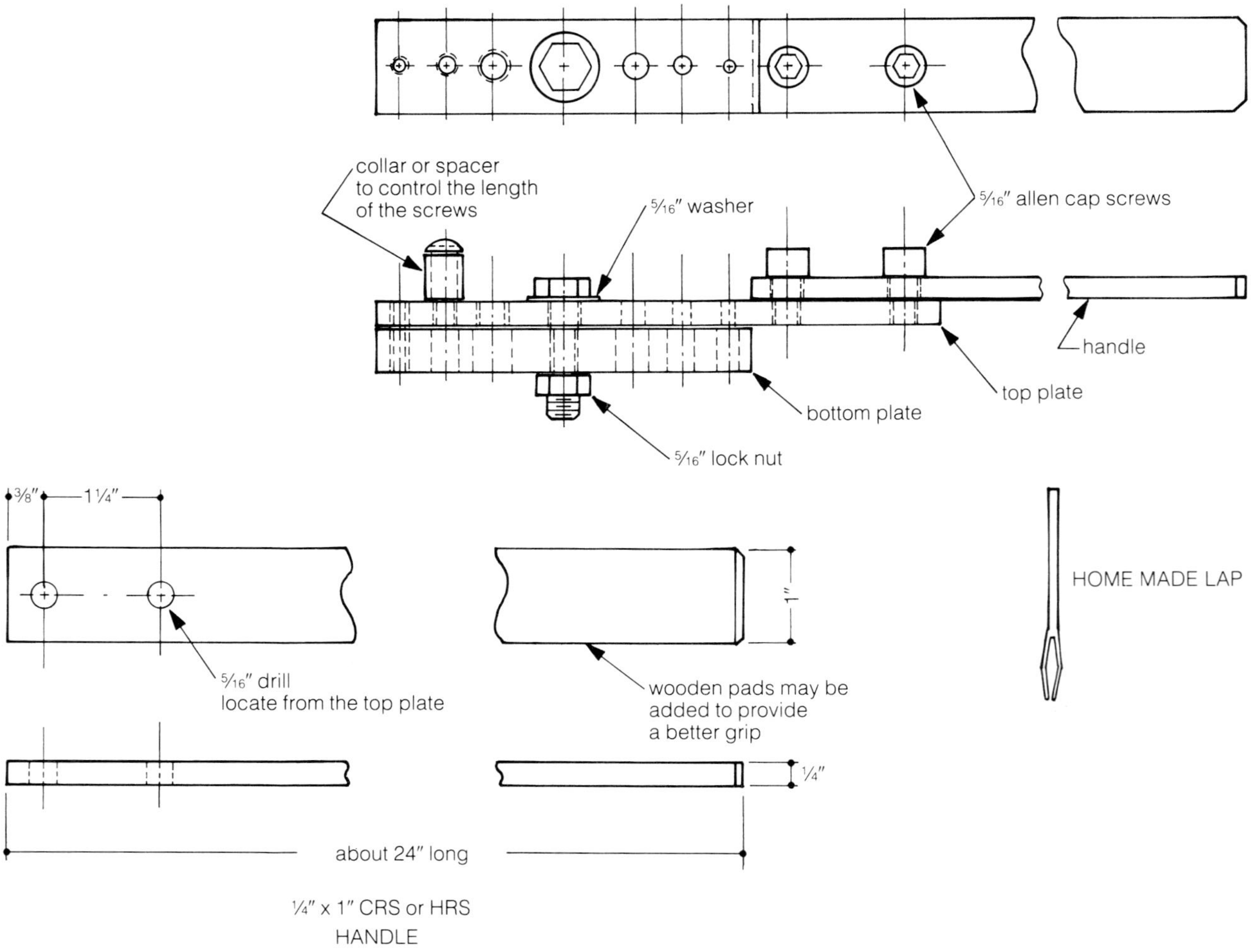

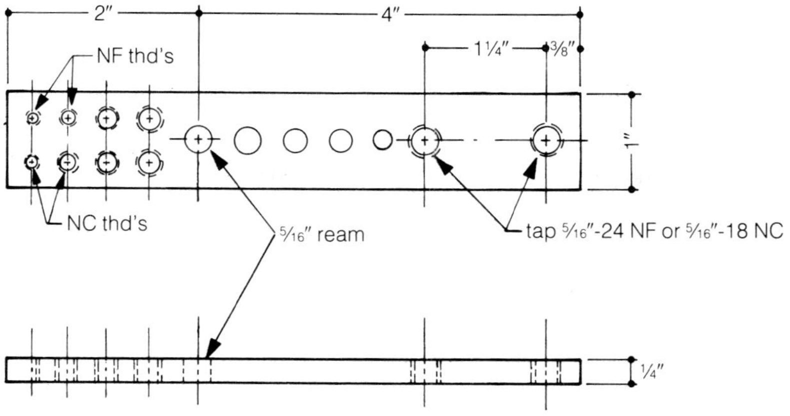

TOP PLATE

Alternate methods of locating the holes shown. Dimensions shown are not important and may be altered to suit the materials available.

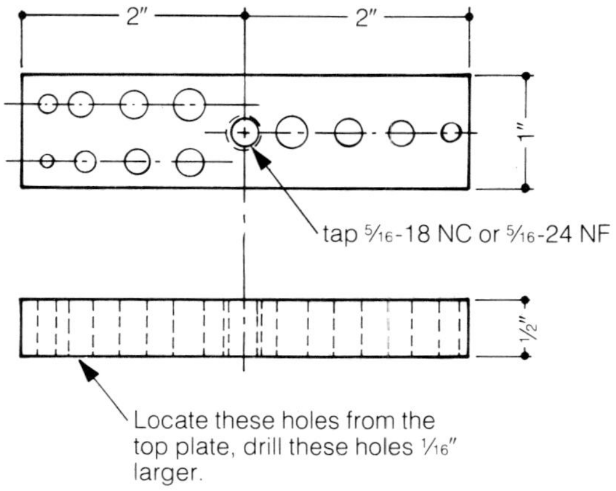

BOTTOM PLATE

An Improved Parallel Clamp

A large pair of clamps is nice to have in a shop. However, most of us put off buying them because they are expensive and not used very often.

These clamps are large enough to use for clamping locomotive cylinders to the frames for drilling and the frames over the stretchers, also. One special feature is that the bottom jaws act as parallels to support the work above the drill press table so you won't make the mistake of drilling through the work into one of the parallels. This is especially distressing if you have bought a set of hardened and ground parallels. There never seems to be enough room for the clamps and the parallels.

To drill locomotive frames, place the frames together and locate the clamps near the ends of the frames in an area where holes will not be drilled.

To drill long work on a drill press, buy a length of wood sold for shelves from a lumberyard. Fasten the board to the drill press table with flathead machine screws, or use wood screws from underneath the table. Now you can slide the frames along the board, instead of trying to support the outboard end on blocks of wood or asking someone to hold the frames. Those readers who have a vertical milling machine with a long table should remove the vise and use the mill as a drill press.

The jaws are made from 1" square cold rolled steel. The threaded rod is available at hardware stores. The retaining clips (Detail 7) are made as a pair and then separated. Locate the 10-32 NF holes in the bottom jaw from the clips during assembly.

The dimensions and materials are not critical and may be altered to suit the requirements or the materials available.

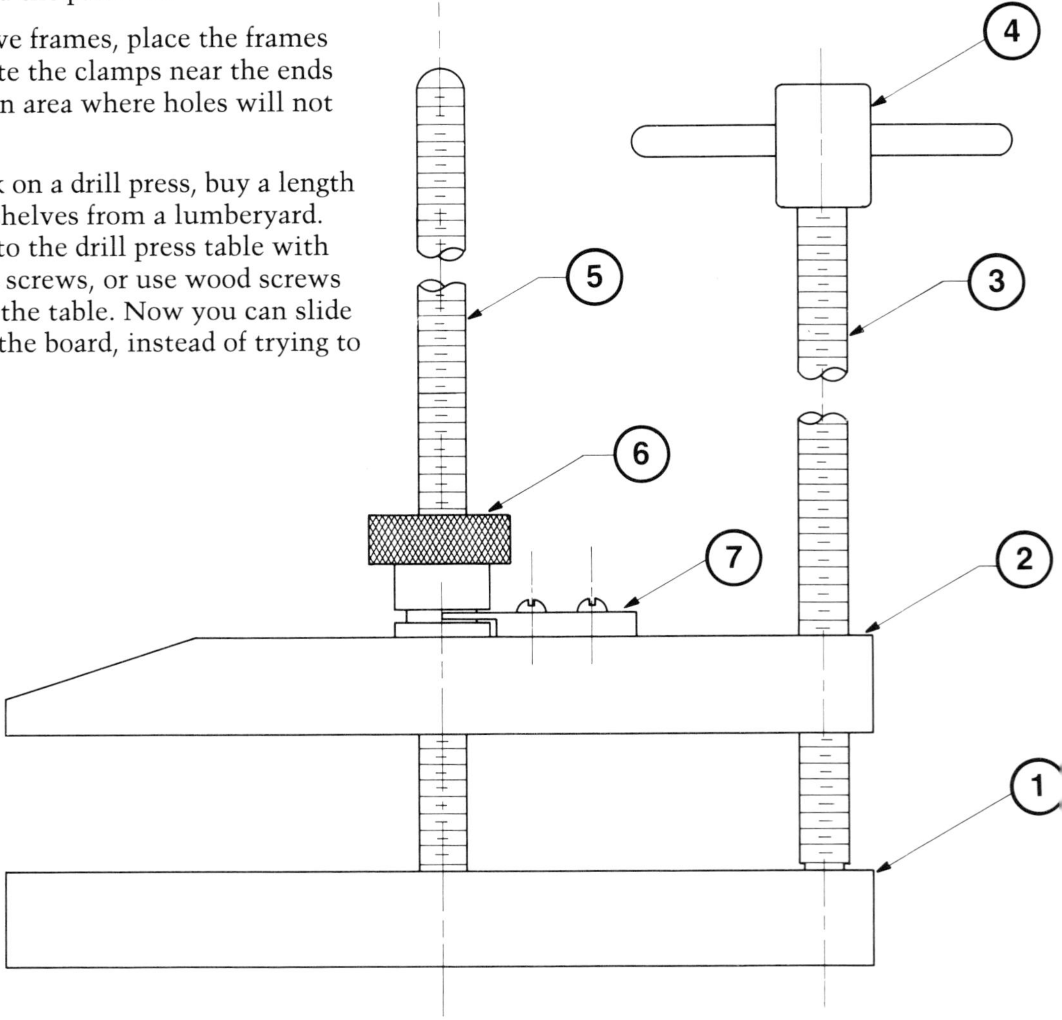

IMPROVED PARALLEL CLAMP

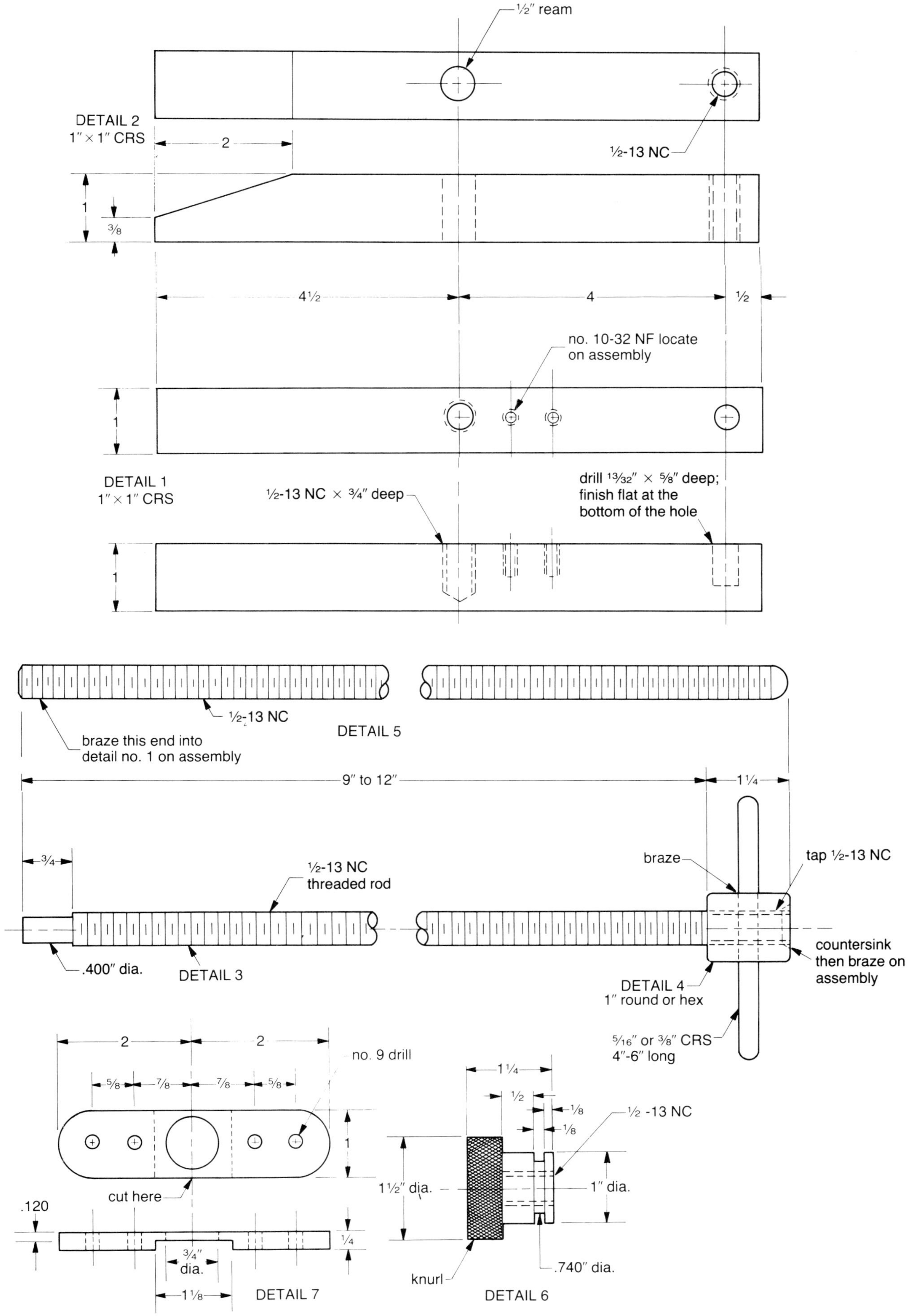

½″ ream
DETAIL 2
1″ × 1″ CRS
2
½-13 NC
1
⅜
4½
4
½
no. 10-32 NF locate
on assembly
1
DETAIL 1
1″ × 1″ CRS
½-13 NC × ¾″ deep
drill 13/32″ × ⅝″ deep;
finish flat at the
bottom of the hole
1
½-13 NC
braze this end into
detail no. 1 on assembly
DETAIL 5
9″ to 12″
1¼
¾
½-13 NC
threaded rod
braze
tap ½-13 NC
.400″ dia.
DETAIL 3
countersink
then braze on
assembly
DETAIL 4
1″ round or hex
5/16″ or ⅜″ CRS
4″-6″ long
2
2
no. 9 drill
⅝
⅞
⅞
⅝
1
cut here
.120
¼
¾″
dia.
1⅛
DETAIL 7
1¼
½
⅛
⅛
½ -13 NC
1½″ dia.
1″ dia.
knurl
.740″ dia.
DETAIL 6

A Plain Surface Gage

This is the type of surface gage that the machinist apprentices used to make for themselves during their first year. The variety of machining operations will be interesting and the surface gage will serve until a more elaborate universal type is made or purchased.

The base is made from an offcut (scrap) from a 3" diameter bar 3/4" to 1" long. Chuck in the 3-jaw, face one side, center drill and then drill right through with a letter "Q" drill. Now use a boring bar or a right-hand turning tool with more clearance ground on the side. Face out a recess 2-1/4" diameter by 1/4" deep. Tap a 3/8-24 NF thread by holding the shank of the tap in the tailstock drill chuck. Pull the belt by hand or use the slowest speed and slide the tailstock along the bed. Use cutting oil and reverse the spindle rotation to back out the tap. This method will ensure that the tapped hole is square with the base. (Both operations must be done at the same setup.)

Reverse the base in the chuck and hold with the chuck jaws inside the recess. Face flat and then take a light cut to just clean up the 3" diameter. Turn a 1" diameter by 1/4" long boss by either facing or plain turning. Now swing the compound slide rest to an angle of 17° as shown on the drawing.

Face the base at this angle by the compound slide and advance the carriage along the bed to set the depth of cut. Break the sharp edges with a file.

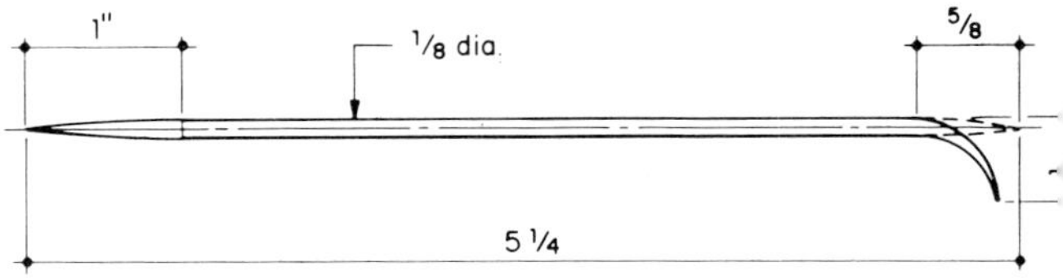

SCRIBER

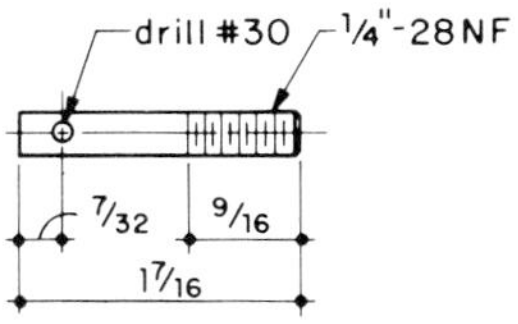

CLAMP SCREW

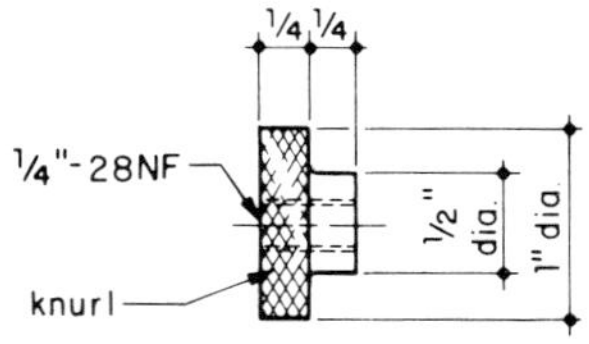

CLAMP NUT

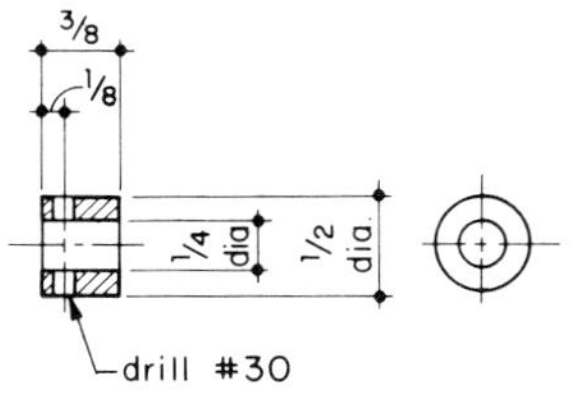

SLEEVE

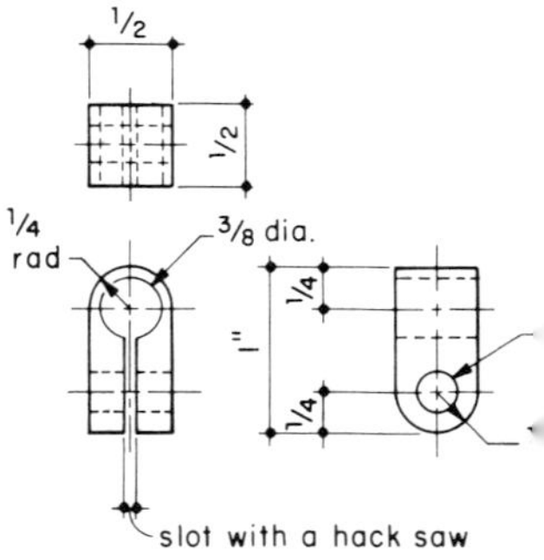

CLAMP

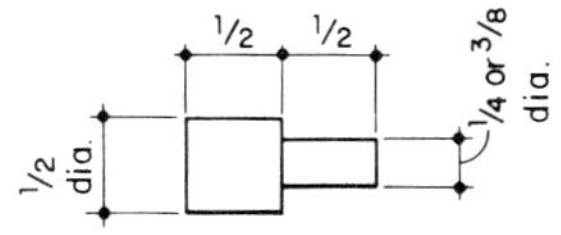

FILING BUTTON GUIDE

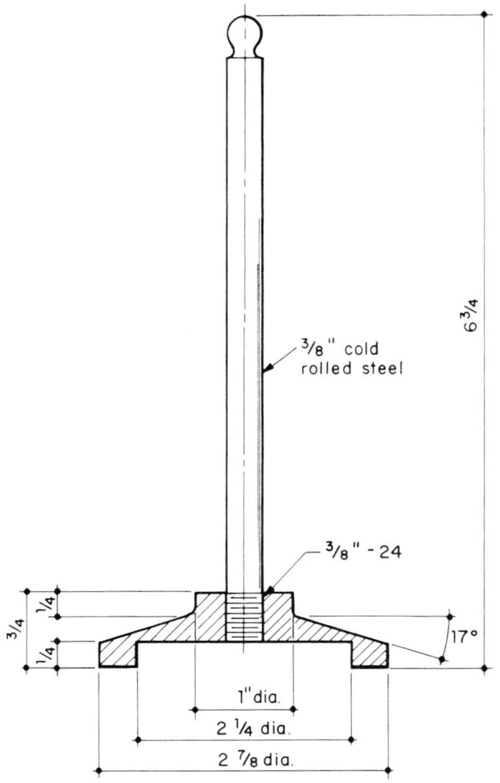

SURFACE GAUGE

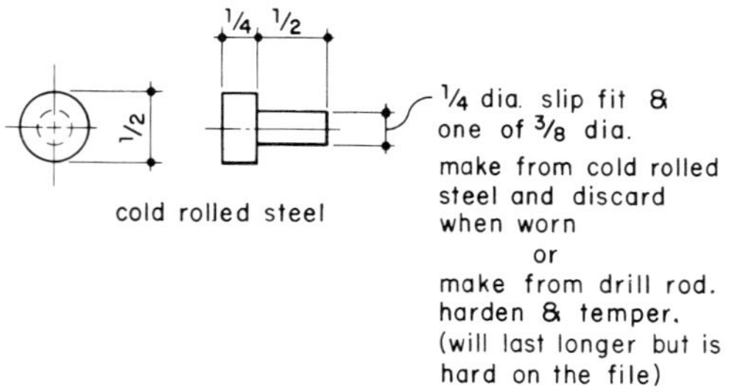

FILING GUIDE BUTTON

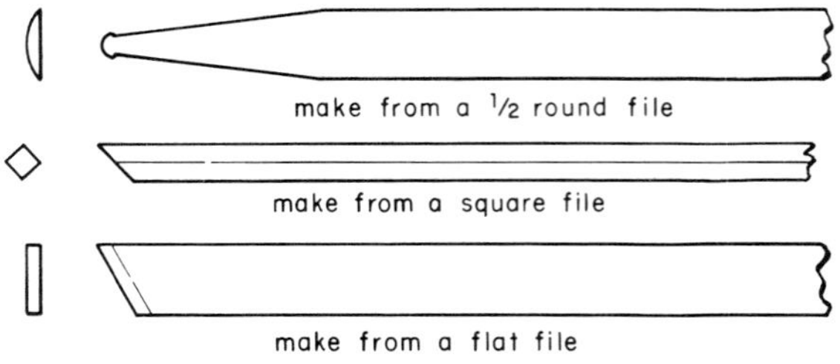

TYPICAL HAND TURNING TOOLS

The post is a piece of 3/8" cold rolled steel. Face, chamfer and thread one end. Turn the top end to any shape you want. Use a long, narrow, round-nose tool bit and manipulate the carriage and cross slide handwheels together to form the shape. Alternately, use a hand-turning tool with the back end of the toolholder or a bar in the tool post as a tool rest.

See the drawing for the shape, ground up from old files. The procedure for turning is similar to turning wood and this will be a good chance to experiment and practice.

The clamp is made from a piece of 1/2" square (key stock) cold rolled steel. Face the ends on the lathe, then lay out and drill the holes. File the ends round with a button guide and then slit with a hacksaw.

The sleeve and the clamp screw are simple parts to make. Drill the No. 30 cross hole with the sleeve and clamp screw together. The knurled thumb nut is made in the same way as the nut on the toolmaker's flat vise but without the spherical face.

The scriber is made from 1/8" drill rod (silver steel). Taper the ends with a single cut file (mill file) on the lathe with the spindle turning at a high rate of speed (about 1000 rpm).

Use the horn of a small anvil to form the carved end. This is bent cold. If you don't have an anvil, hold a 1" diameter bar in the bench vise and use the bar as an anvil.

Harden and temper each end separately with a small propane torch. Temper to a dark straw color.

An Easy-to-Make Vernier Height Gage

Most home workshops have a 6" vernier caliper, but few have a vernier height gage, largely because of the price of the latter. This lack can be solved easily. Two or three evenings spent making these attachments will convert your caliper into a height gage when one is needed, without spoiling it for its original purpose.

When making the base, the slot is most easily made with one of the screw-slotting cutters. These are available in a range of thicknesses from .020" up and are relatively inexpensive. They are 2-3/4" diameter with a 1" bore. For those without a milling machine, make up a mandrel for the lathe and use the milling slide.

To sharpen the scriber after hardening and tempering, assemble the height gage and place it on the surface plate or drill press table. Place a piece of emery cloth flat on the surface plate in front of the height gage. Push the vernier slide down until the scriber touches the emery cloth, then slide the scriber back and forth until it is sharp and the underside surface is dead parallel with the surface plate. Any time the scriber becomes dulled from use, it should be resharpened by this same method.

To set the scriber in position, adjust the vernier to read .100" and lock the slide. Loosen the scriber setscrew and bring the scriber down to touch the surface plate and retighten the setscrew.

Just remember that your base or zero reading is .100" instead of 0" and you will have a very simple and useful vernier height gage while still retaining the usefulness of your vernier caliper.

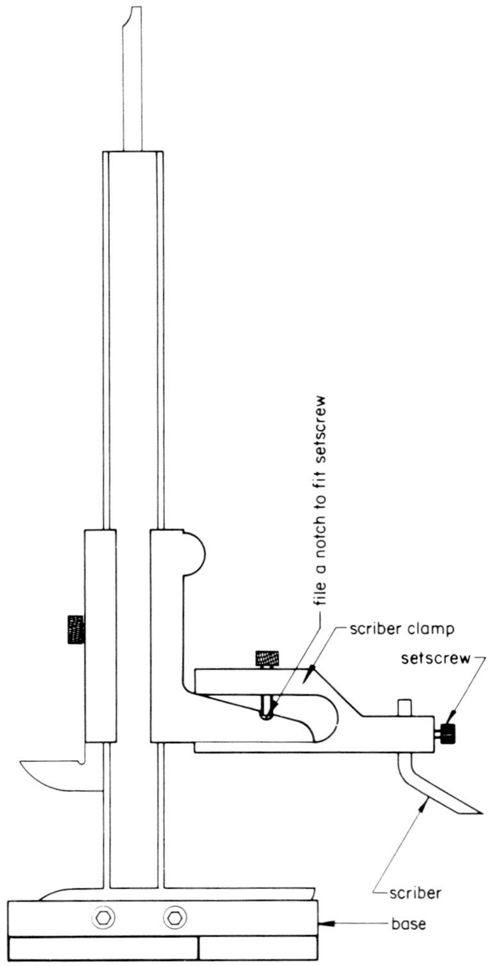

SIMPLE HEIGHT GAUGE

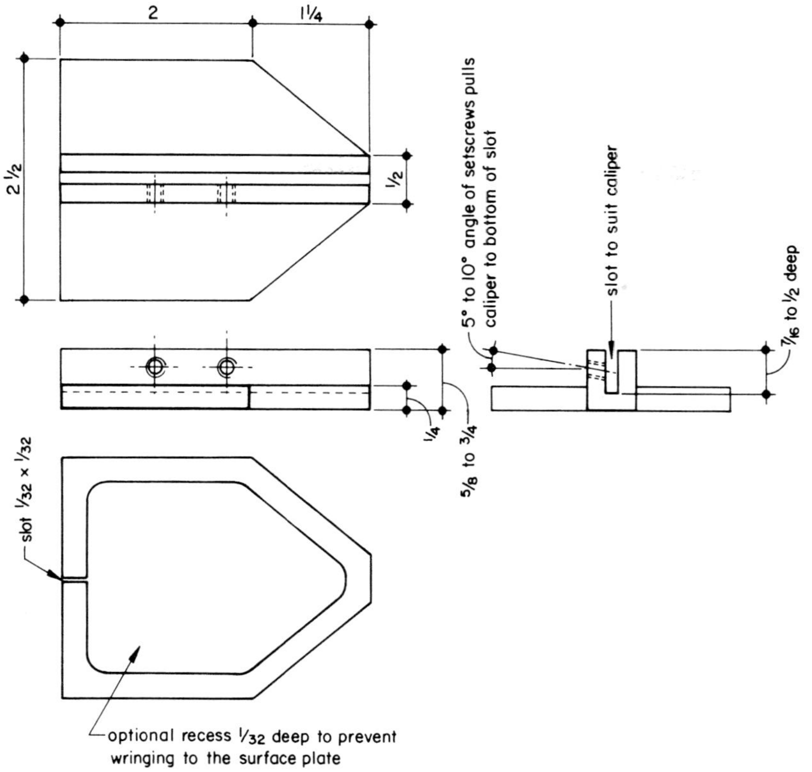

BASE
Cold-rolled Steel, 1 required

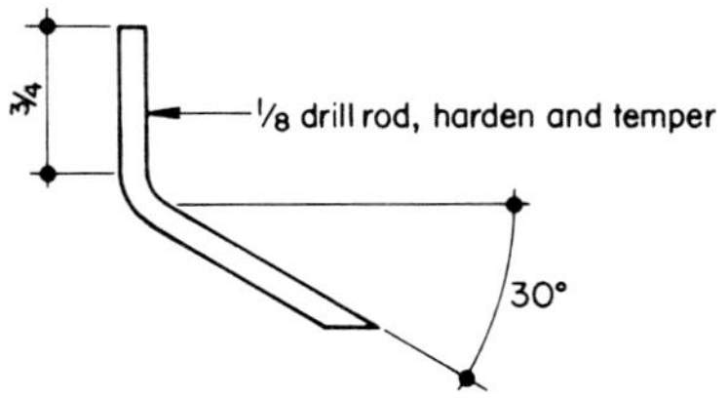

SCRIBER
Drill Rod, 1 required

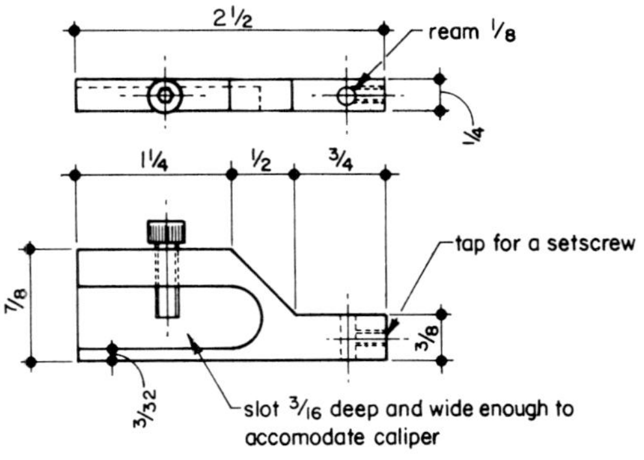

SCRIBER CLAMP
Cold-rolled Steel, 1 required

A Marking Gage

This little tool will be useful when marking out sheet metal, plates and carpentry work. The time required will be one or two evenings and the turning is not particularly difficult. Whatever material that can be measured with a rule and is pleasing to the eye will be suitable.

To make the spindle, cut off a 6" length of 5/16" diameter cold rolled steel rod. Face one end flat. Now swing the toolholder around so the back end points toward the lathe chuck and the shank of the toolholder is parallel with the lathe axis. The toolholder shank will serve as a tool rest for turning the back end of the spindle. The rod should project from 1" to 1-1/2" beyond the lathe chuck jaws. Use a round-end hand turning tool ground from an old flat file. Use a lathe speed of about 300 rpm. The toolholder should be about 1/16" to 1/8" from the spindle. If the toolholder is too far from the work, a "dig in" is likely. Use the hand tool in the same manner as a chisel when turning wood. This is fun, and when the technique has been mastered, the turning of ball handles and other odd shapes is easily done.

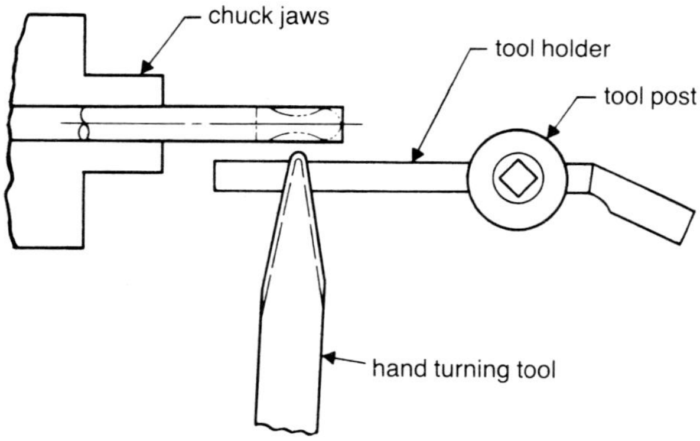

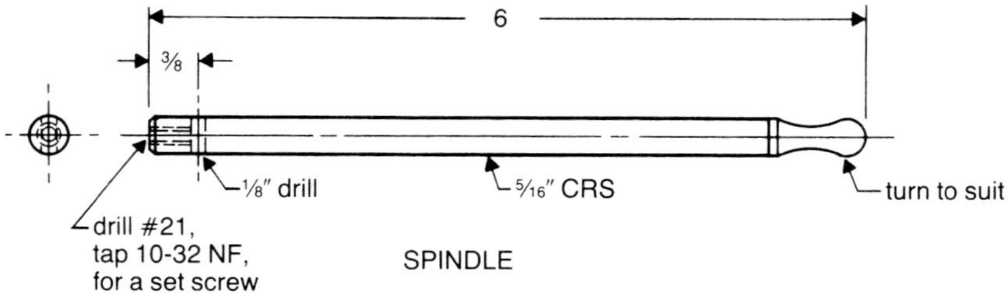

SPINDLE

Now to make the other end of the spindle. Face the end and then mark a line 3/8" from the end. Drill 1/8" through the spindle 3/8" from the end. Either locate the drill by eye to drill the hole or use a flat vise. When the 1/8" cross hole has been drilled, return the spindle to the lathe. With the lathe turning about 1000 rpm, center drill and then use a No. 21 drill. Drill the hole about 3/4" deep. This extra depth beyond the 1/8" cross hole will accommodate the taper at the end of the tap, so there will be a full thread up to the 1/8" hole. The tapping operation takes place on the lathe. Hold the round part of the shank of the No. 10-32 NF tap in the drill chuck in the tailstock. Do not clamp the tailstock to the bed, for the tailstock must be free to slide along the bed. Rotate the lathe chuck with your left hand and push the tailstock along the bed with your right hand. If the ordinary tapering or plug tap is used, reverse the direction of rotation to back the tap out and remove the shavings after every two or three revolutions. Stop threading when the taper part of the tap has passed by the 1/8" hole. Look in the 1/8" hole to see this. If a modern spiral flute tap for blind holes is used, the tapping operation can be carried out without reversing until the full depth is reached, since the shavings will feed back along the flutes just like a twist drill. With a spiral flute tap, I would not bother to rotate the chuck by hand, but would set the lathe speed to about 50 rpm and reverse the motor when the full depth is reached.

Don't forget to use cutting oil or one of the special preparations for this operation.

Now to make the gage head. Chuck a piece of 1-1/4" to 1-1/2" diameter bar of mild steel in the lathe with about 2" projecting beyond the jaws. Face the end and then turn to 5/8" diameter for 1-3/8" length. Use a speed of about 300 rpm and a feed of about .005" per revolution. Round off the point of the lathe tool bit with an oil stone for the final cut to have a fine finish. Now knurl in two places as shown on the drawing. Use about 90-100 rpm and don't forget to use some cutting oil.

Now center drill, drill 1/4" to a depth of 19/64", and ream 5/16" or use a 5/16" drill as a reamer. To do this, round off the lips of the drill slightly with an oilstone as shown in the drawing.

Clean up the (1-1/4 to 1-1/2") diameter of the flange. Then round off the edge slightly with a single cut mill file and do the same at the end. Bring a countersink mounted in the tailstock chuck up to take a slight cut to remove the sharp edge from the 5/16" hole.

Now part off, leaving the flange width about 1/4" for a facing operation. If you are not confident about this operation, remove the work from the lathe and saw off at the bench vise or use a power hacksaw.

Reverse the gage head in the lathe chuck and grip lightly on the 5/8" diameter. If the flange is touching the chuck jaws, there is less chance of the work slipping in the chuck. Face to 5/32" to 3/16" thick, break the sharp edge with a file, and lightly countersink the 5/16" hole.

To drill and tap the No. 10-32 NF hole, grip the gage head by the ends in the drill press vise. This will avoid squashing the gage head and distorting the bore. Center drill, drill No. 21 and then tap by holding the tap in the drill chuck and rotating the drill chuck with your left hand while pulling down on the feed lever with your right hand. Reverse the direction of rotation to remove the tap and lift up slightly on the feed lever. If the vise has been clamped in place on the drill press table, the tap will follow exactly concentric and straight in the No. 21 hole.

The scriber is made from 1/8" drill rod about 7/8" long. Hold in the lathe chuck with about 1/4" projecting. Run the lathe about 700 to 1000 rpm. Round off one end with a single-cut file. Reverse the scriber in the chuck with about 1/2" projecting. Now use a good single cut file to file a long sharp point.

Hold the round end with an old set of pliers for heating and quenching to harden the point.

The knurled thumbscrew is fun to make. Chuck a piece of 1/2" cold rolled steel rod with about 1" projecting. Knurl about 3/8" to 1/2" from the end. Now face the end, and turn to .185" to .190" diameter. Round off the end and chamfer the knurl with a file.

To thread the screw, a tailstock die holder will be used. Turn the lathe chuck by hand or use a low speed (50 rpm). Hold the die holder with your right hand and let go when the die reaches the screw head. Reverse the spindle rotation to remove the die.

A standard die stock can be used, but you will have to be careful to start the die square with the work. Remove the drill chuck or center from the tailstock spindle. Place the start side of the die against the screw and bring the tailstock up to press on the back side of the die. Place one end of the die stock on top of the compound slide rest or on top of the toolholder set parallel with the lathe axis. Rotate the chuck with your left hand and turn the tailstock handwheel with your right hand until the die is started. Continue by hand until the die reaches the screw head, etc.

Now part off the screw and reverse the screw in the 3-jaw chuck. Grip the screw thread lightly for facing, or hold the spindle in the 3-jaw chuck and screw in the thumbscrew. After facing, use a pair of pliers on the knurling to remove the screw.

To use the gage, measure from the scriber point to the gage face with a steel rule. This gage will be more convenient to use than a combination square and a scriber for drawing lines parallel to an edge.

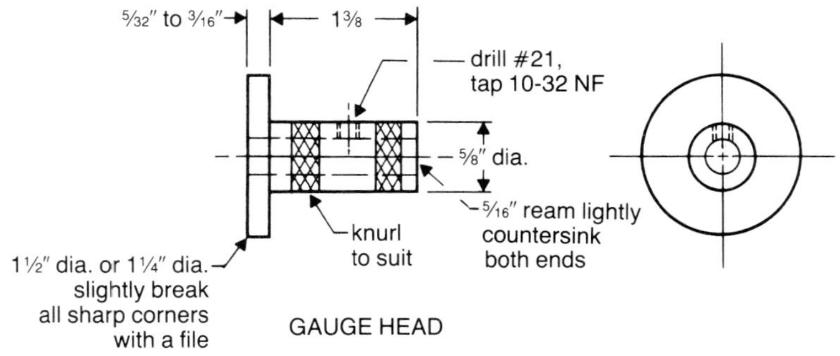

GAUGE HEAD

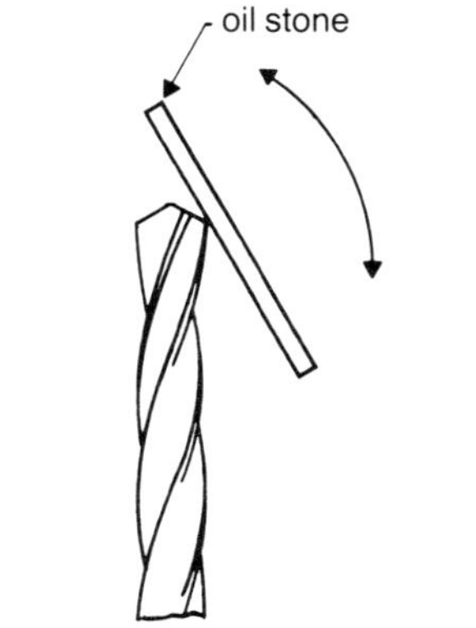

HOW TO ROUND OFF THE LIPS OF A DRILL TO USE THE DRILL AS A REAMER

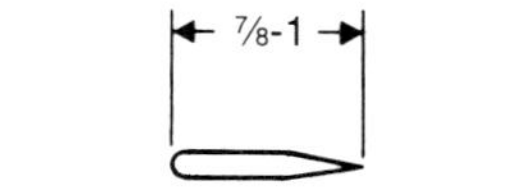

1/8" DRILL ROD SCRIBER

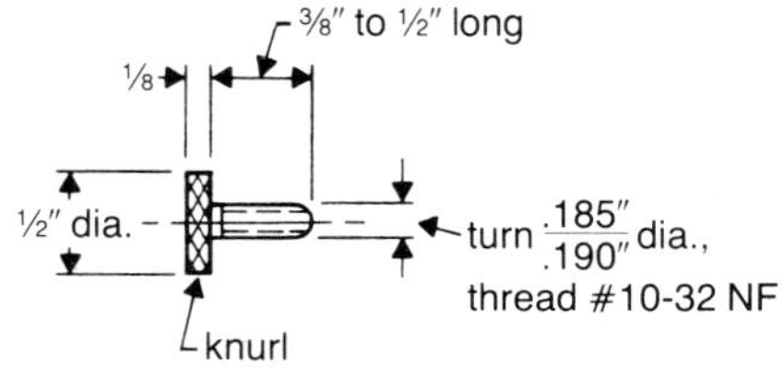

KNURLED THUMB SCREW

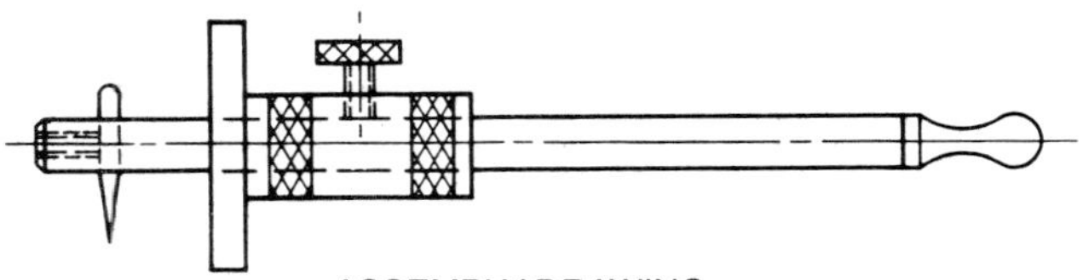

ASSEMBLY DRAWING

A Universal Surface Gage

This gage will be useful in the home shop for marking out castings and plates, but it can also be used to locate the work accurately from the layout lines when setting up the shaper or milling machine, and for locating work in the four-jaw lathe chuck and the faceplate. If the design is altered to hold a small dial indicator, its range of uses will be increased.

The dimensions are not critical and may be changed to suit the material available or to make a larger or smaller gage. The vee in the base is optional. The purpose of the vee was to allow the gage to be used on the boring bar of a horizontal boring machine to set up large castings from the scribed circle around a cored hole (after the wooden center plug had been removed). I have omitted the four vertical pins usually fitted in the corners of the base, because they are seldom used and are difficult to fit accurately enough for a nice "push fit." The pins are used to guide the gage along the edge of a plate when scribing lines parallel to the edge. For most small work the gage will be used on a surface plate.

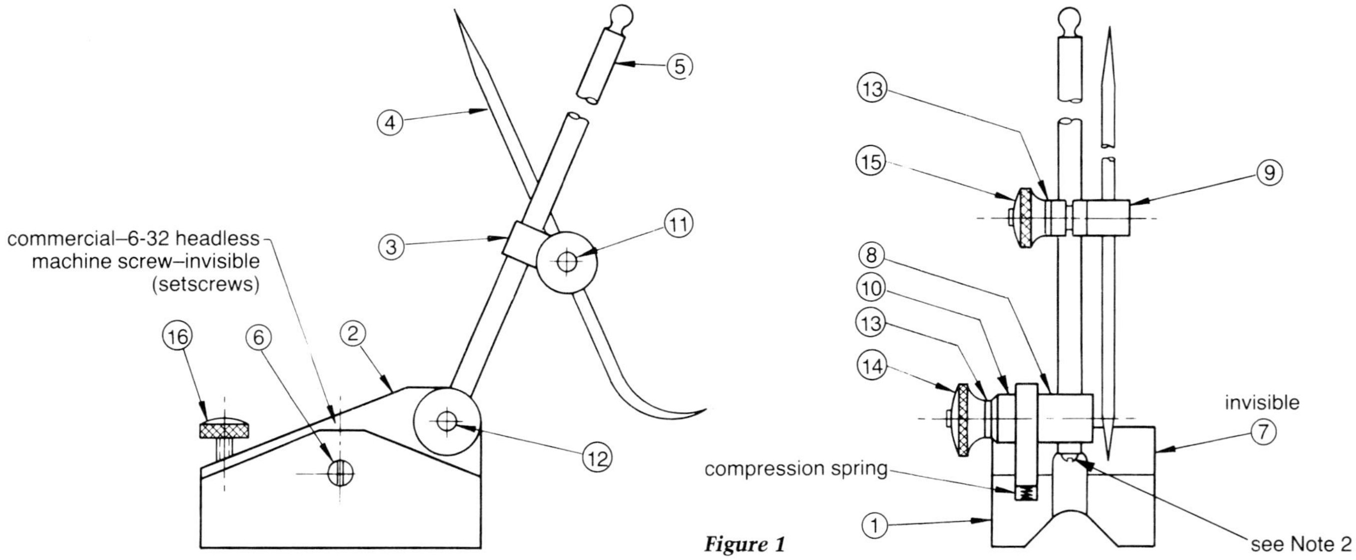

Figure 1

Setting the scriber on top of the bar is the first step in locating the center of the bar accurately.

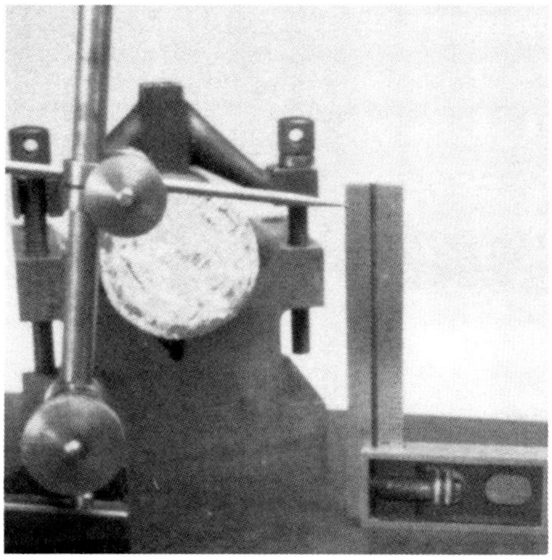

The measurement is transferred from the top of the bar to a scale.

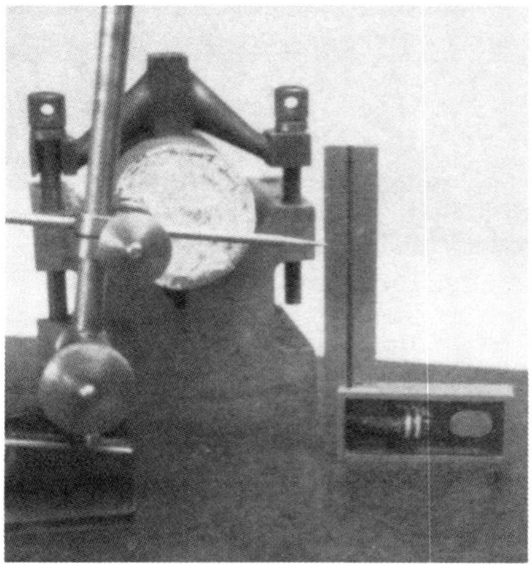

The scriber is located on the scale at a distance from the previous setting equal to the radius of the bar prior to scribing the center lines.

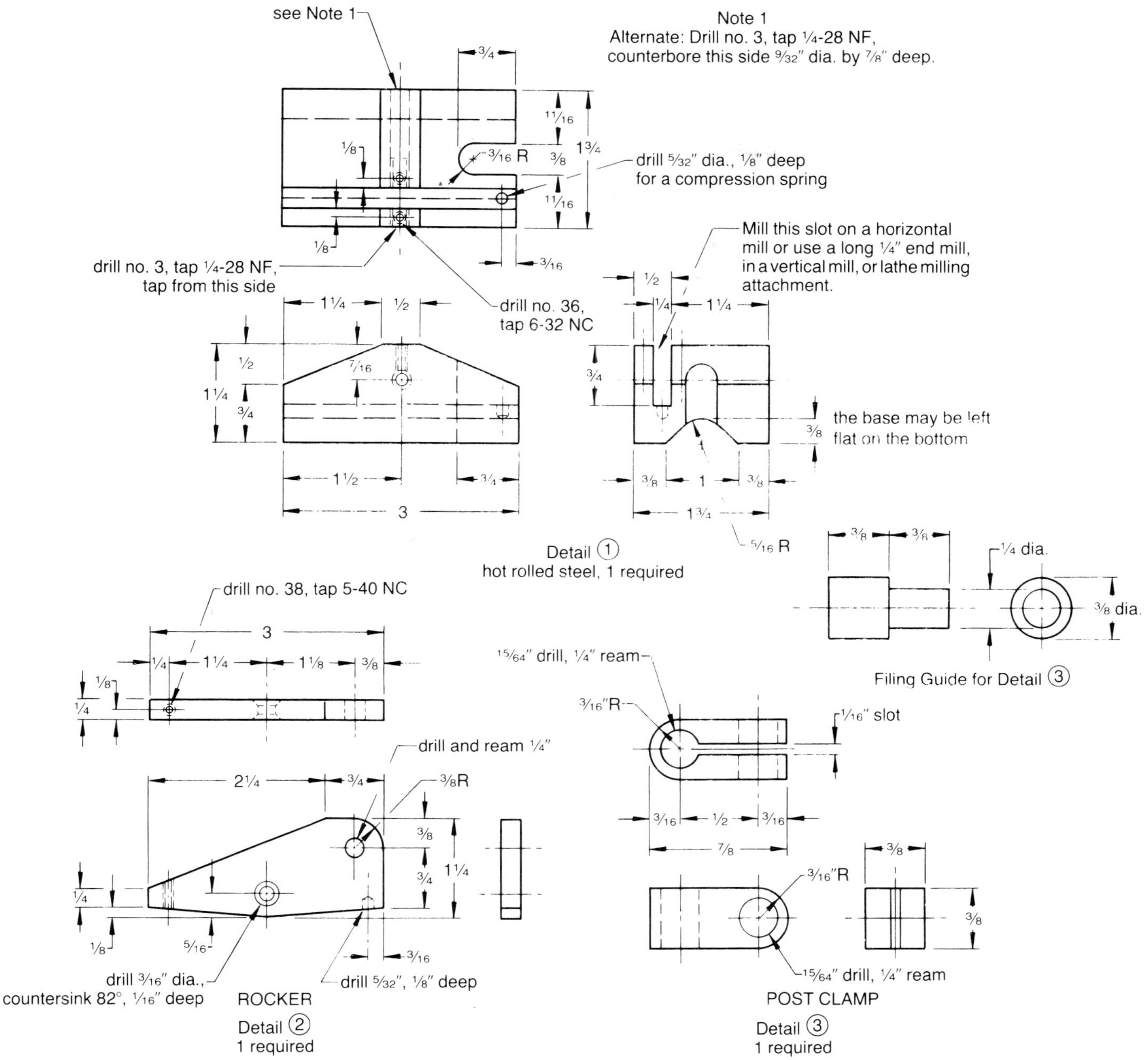

A Line is scribed on the end of the bar at the center height.

Rotate the bar 90° and align the center line with the square. Then scribe a second center line at 90° to the previously scribed center line.

A casting is leveled in the machine vise using the scriber to indicate when the largest layout line is level.

The base can be made from cold or hot rolled mild steel, or a piece of tool steel if it is available. Machine the block on all six surfaces and then mark the location of the 1/4"-wide slot and the holes. Drill and tap the 1/4-28 NF thread first. Then mill the 1/4"-wide slot. The sloping top surfaces are made by tipping the block in the milling machine vise so the layout line is parallel with the top of the vise jaws and from 1/16"-1/8" above the top of the jaws. To smooth all the flat surfaces after the base is finished, place a sheet of emery cloth on a surface plate with the abrasive side up. Slide the base back and forth a few strokes on each face to remove any machining marks. Do not use a circular motion. Instead of using a surface plate, the emery or aluminum oxide cloth could be glued to a piece of 3/4"-thick plywood or particle board. This makes an effective "poor man's" surface grinder. The 5/32" diameter hole in the bottom of the 1/4"-wide slot is to hold a small compression spring to make the rocker (Detail 2) rock against the adjusting screw (Detail 16). Alter the diameter of the

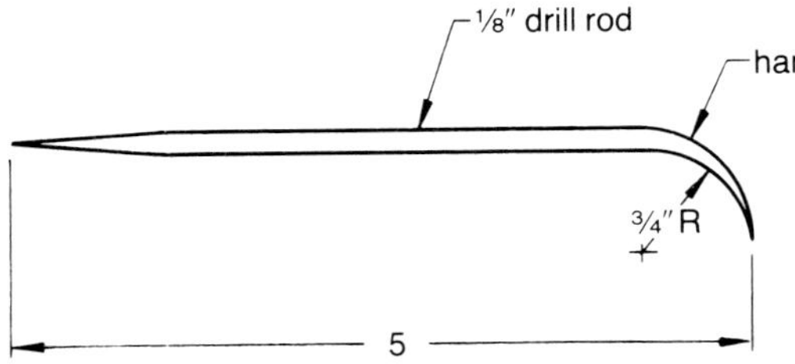

SCRIBER
Detail ④
1 required

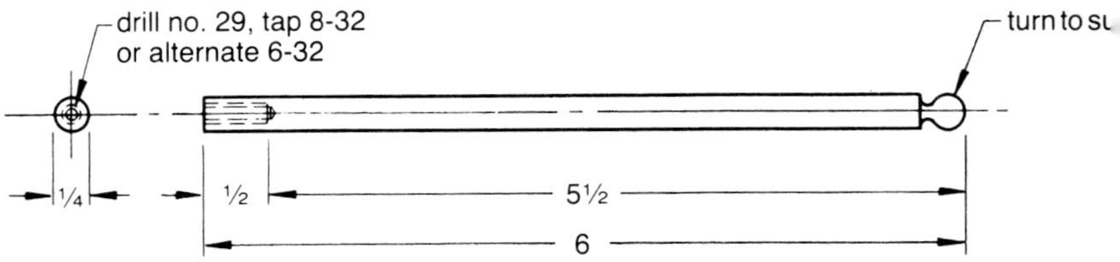

POST
Detail ⑤
1 required

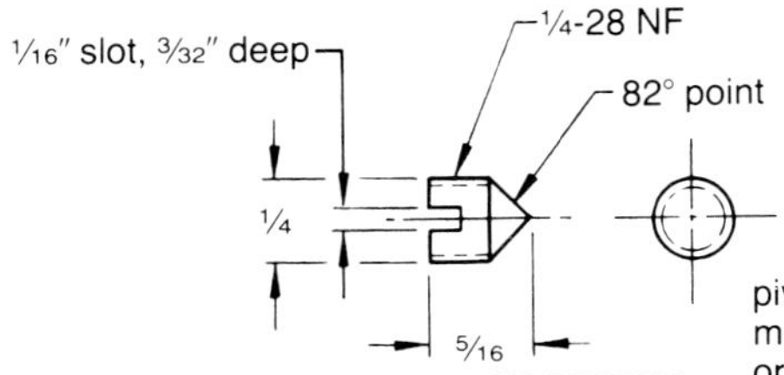

PIVOT SCREW
Detail ⑥
1 required

pivot screws
made from CRS
or drill rod

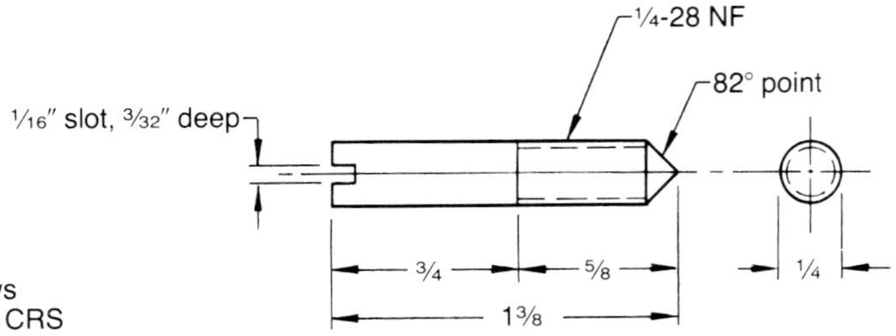

PIVOT SCREW
Detail ⑦
1 required

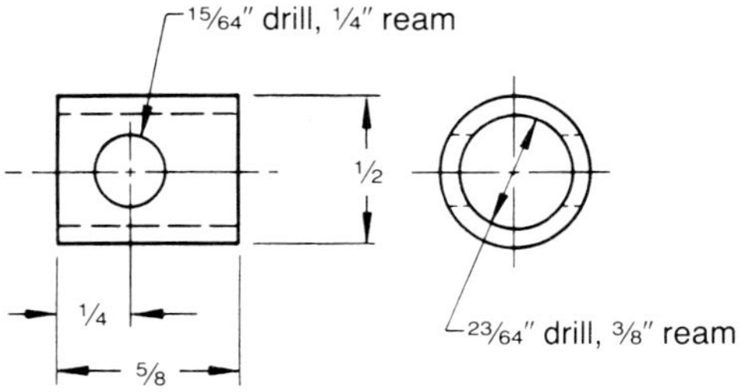

SLEEVE

Detail ⑧
1 required CRS

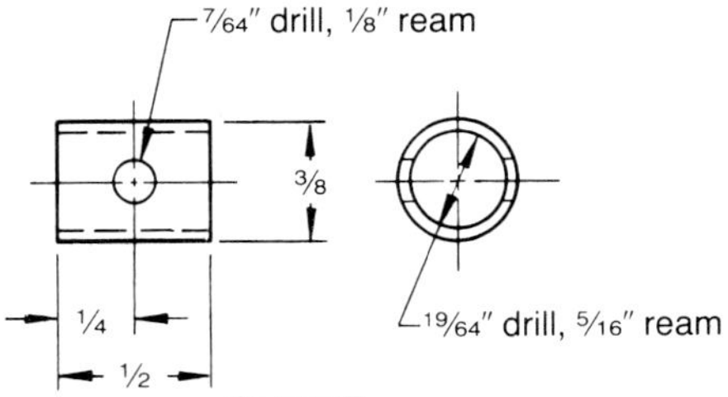

SLEEVE

Detail ⑨
1 required CRS

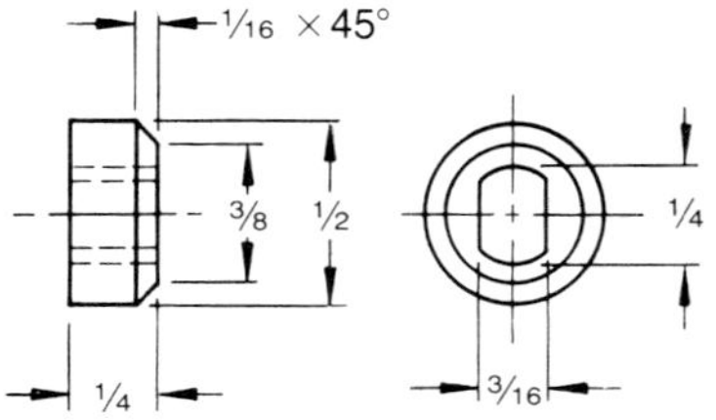

SPECIAL WASHER

Detail ⑩
1 required CRS

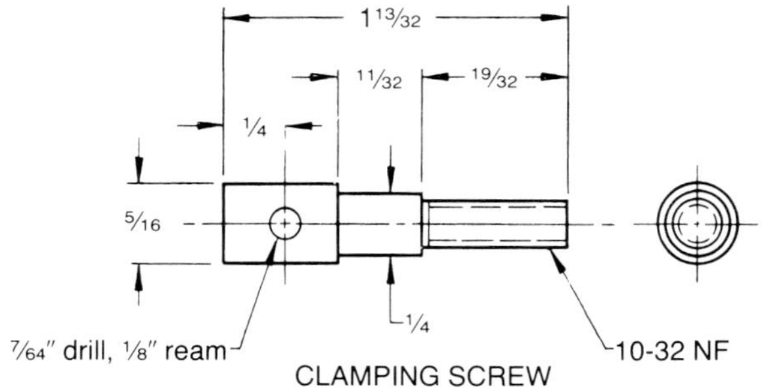

CLAMPING SCREW

Detail ⑪
1 required CRS

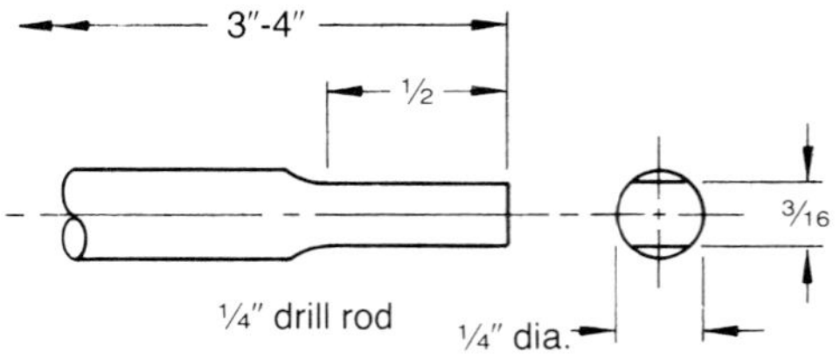

Broach for Detail 10
harden and temper.

Details 9 and 11 may be increased in diameter from 3/8" dia. to 5/8" to permit a 1/4" hole or one to suit the round shank on the back of a small dial indicator.

hole to suit the spring. Try a spring from a ballpoint pen or a discarded light switch, if your local hardware store has nothing suitable.

The scriber (Detail 4) is made by filing the points with a lathe file while the lathe is turning about 1,000 rpm. Bend one end around a piece of 1" diameter bar held in a bench vise, or use an anvil. Harden in oil, but do not temper if oil hardening drill rod is used. Sharpen the points on an oil stone.

The knurled thumb nuts (Details 14 and 15) are made by placing a piece of round bar, preferably screw stock, in the three-jaw chuck with 1"-1½" projecting beyond the jaws. Face the end flat and take a light cut along the bar to make sure it is running true. Now knurl the place on the bar where the knurl is required. Use a speed of about 100 rpm. Remember, this is similar to a swaging operation, so the knurls are forced into the work by the cross slide screw. This operation is very hard on the lathe.

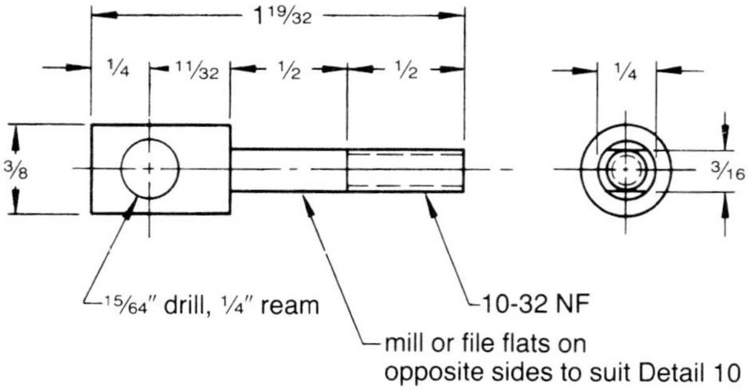

CLAMPING SCREW
Detail ⑫
1 required, CRS

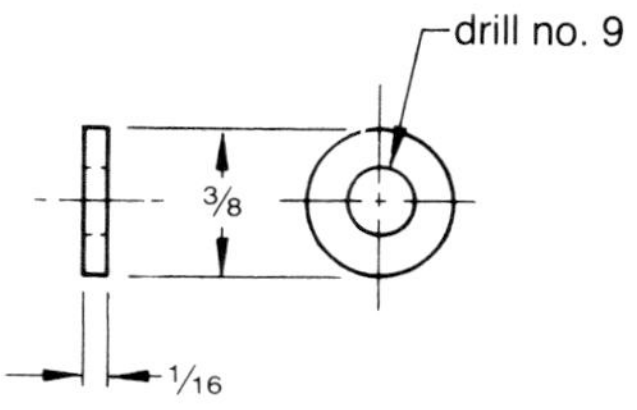

WASHER
Detail ⑬
2 required, CRS

Where CRS is specified for turned parts leaded screw stock may be used. Note – CRS means cold rolled steel.

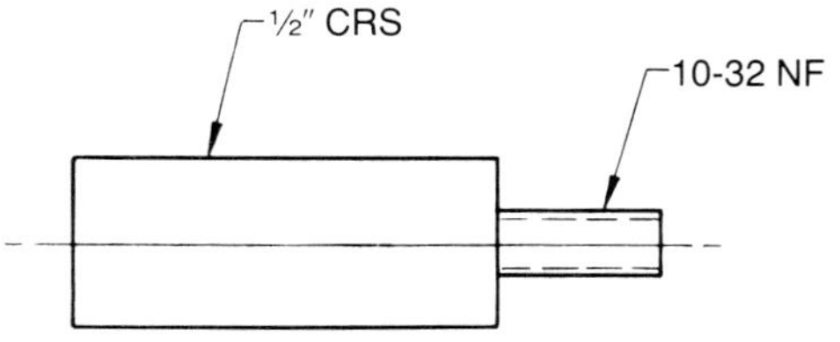

Arbor for Facing Detail 14 and Detail 15

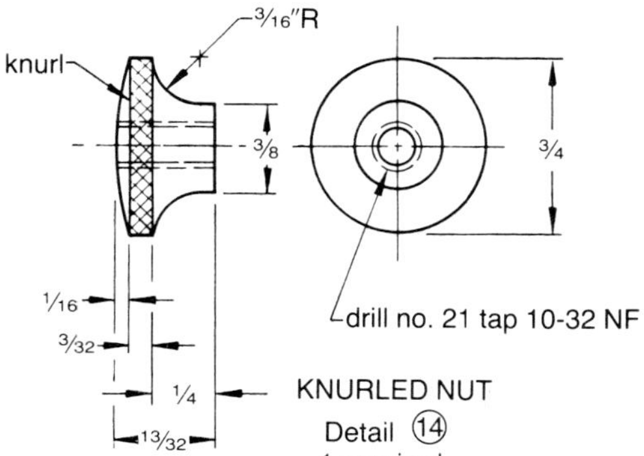

KNURLED NUT
Detail ⑭
1 required

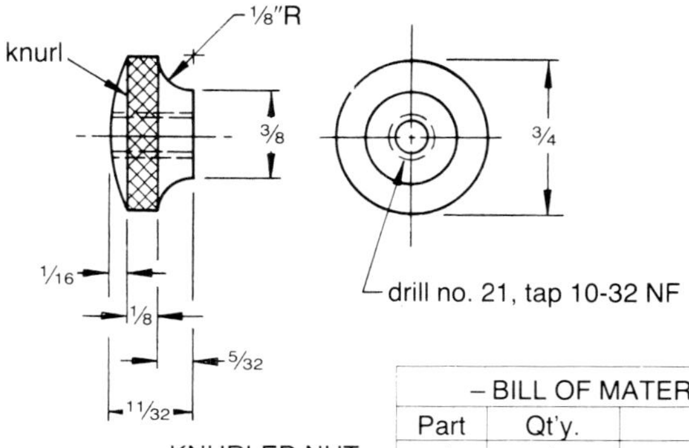

KNURLED NUT
Detail ⑮
1 required

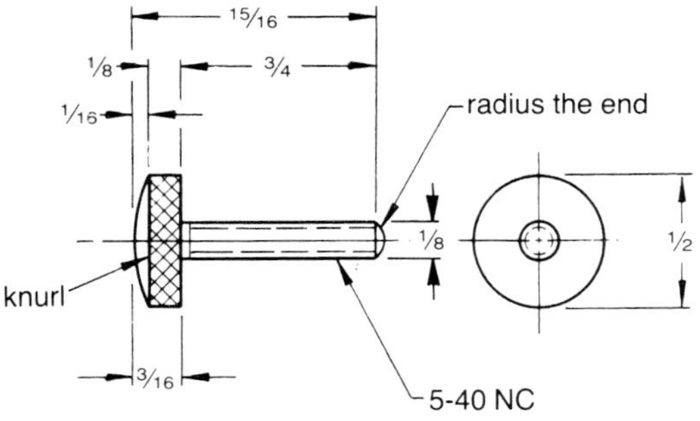

KNURLED THUMB SCREW
Detail ⑯
1 required

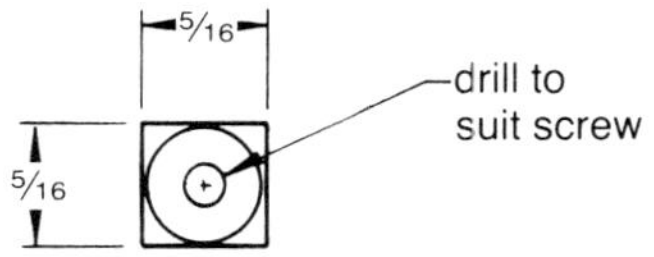

SQUARE SCRIBER
for the bottom of the post

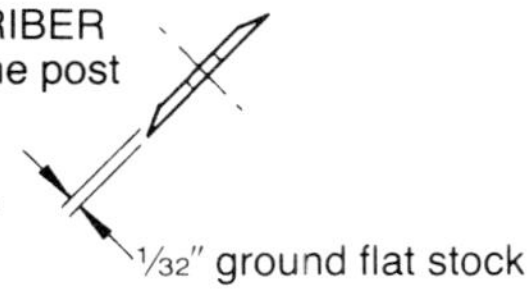

Quench in oil – do not temper.
Bevel the corners by turning.

– BILL OF MATERIALS –		
Part	Qt'y.	Mat'l.
1	1	CRS
2	1	CRS
3	1	CRS
4	1	1/8" drill rod
5	1	1/4" drill rod or CRS
6	1	1/4" drill rod
7	1	1/4" drill rod
8	1	CRS
9	1	CRS
10	1	CRS
11	1	CRS
12	1	CRS
13	2	CRS
14	1	CRS
15	1	CRS
16	1	CRS

Next, form the radius with a round nose tool bit. Then drill, tap and part off.

To finish the ends of the nuts, make up the threaded arbor shown in the drawings. Place the arbor in the three-jaw chuck with the shoulder just inside the jaws. This is done so when the chuck is released, the nut can be easily removed from the arbor. The nut may be faced flat with a small chamfer at the edge, if you feel you will have difficulty forming the curved surface.

To radius the back of the nut, use a round-nose tool bit. Move the apron handwheel slightly with your left hand, and at the same time turn the cross slide handle at a faster rate with your right hand. Actually, this is more difficult to explain than to do on the lathe. Finish the curved surface with a single cut mill or lathe

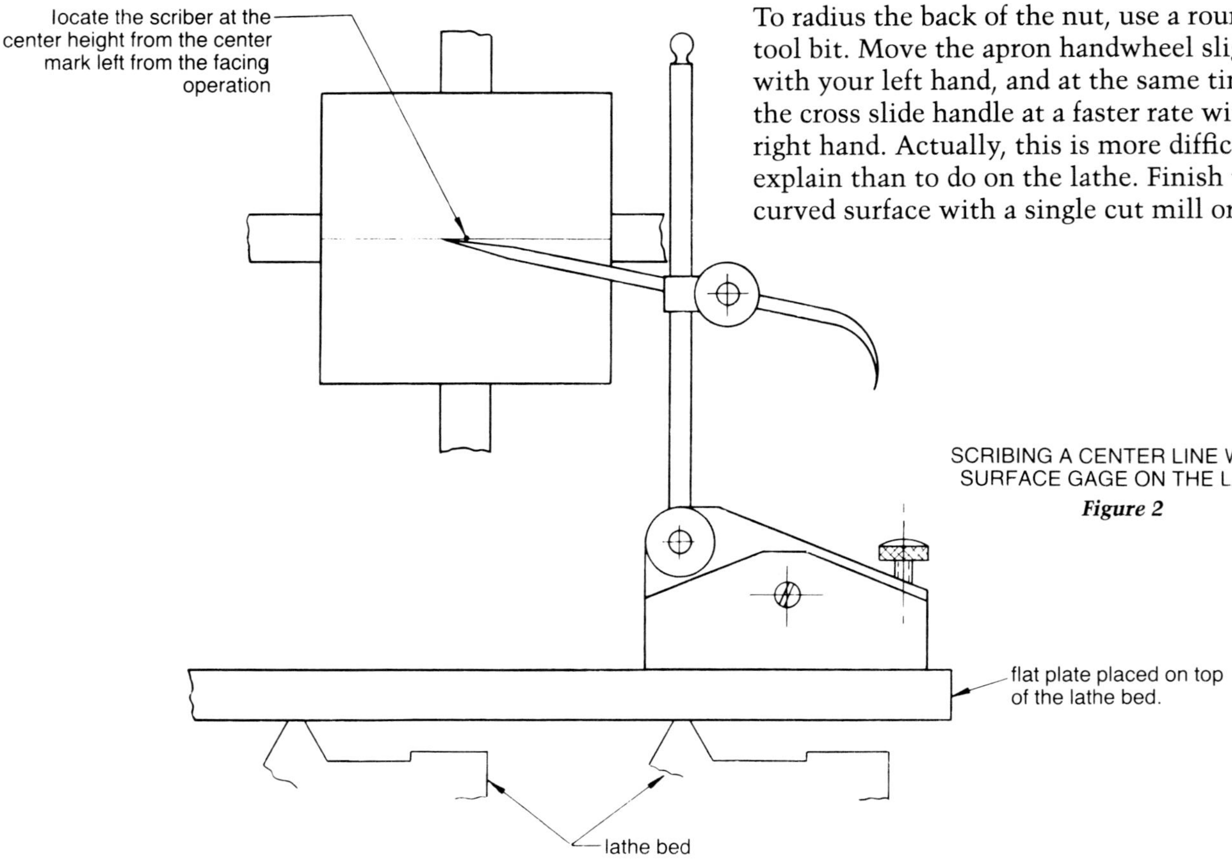

SCRIBING A CENTER LINE WITH A SURFACE GAGE ON THE LATHE
Figure 2

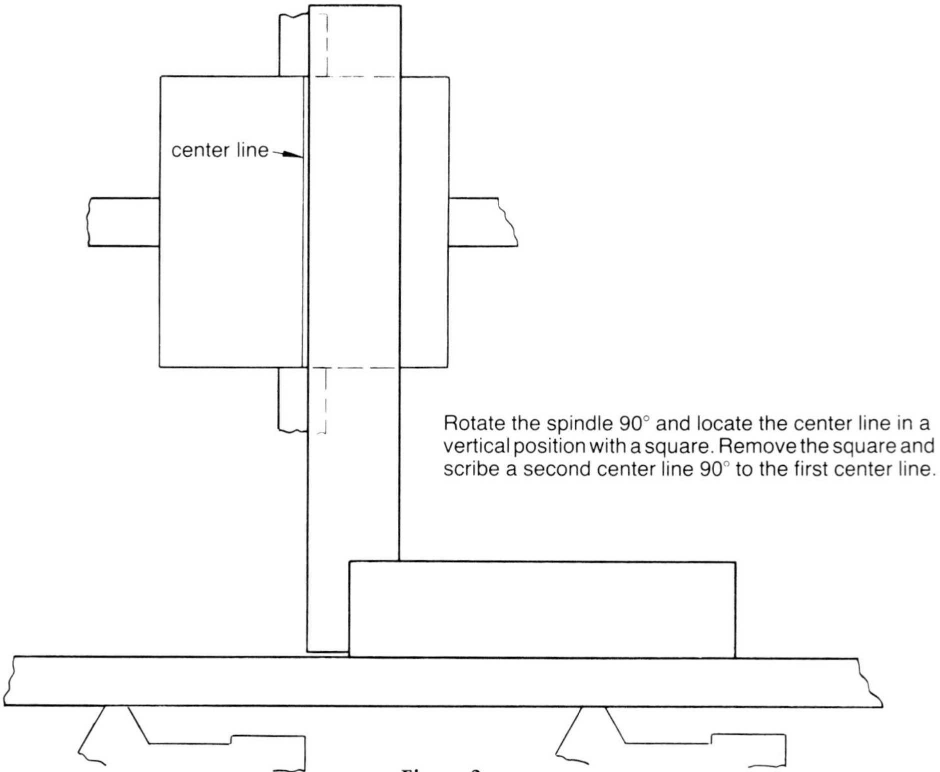

Figure 3

The bent end of the scriber is used to level a casting in the milling machine.

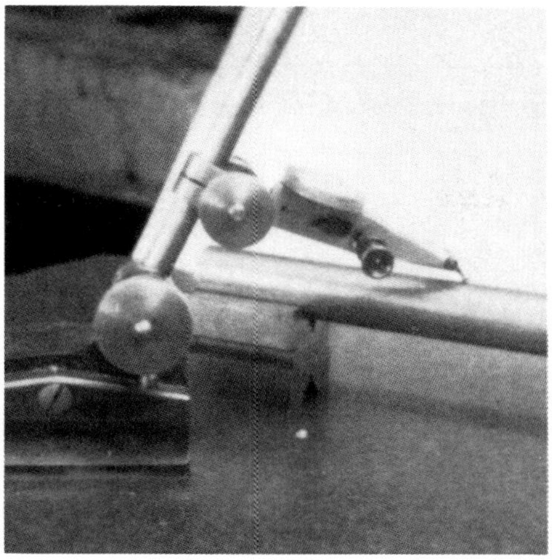

A shaft is tested for straightness by use of a dial indicator in the surface gage.

A dial indicator is used in the surface gage to locate work accurately in a four-jaw chuck.

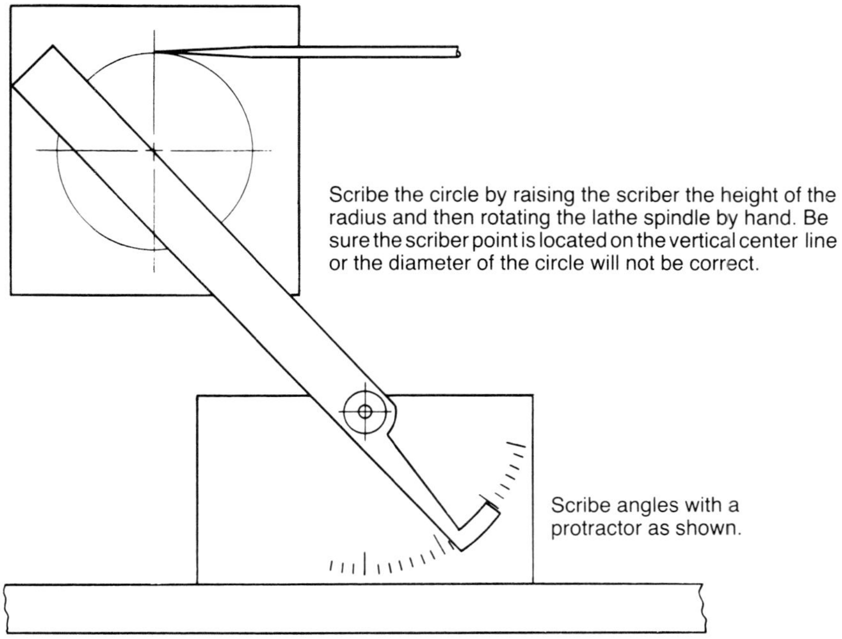

Figure 4

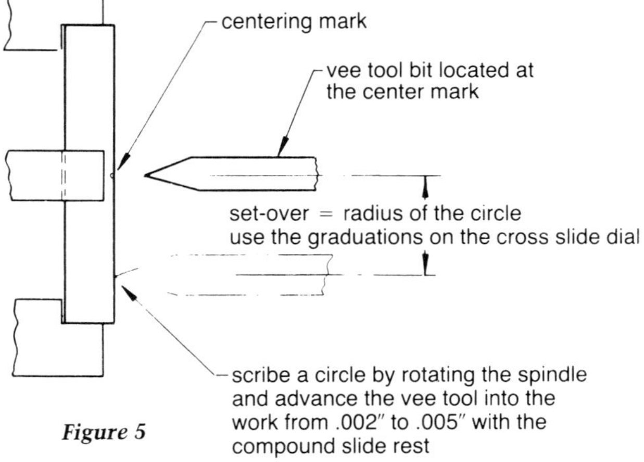

Figure 5

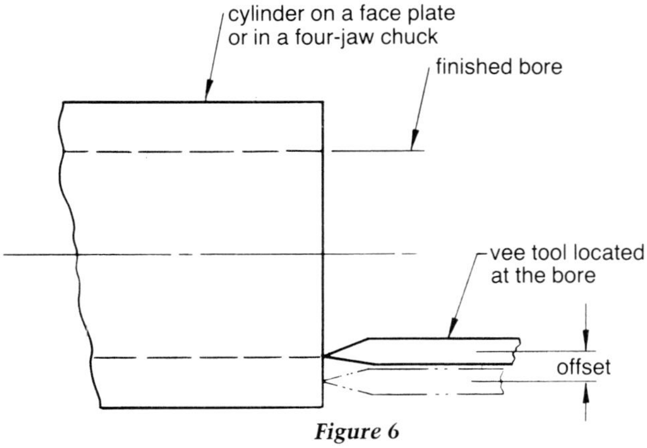

Figure 6

file using long strokes and rotating the file over the curved surface. The curved surface can also be made by hand turning using a turning tool ground up from an old file and the back of the tool holder as a tool rest.

The post clamp (Detail 3) is made from a 1" length of 3/8" square cold rolled steel. Lay out, drill and ream the holes carefully. Make up the filing guide button shown on the drawings.

Place the filing guide in one of the holes, hold the block in the bench vise and file the 3/16" radius. Repeat for the other end. Hold the block vertically in the bench vise to hacksaw the slot.

The special washer (Detail 10) has a 3/16" diameter hole in the center. This hole is filed out nearly to the layout lines and then is finished to size by forcing the special broach through the washer in an arbor press or by hammering with a brass hammer. Use a 3/16" diameter brass rod to drive the broach out of the washer.

The clamping assembly holding the scriber (Details 9 and 11) may be increased in size, or a larger hole may be drilled at 90° to the 1/8" scriber hole to suit a small dial indicator.

A universal surface gage can be easily purchased, for they are not too expensive. However, the gage offers some interesting machining and bench operations and will be a source of pride in many years of service. I have one larger and one considerably smaller than this one, and I find them both useful.

I sometimes use them on the lathe by placing a flat plate on top of the lathe bed as a surface plate. If the center mark left from facing is used as a reference, many times the work can be marked out directly on the lathe. If there is a cored hole in a casting, then the center height can be taken from the tailstock center. When one line has been scribed at the center height across the work, rotate the lathe spindle 90° and line up the scribed line with a precision square set on the surface plate. If a protractor is used, it will be a simple matter to lay out bolt holes in a circle. Scribe the circle with a 60° vee threading tool. Use the cross slide graduations to locate the radius of the bolt circle from the center or from the edge of the bore (see drawings).

A Chip Catcher for the Shaper

The best way to prevent chips (shavings or swarf) from being tracked through the house is to keep them off the shop floor in the first place. If your wife steps on a sickle-shaped shaving with her bare foot during the night, she may have some harsh things to say about you and your workshop. In the interest of harmonious domestic relations, one evening devoted to making this chip catcher will be time well spent.

This chip catcher is dimensioned to fit an Atlas 7" shaper, but any shaper can be fitted with one. Large shapers in commercial machine shops often have a bin or barrel placed in front of the table for the shavings to fall into. Our small shapers take lighter cuts at higher speeds, so the shavings tend to fly. We need a chip catcher that is similar to the old grass catchers that were attached to the reel-type lawnmowers.

I used an old, square, 5-gallon can (equal to 6 gallons in the US) to save time, but the catcher can be made of sheet metal or plywood. The side of the can facing the shaper is cut out. There are two pieces of 1/8 × 1/2 × 1/2" angle iron bolted underneath the can. These angles have two 1/4" wide slots to slip over the ends of two 1/4" threaded rods that pass through the top slot in the sides of the table. These rods have a nut and washer at each end, with the thread extending 3/8" to 1/2" beyond the nuts.

The chip catcher is slipped into place after the machine is set to make a cut. The shavings are emptied into a garbage can at the end of the evening.

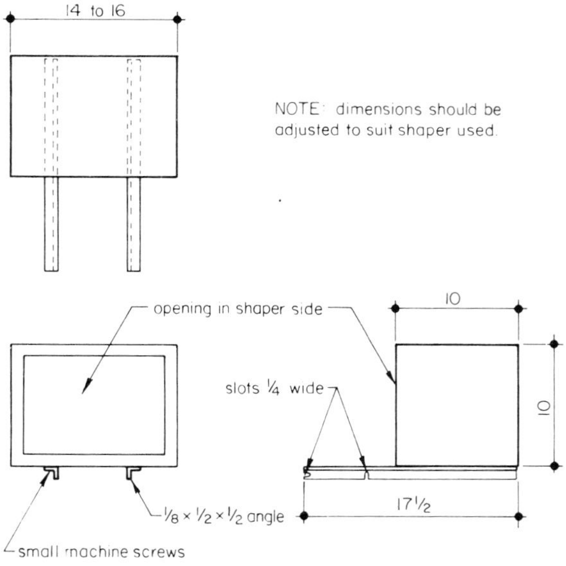

SHAPER CHIP CATCHER

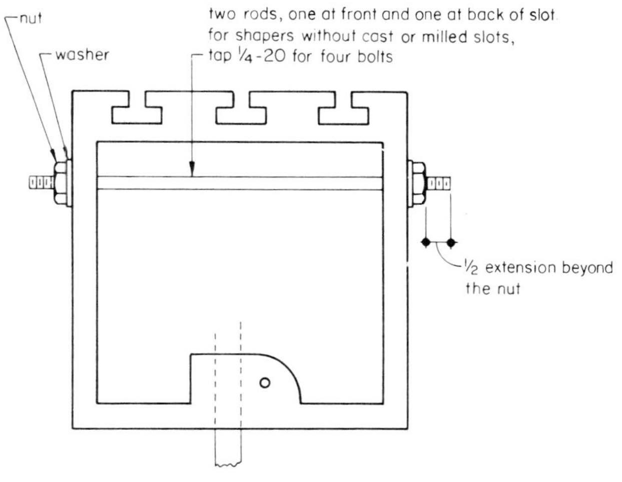

FRONT VIEW OF SHAPER TABLE

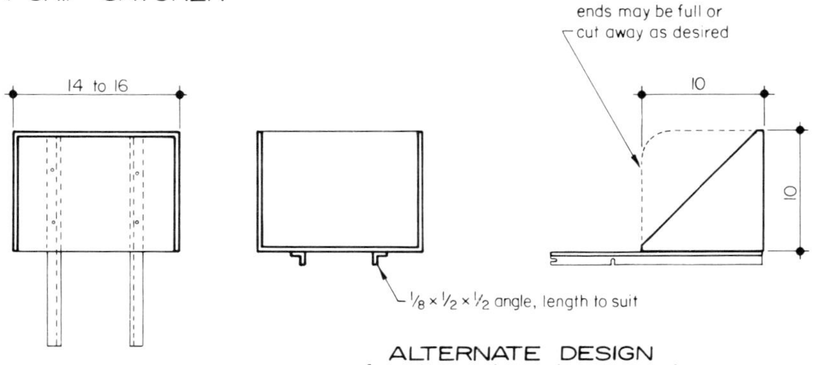

ALTERNATE DESIGN
for plywood or sheet metal

A Depth Gage Attachment

Recently, I was working on a casting that had been machined on the bottom, and was to be machined in a recess in the center of the casting. The walls on the sides of the casting were too high to permit measurement with an ordinary micrometer or a vernier caliper and the tops of the sides were too rough "as cast" to permit the use of a depth micrometer. I remembered when I was young, Starrett used to supply a depth attachment with their vernier height gages. My own vernier height gage is a 10" Mitutoyo, and I'm not sure if a depth attachment is available; however, my need was immediate, so I made one in a few minutes.

I used a piece of cold rolled steel 1/4 x 1/2 x 3-1/4" long. Ground flat stock (gage plate) would be preferable, but I couldn't find any in my tool steel storage box. The CRS was a little rusty, but this was soon corrected on the "armstrong surface grinder." That is a piece of abrasive cloth placed on a surface plate; the work is rubbed back and forth a few strokes on each face. Be sure to use linear strokes, not circular ones.

I marked out the location for the depth rod and the clamping screw with the vernier height gage, and then used the wiggler on the vertical mill to locate the positions accurately for drilling. Other methods may be used, but this method is quick and accurate.

The Starrett depth gage attachment used a collet-type chuck to clamp the depth rod in position. I was in a hurry, so I used a simple clamping screw. In fact, I used a 4-40 Allen cap screw, but this will be replaced with a knurled brass clamping screw very soon.

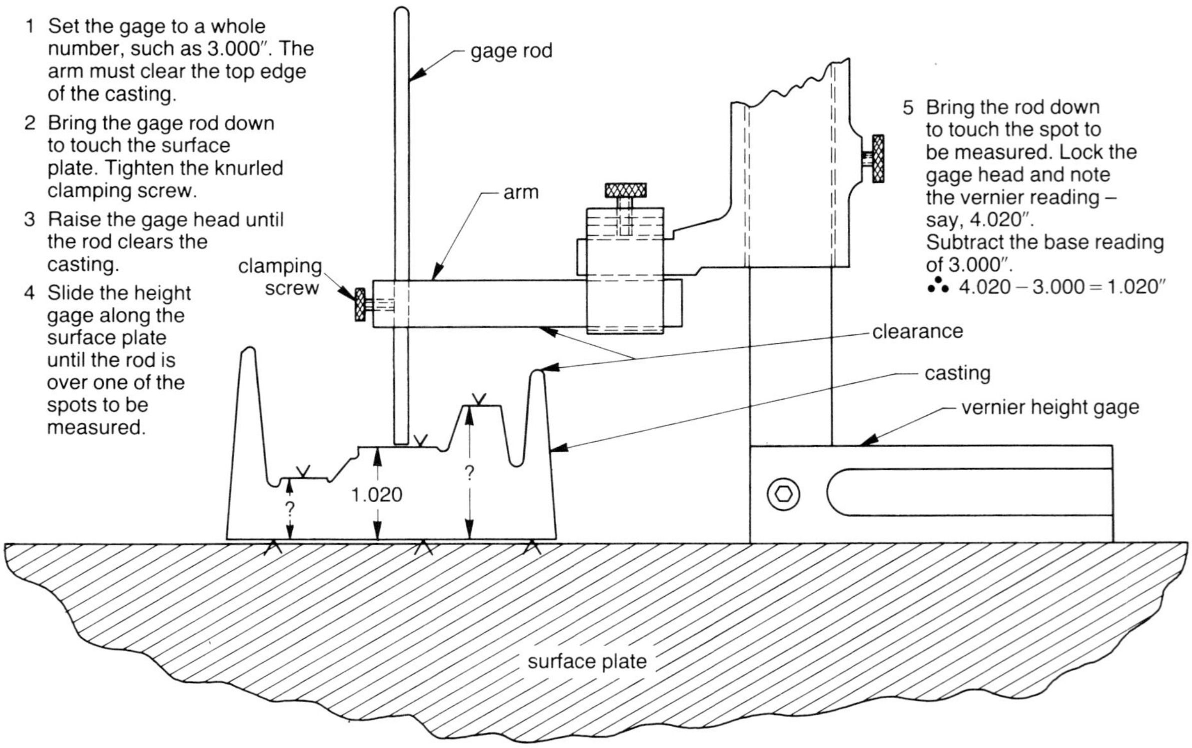

PROCEDURE TO USE A VERNIER HEIGHT GAGE DEPTH ATTACHMENT

The vernier height gage.

The depth gage in place and set to 1.000".

Measuring the depth inside a casting.

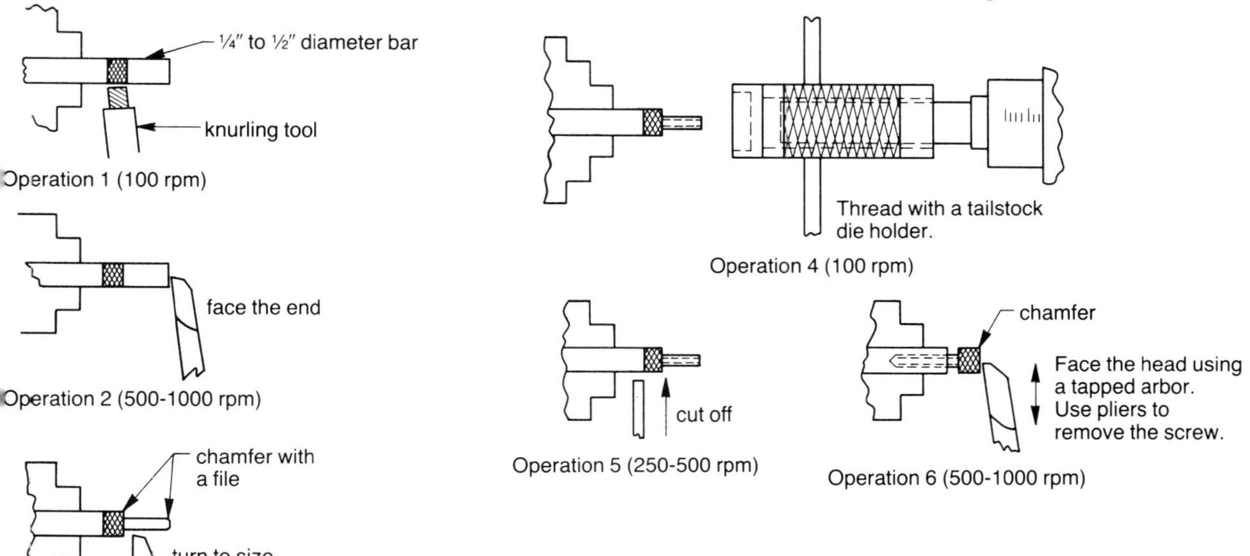

PROCEDURE TO FOLLOW TO MAKE
A KNURLED THUMB SCREW
Machining time: 10-15 minutes

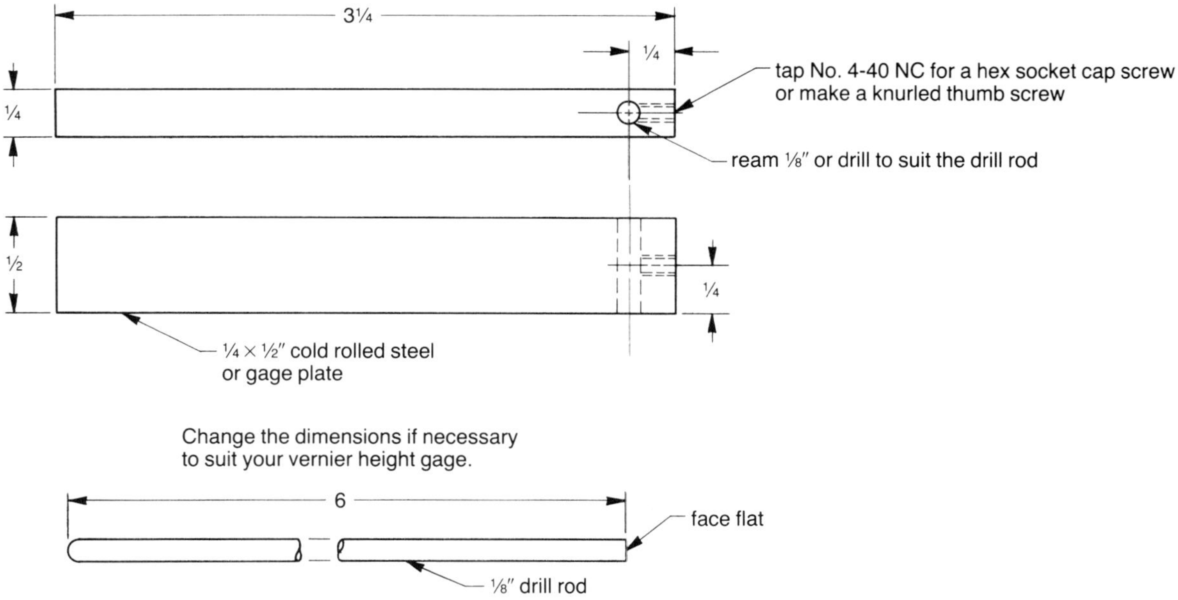

A Toolmaker's Flat Vise

The name "Toolmaker's Flat Vise" is really a misnomer; the term denotes a fine craftsman, or an object made with great precision. This little vise lives up to the name of this column, since everything came from the scrapbox, and there's nothing precise about it. However, it's perfect for holding two pieces together for drilling and tapping holes concentrically and for odd parts that cannot be held easily in the usual drill press vise. The dimensions can be varied to suit the material available or your personal whims.

These vises were usually made to suit the requirements of the individual machinist. They are not generally available from tool suppliers, nor would the machinists have bought them back in the "good old days" because, as a rule, those guys were a crusty bunch of eccentrics set in their ways.

Most flat vises were short and held on the drill press table by hand. This vise has been extended to accommodate a 3/8" wide slot into which the square under a carriage bolt head is placed. The bolt must be long enough to pass through the vise and the drill press table, and enter the tee nut for 1/2". A commercial knob could be used, but I prefer an extended tee nut, as shown on the drawings. Then my knuckles will not hit the ribs under the drill press table.

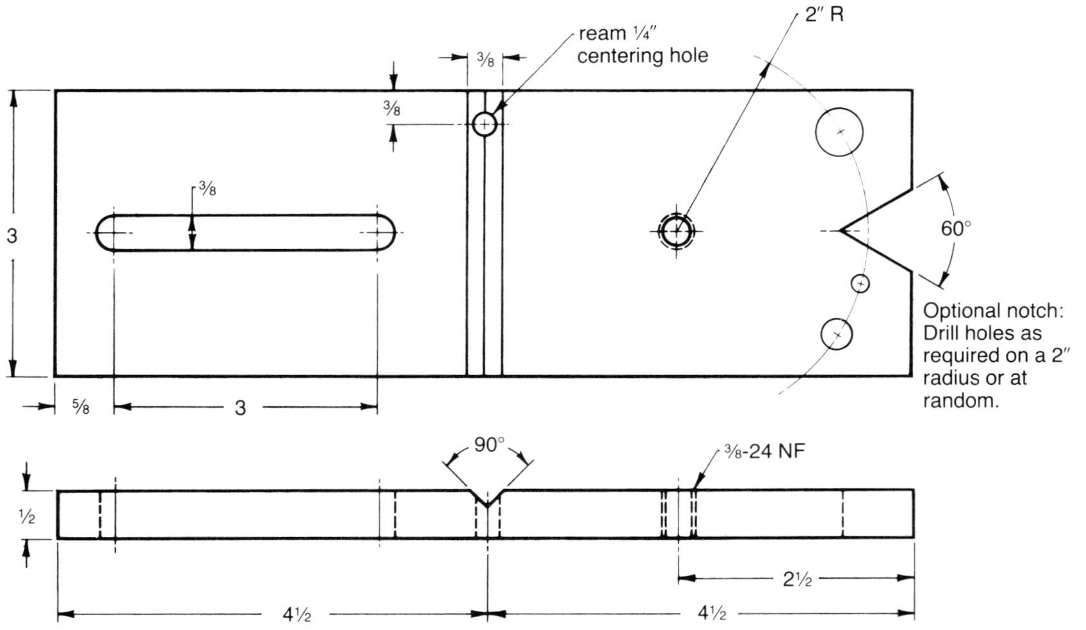

BASEPLATE
1/2 × 3 × 9" crs

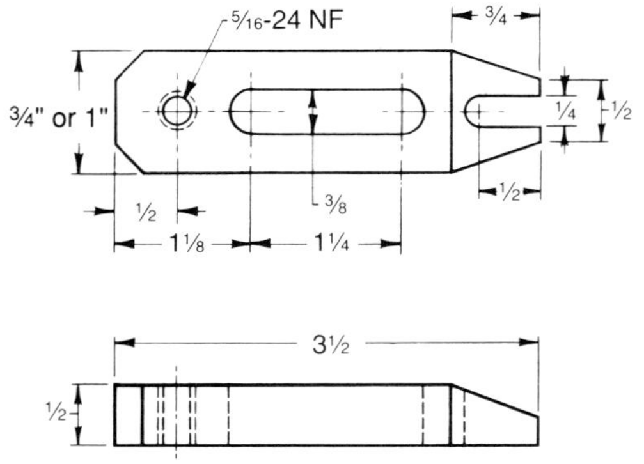

STRAP
1/2 × 1 × 3 1/2" crs

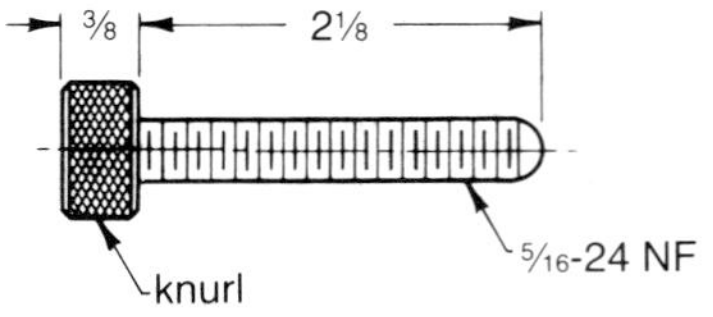

CLAMPING SCREW
3/4" dia. C12L14

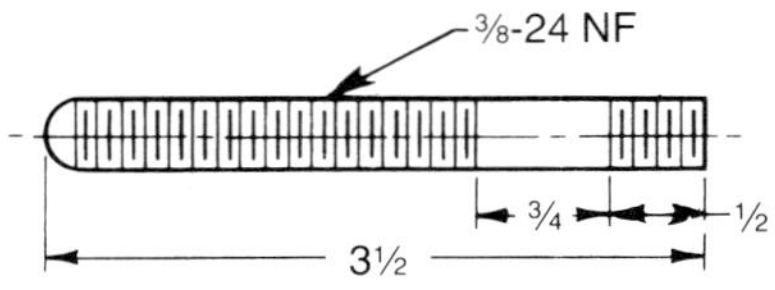

POST
3/8" C12L14

The base is made from cold rolled or hot rolled mild steel. Mine has been surface-ground, but this is certainly not necessary except my vise was badly rusted from being stored in the garage.

The 90° vee groove is optional, and will be used only to drill cross holes in a shaft exactly on center.

The most common method of machining a vee groove like this would be to use a 90° vee cutter on a horizontal milling machine. Most readers will have the usual mill/drill type of vertical milling machine, so this is how they can proceed.

Hold the vise in the milling machine vise at an angle of 45°. Locate the base by placing a protractor or a combination square on top of the vise jaws and align the vise base with the rule. Then mill out the groove with an ordinary end mill. Take several light cuts and do not feed too quickly.

The 1/4" hole in the vee is used to align the vee under the spindle axis of the drill press. Place a 1/4" pin in the drill chuck and then lower the spindle until the pin enters the 1/4" hole. Clamp the vise to the drill press table with the carriage bolt and tee nut. Now place the shaft to be drilled in the vee groove and clamp the shaft in the vee groove with the "vise strap" or clamping bar. Start the hole with a center drill and then drill to size.

To make the 1/4" alignment hole in the vee groove, be sure to use the 1/2" diameter pin with the 90° conical end to align the vee groove under the drill press spindle for drilling and reaming the 1/4" hole.

Make the spherical washer and nut at the same time. Make the washer first and form the concave surface with a tool bit that has been ground to this radius. The radius does not have to be exact. Form the matching curve on the nut with a hard turning tool made from an old file, or file the nut to match the curve in the washer.

Drill the clearance holes in the base, as shown in the drawing, when they are required. Do not be afraid to drill extra holes when they are needed. When your base looks like a piece of Swiss cheese, throw it away and make another.

Finger Plate Vise

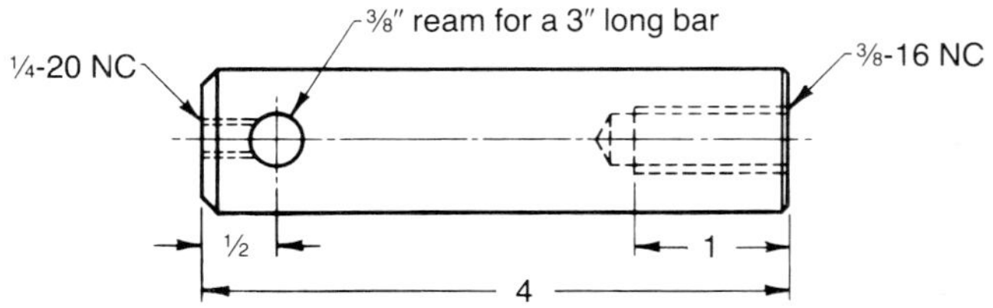

TEE NUT
1" dia. C12L14
use with a 3/8 × 1 1/2" carriage bolt

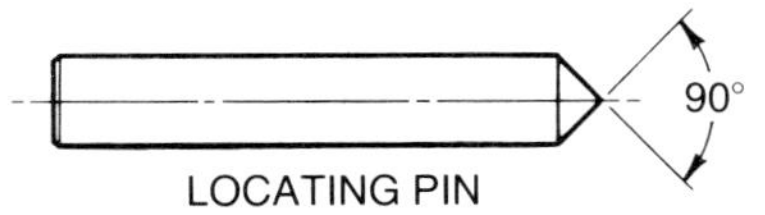

LOCATING PIN
1/2" crs

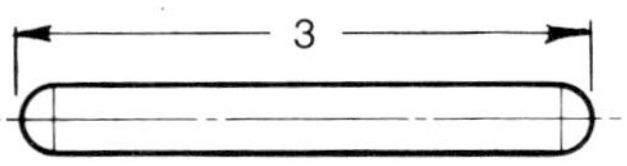

TEE NUT BAR
3/8" dia. crs or drill rod

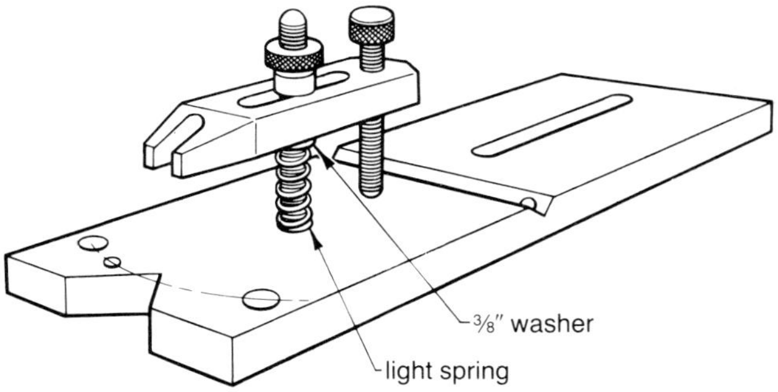

ASSEMBLY

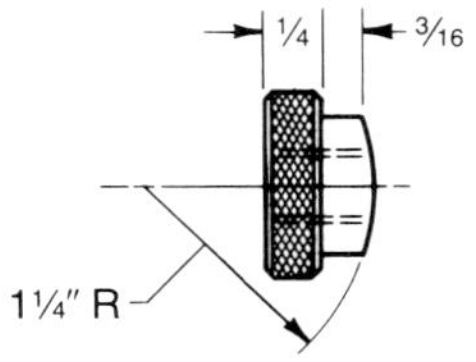

CLAMPING NUT
1" dia. C12L14

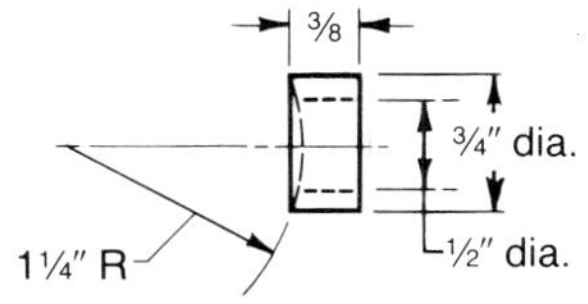

SPHERICAL WASHER
3/4" dia. C12L14

A Small Machine Vise

The design of this useful drill press vise for small parts calls for a piece of 1-1/2 × 1-1/2 × 4" cold rolled steel for the main body of the vise. The dimensions may be altered to suit the material available. For example, if you are going to use 1-1/2" square hot rolled mild steel, simply make the vise 1-7/16" square. The dimensions could also be doubled in size to make a 3" drill vise.

Make the main body (Detail 1) of the vise first. If you are using cold rolled steel, face the ends of the block square in the four-jaw lathe chuck. Hot rolled mild steel will have to be machined all over. Follow the procedure shown in the drawings. The 7/8 × 2-7/8" opening should be made on a vertical or horizontal milling machine, or on a lathe milling attachment. To reduce the amount of machining, chain drill a row of holes along the bottom, saw down the ends and remove the section with a cold chisel. Finish the bottom and sides of the opening by milling.

The next operation will be milling the 1/2" and 11/16" wide slots in the bottom. Note that the cutter will break through into the vise seat. *Do not* perform this operation first, for you will probably squeeze the sides and bend them while holding the work in a milling machine vise to mill the 7/8 × 2-7/8" opening.

The moving jaw (Detail 2) will be the most difficult part to lay out and make. Clamp the moving jaw in place in the main body when drilling and tapping the 3/8-24 NF thread for the vise screw (Detail 5). Be sure to set the body (Detail 1) vertical in the drill press vise. The tapped hole must be "square" with the sides and base of the vise body, or the end at the vise screw will bind in the moving jaw.

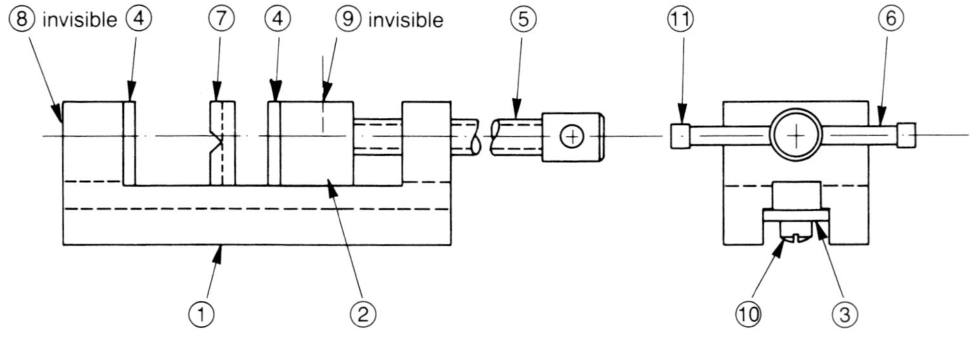

VISE ASSEMBLY

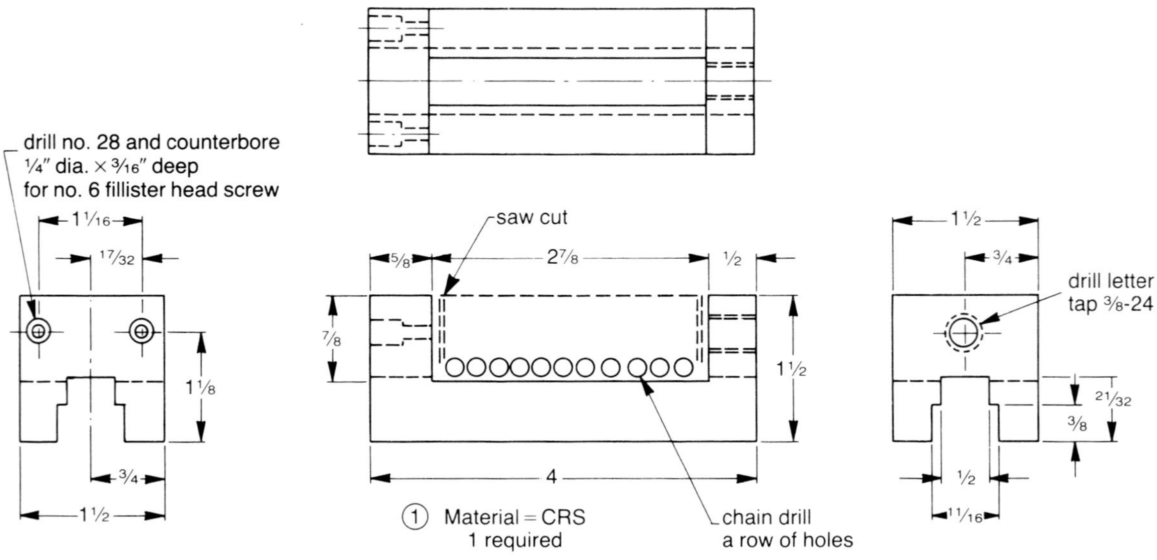

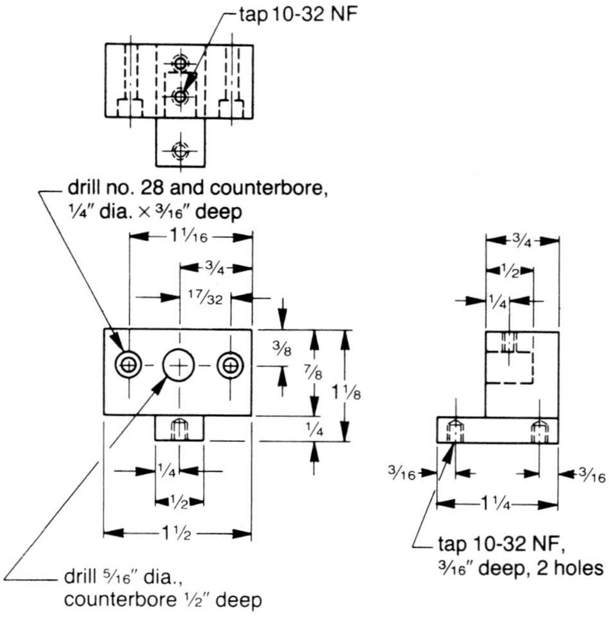

② Material = CRS
1 required

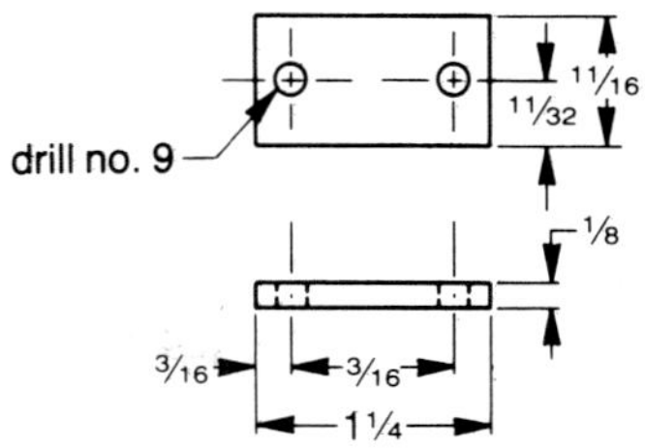

③ Material = CRS
1 required

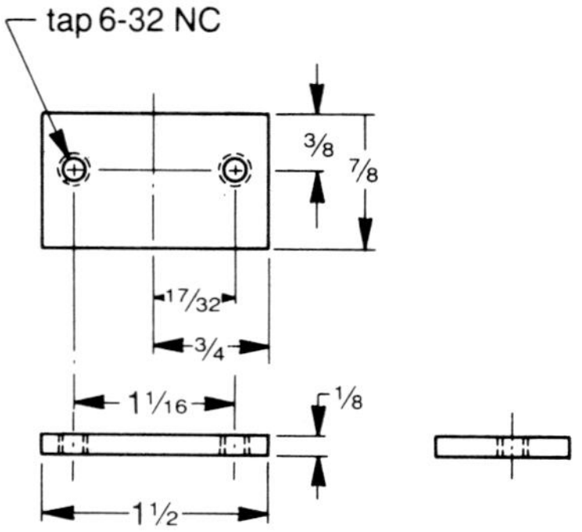

④ Material = CRS
2 required

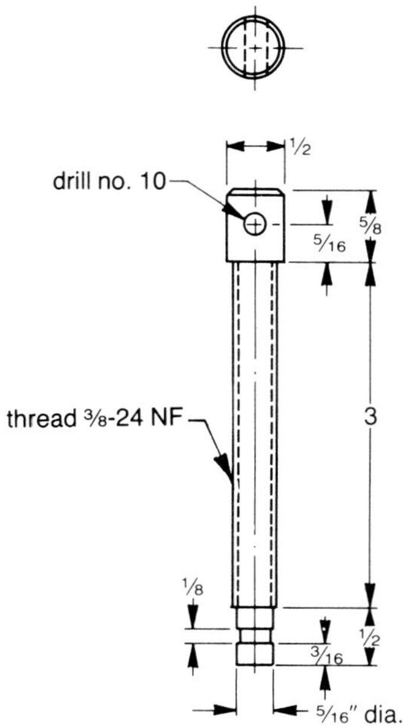

⑤ Material = 1/2" CRS
1 required

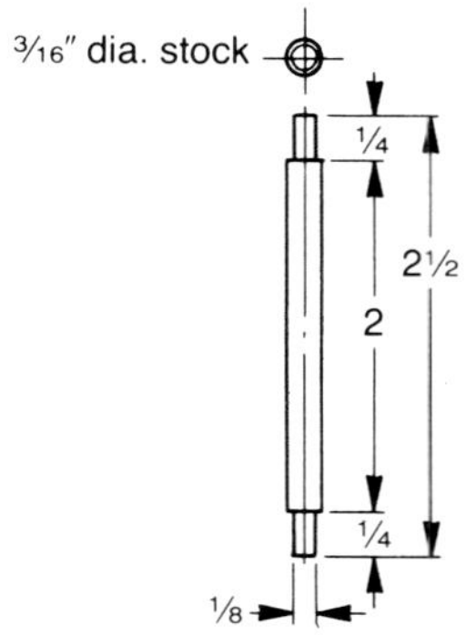

⑥ Material = CRS
1 required

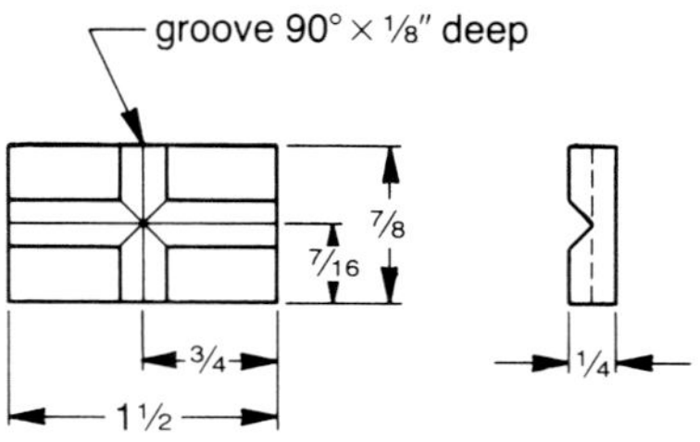

⑦ Material = CRS
1 required

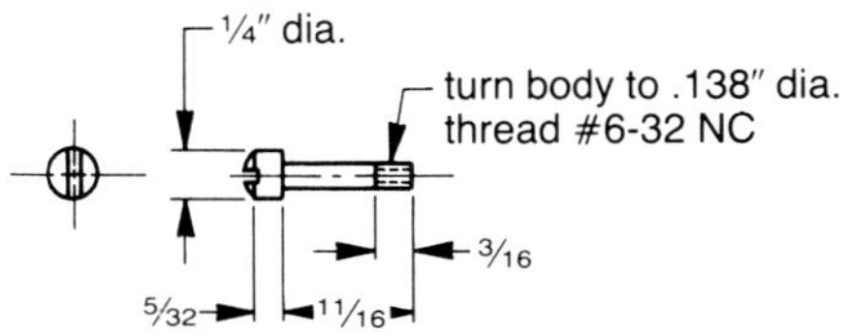

⑧ Material = CRS
4 required

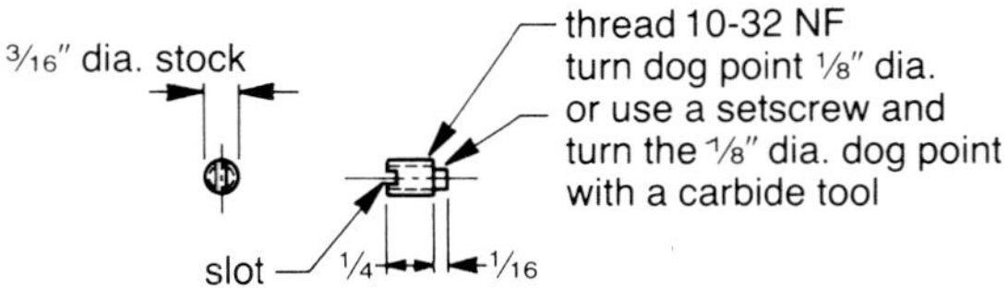

⑨ Material = CRS
1 required

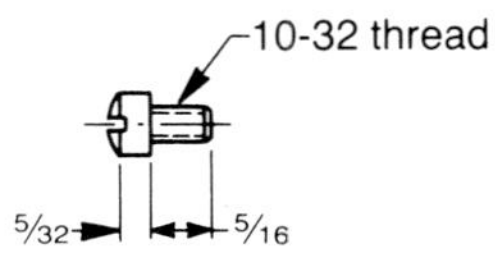

⑩ 5/16" dia. CRS
2 required

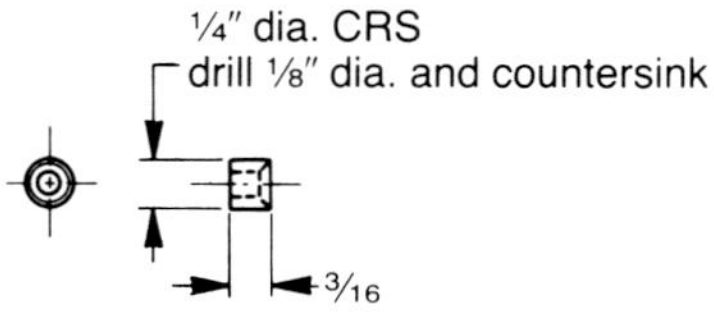

⑪ Material = CRS
2 required

When drilling the vise body with the "Q" drill, run the drill into the moving jaw to form a "dimple." Then drill the 5/16" hole about 3/8" deep. The hole can be finished flat on the bottom later with a counterbore, a 5/16" drill ground flat, or a homemade "D" drill (see drawing).

The thread on the vise screw should be cut on the lathe. Cutting a long thread like this with a die often results in a "drunken" thread. Locate the groove on the end of the vise screw for the dog point by running a no. 21 drill down the tapped hole for the dog point setscrew (Detail 9) and making a mark on the vise screw. Use a 1/8" wide parting tool to form the recess or undercut.

The ends of the sliding handle (Details 11 and 6) are held in place by riveting over the end of the bar. One end can be done in a bench vise, but don't forget to put the bar in place on the vise screw before riveting the second cap in place.

The vise inserts and the vee block are optional. I have given the dimensions in case you want to make the fillister head screws, which take only a few minutes to make. Turn to size, thread with a die, part off and face the head. The slots are made with a hacksaw.

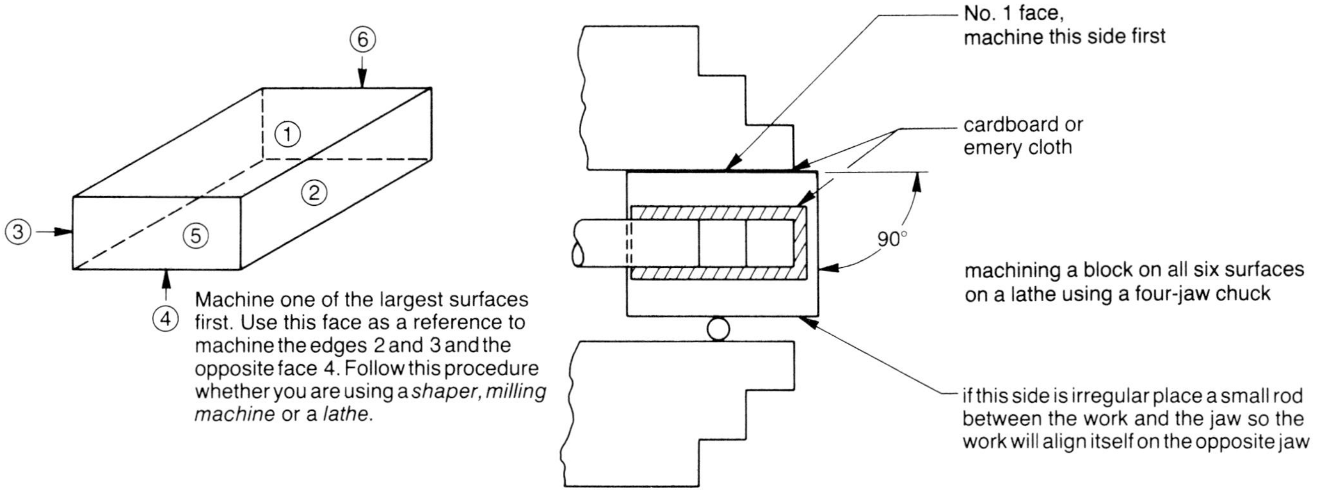

HOW TO MACHINE A BLOCK ON ALL SIX SURFACES

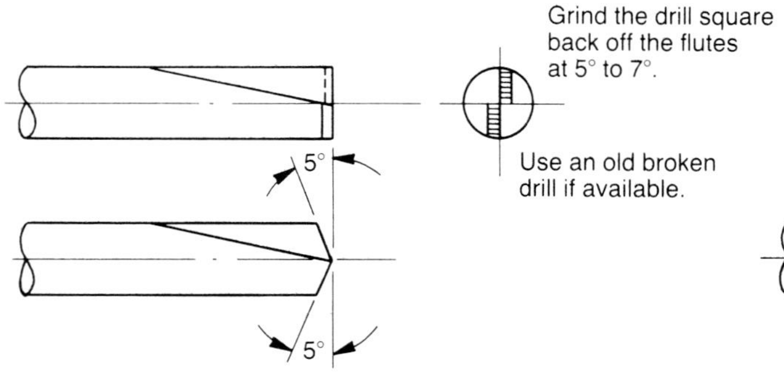

TWIST DRILL CONVERTED TO A COUNTERBORE OR FLAT BOTTOM DRILL

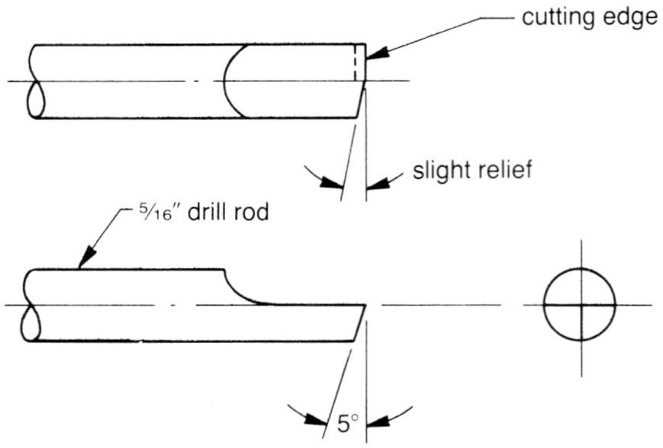

HOMEMADE COUNTERBORE OR "D" BIT

Make from drill rod. File or mill flat to the center line. Harden in oil. Do not temper.

A Band Saw Slow Speed Attachment

I had been wanting a metal cutting band saw for many years, but the cost seemed hard to justify for a machine that will be used only occasionally. I have a homemade power hacksaw and one of the small vertical/horizontal cutoff band saws that are so popular.

When a friend offered me a 14" Taiwanese band saw for a price I couldn't refuse, I decided to design and make a slow speed attachment using Boston change gears available everywhere, and for a moderate cost.

My band saw is similar to the 14" Delta, and is called a "King"; however, all the machines

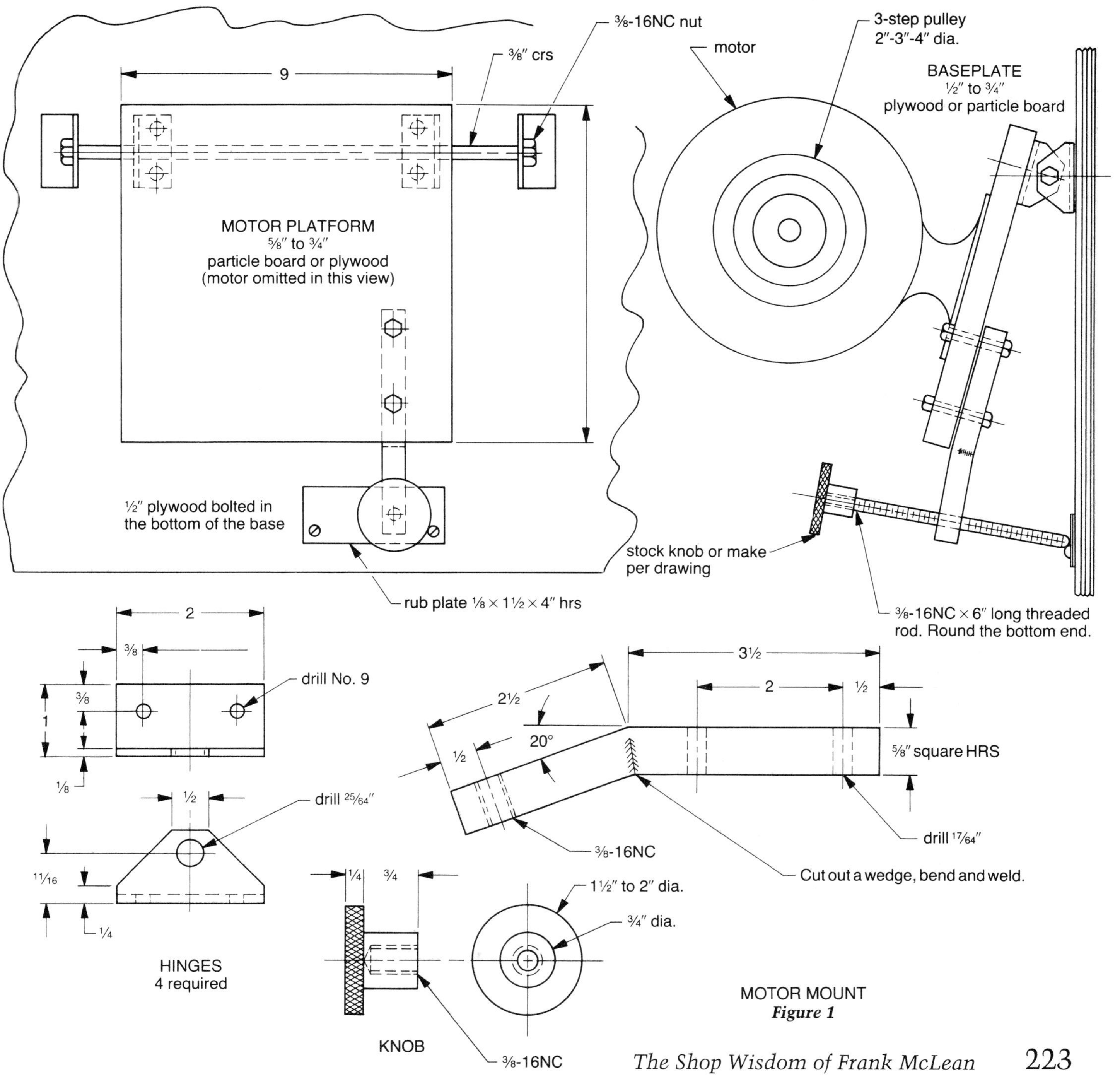

MOTOR MOUNT
Figure 1

advertised in *The Home Shop Machinist* seem to be similar.

When buying a band saw, make sure the wheels run true and in line when a straight edge is held vertically against both wheels. Some machines are available with step pulleys and others have a gear box for reducing the speeds for metal cutting. Usually, the speeds are too fast for sawing thick steel and particularly alloy and tool steels. I took a single speed machine, but I put a three-step pulley on the motor.

The adjustable motor mount permits the motor to move over to bring each step of the motor pulley in line with the 6" driven pulley when it is located on the wheel shaft for cutting wood or on the speed reducer for metal cutting.

I brought my band saw into the basement by sliding the box down the basement stairs with me at the bottom controlling the rate of descent. The box weighed 76 kilograms. I should have opened the box outside and carried the machine into the basement in pieces.

I assembled the base, or cabinet, first. A piece of 1/2" plywood was fitted inside under the top to add stiffness for the machine is quite heavy. The plywood was cut to fit and the opening to clear the vee belt was cut with a hand jigsaw. The plywood is held in place by the machine mounting bolts and No. 10 machine screws in the corners.

The instruction sheet calls for the motor to be bolted to a vertical channel in the base. The one horsepower motor is a heavy "lump" and would be difficult to hold in place while putting in the bolts and adjusting the alignment.

I designed a different motor mount (Figure 1 and Photo 2). I bolted a piece of 5/8" particle board in the base and made a plate of the same material to hold the motor. The hinges are made from 1/8 x 1" angle, and the motor plate slides on a 3/8" diameter rod to align the pulleys.

The belt tension is controlled by a piece of 3/8-16 threaded rod resting on a rub plate. A commercial or a homemade knob is screwed tightly on top of the rod. Use *Loctite* or hold the rod in the bench vise with soft jaws (brass, aluminum or *Masonite*) and screw the knob until it jams on the thread.

I calculated the RPM of the wheel shaft and the cutting speeds in feet per minute for wood and metal cutting prior to any major design work to prevent serious mistakes. The RPM of the motor is 1720, and the motor pulley has three steps of 2", 3" and 4" diameter. The pulley on the wheel shaft is 6" in diameter.

2 ***A view of the adjustable motor mount.***

Wheel shaft RPM:

2" motor pulley = $\frac{1720 \times 2}{6}$ = 573 rpm

3" motor pulley = $\frac{1720 \times 3}{6}$ = 860 rpm

4" motor pulley = $\frac{1720 \times 4}{6}$ = 1146 rpm

The 14" diameter wheels have a circumference of 3.66 feet.

Cutting Speeds for Wood:

2" motor pulley = 573 rpm x 3.66 ft. = 2097 fpm
3" motor pulley = 860 rpm x 3.66 ft. = 3147 fpm
4" motor pulley = 1146 rpm x 3.66 ft. = 4194 fpm

I decided to use a gear reduction of 1:36 to have cutting speeds available for various metals, and not just non-ferrous metals like many band saws. This great gear ratio is possible by using compound gearing similar to the change gears on a lathe. In fact, I used lathe change gears from Boston Gear, because they fitted perfectly into the design without any modification beyond an added hub.

Cutting Speeds for Metal:

2" motor pulley = 2097 fpm ÷ 36 = 58 fpm
This speed will be suitable for tool steel, alloy steels, stainless steels, etc.

3" motor pulley = 3147 fpm ÷ 36 = 87 fpm
This is an ideal speed for cutting hot and cold rolled mild steel, cast iron, tough bronze, etc.

4" motor pulley = 4194 fpm ÷ 36 = 116 fpm

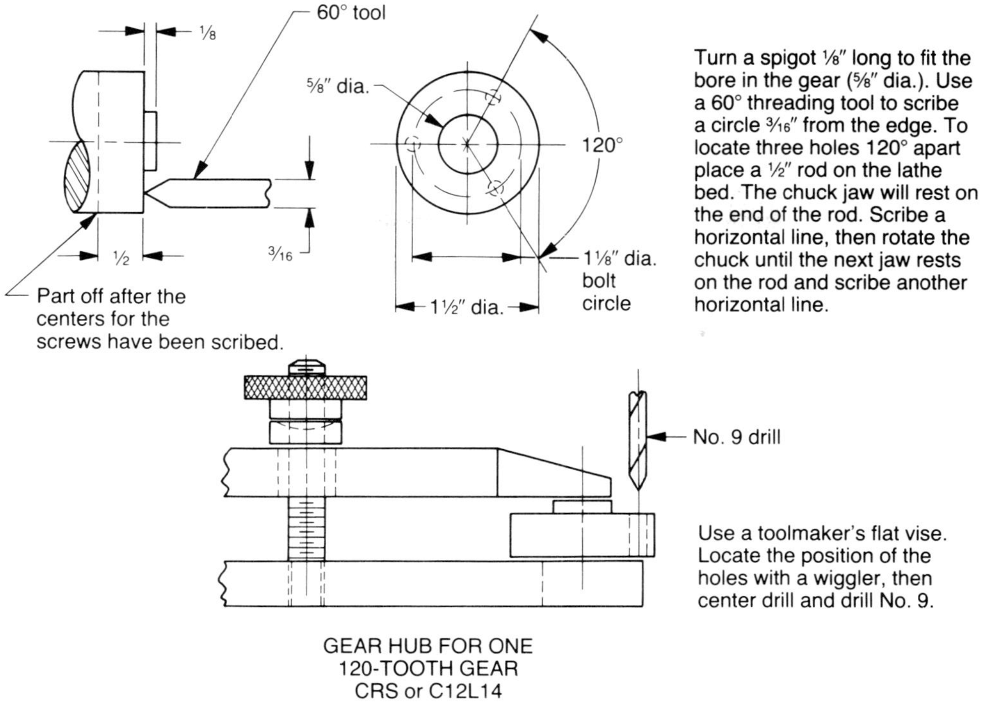

Figure 3

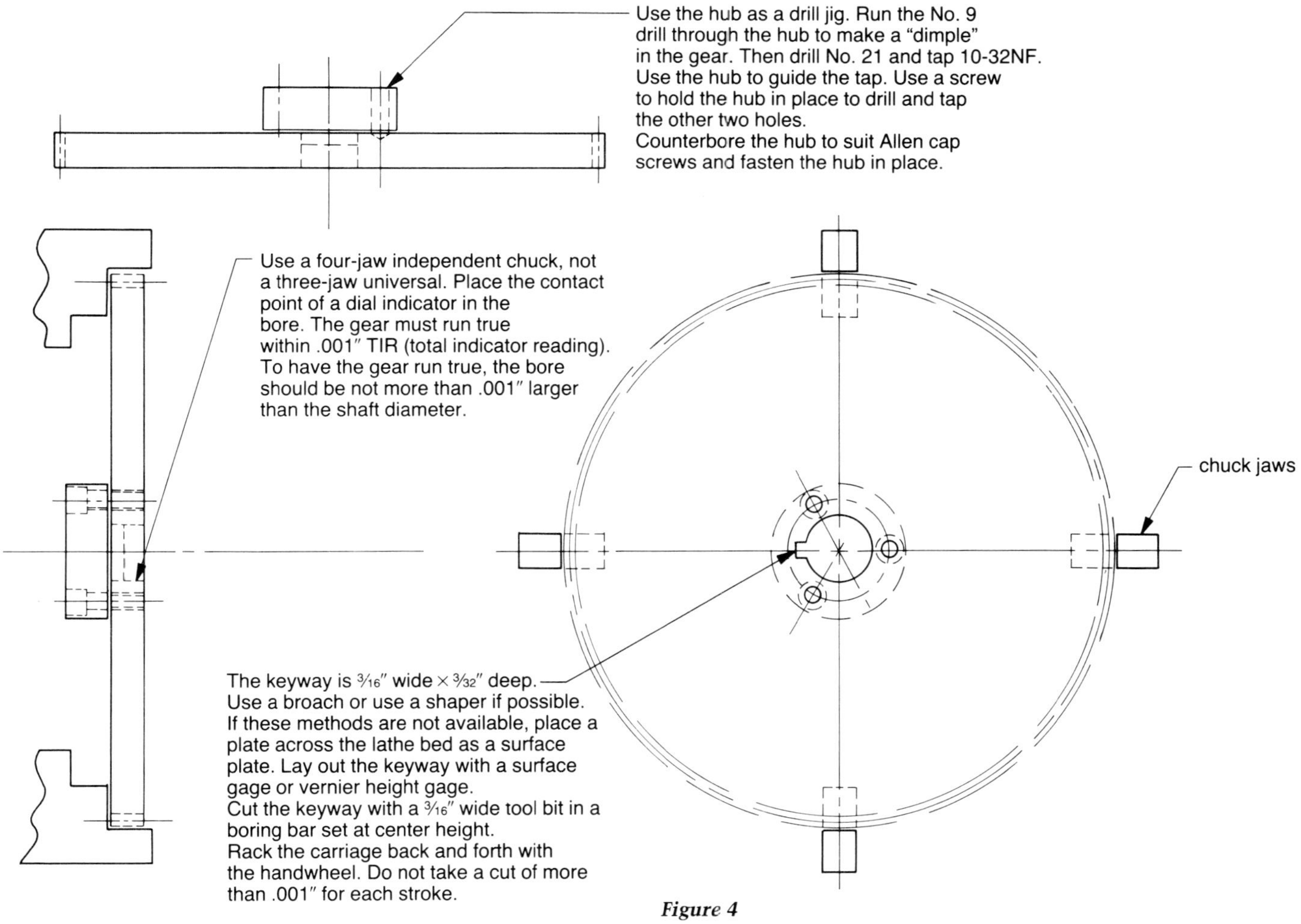

Figure 4

This is the speed to use for cutting aluminum, brass, copper, etc.

I have assumed that carbon steel blades will be used rather than high speed steel blades for economy. I recommend DoAll *Dart* band saw blades.

The gear speed reducer is very quiet in operation. In fact, it works better than I had anticipated.

5 ***The main driven gear with its hub bolted in place. Note the 3/16" wide keyway. This gear goes on the lower wheel shaft of the band saw.***

For once, anticipation was not greater than realization.

Changing over from wood to metal cutting takes about three minutes. Loosen the clamping bolt holding the adapter in place by passing a 9/16" socket wrench through one of the lightening holes in the wheel. Slide the adapter up until the gear teeth are in mesh. Then tighten the bolt. Loosen the setscrew holding the 6" vee pulley on the wheel shaft and transfer the pulley to the sleeve on the lower drive shaft "B" and tighten the setscrew on the key. Install the shorter vee belt (you will need two), and align the motor pulley with the driven pulley visually by sliding the motor along the 3/8" shaft.

The belt tension is adjusted by turning the knob on top of the 3/8" threaded rod. The bottom of the rod is rounded off and slides on a rub plate. There should be about 1" of movement of the belt using moderate pressure about midway between the pulleys.

Changing band saw blades takes longer and must be done with care. I recommend the book put out by Delta, called *Getting the Most Out of Your Band Saw.*

The design of the slow speed attachment or adapter may have to be altered to suit your machine. My band saw had a 3/4" wide slot in the casting for a slow speed attachment, so this made the designing easy.

A few minutes with a file removed the rough edges from the casting. I filed until a piece of 3/4" wide CRS fitted easily in the slot. My machine has ribs cast on the sides of the slot

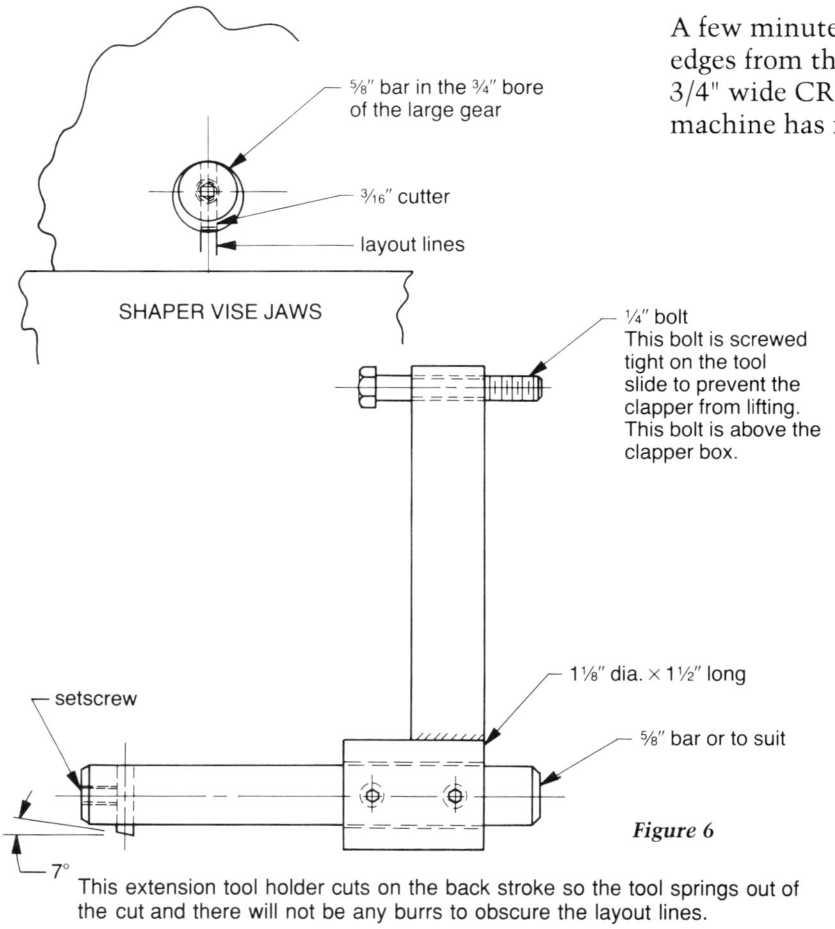

Figure 6

SHAPER EXTENSION TOOL HOLDER
for small shapers such as Atlas, South Bend, Delta, etc.

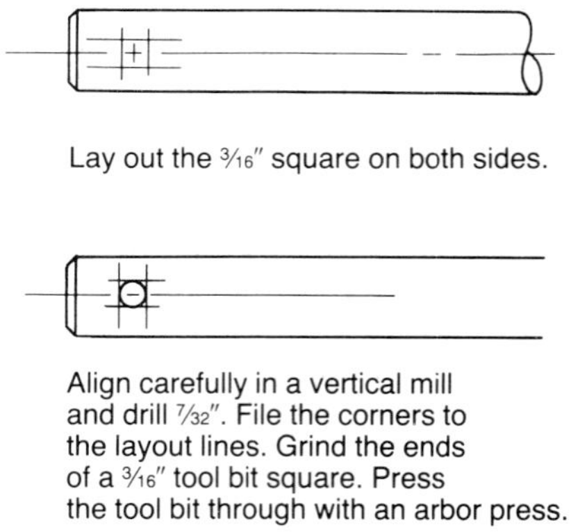

A SIMPLE METHOD OF BROACHING SQUARE HOLES
for tool bits in boring bars, extension, and other special tool holders

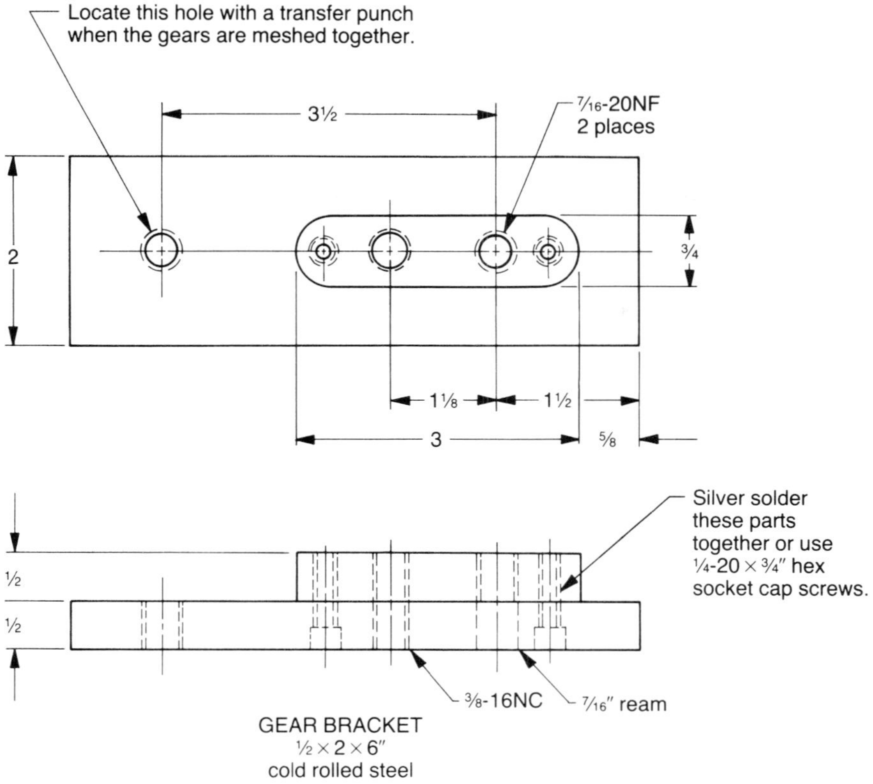

Figure 7

on the inside or wheel side so the total thickness over the ribs is 5/8". You will notice the tongue fitting in the slot is only 1/2" thick, leaving 1/8" between the tongue and the keeper plate.

For those machines without a slot, there are two solutions. A 3/4" wide slot could be made by marking out, chain drilling, and then filing to shape along the layout lines. Alternately, a plate similar to Figure 9 could be used.

Be sure to measure and plan carefully before starting to cut any metal or ordering the gears. Some careful workmanship will be required if the slow speed attachment is to be satisfactory.

Start by making the hub for the large gear that goes on the wheel shaft. Follow the steps shown in Figures 3 and 4. The 1/8" long projection shown in Figure 3 is only to locate the hub in the bore of the gear so it will be concentric with the gear for drilling and tapping the holes for the machine screws.

Photo 5 shows the main driven gear with its hub bolted in place.

Place the gear in the four-jaw chuck with the hub inside the chuck. Place the contact point of the dial indicator in the bore and adjust the chuck jaws until the total indicator reading "TIR" is less than .001". The usual three-jaw chuck will not be accurate enough for this if you want the gears to run quietly. The final size of the bore should not be more than .001" larger than the wheel shaft diameter. Bore to size for accuracy.

The 3/16" keyway should be broached, if possible, either by you or at a local machine shop. Alternately, the keyway will have to be machined on a shaper or done on the lathe by planing. Mine was cut on a shaper using the extension tool holder shown in Figure 6.

Before cutting a keyway on a shaper or a lathe, a layout should be made while the gear is in the lathe chuck. Some lathes have a cross slide than has been machined on top, and this surface can be used. I place a piece of CRS plate on top of the lathe bed to act as a surface plate. Use a vernier height gage or a universal surface gage.

Bring the scriber to touch the bottom of the bore and note the reading. Raise the scriber half the diameter of the bore and scribe a line right across the gear. Now lay out the sides of the keyway .093" (3/32") above and below the center line.

8 ***The gear bracket.***

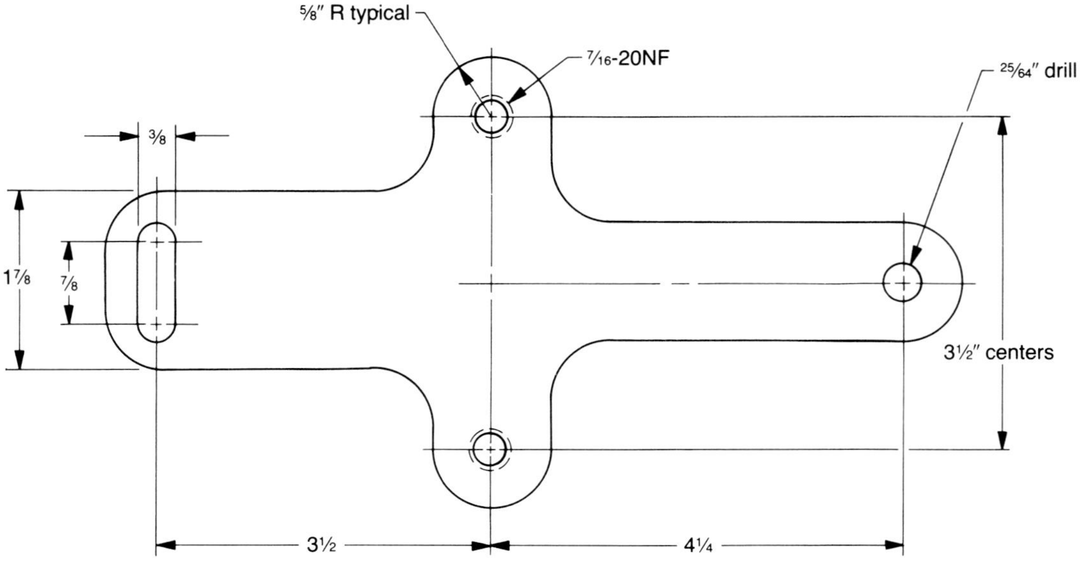

ALTERNATE GEAR BRACKET
for machines without a 3/4" wide
slot in the casting
1/2" to 3/4" HRS

Figure 9

For planing the keyway in the lathe, adjust the height of the boring bar until the cutter matches the layout lines. Bring the cutter to touch the bore and set the cross slide dial to zero or note the reading. Now rack the carriage back and forth with the apron handwheel, taking very shallow cuts – not more than .001". The theoretical depth is 3/32", but make the depth .100" for a little clearance over the key (see the November/December '89 issue of *HSM*).

The setup for the shaper will require some care. The center line scribed across the gear will be set vertically and aligned with a square placed on top of the vise jaws.

Locate the edges of the cutter with the layout lines and bring the cutter to touch the bore. Note the reading on the tool slide dial. Be sure to turn the pulley by hand to operate the shaper through one full stroke to be sure everything clears before turning on the power.

The procedure for drilling and then filing a square hole for a 3/16" square tool bit is shown in Figure 6.

The keyway may be a bit too tight for the key, but this can be easily corrected by a little careful filing. Place the gear in the bench vise with the keyway in a horizontal position as when marking out. Choose a narrow pillar file with a safe edge (no teeth).

Sit on a low stool so your eyes are nearly level with the bore and take a few careful strokes with the file. Turn the gear over in the vise and file the other side of the keyway, and then try the key for fit. Repeat the filing until the key enters the keyway when pushed but without any "shake."

Some may not like this position for filing, so they should stand up and file as usual if this is more comfortable.

Drill and tap a hole for a 10-32 NF setscrew over the keyway so the setscrew will touch the key rather than score the shaft.

Make the gear bracket (Figure 7 and Photo 8) from CRS, if possible. I used a piece of 1/2 x 2" hot rolled mild steel because it was in the scrap box. Bolt or silver solder the pieces together. Scribe a center line on both sides of the plate. Lay out, drill, and tap the 3/8-16 NC thread. The top hole for the 7/16-20 NF thread calls for a reamed hole. This will have to be done at one setup so the hole will be concentric with the thread. The lower 7/16-20 NF thread will be done later during the assembly. The keeper plate (Figure 10) can be made at this time so you can try bolting the adapter in position.

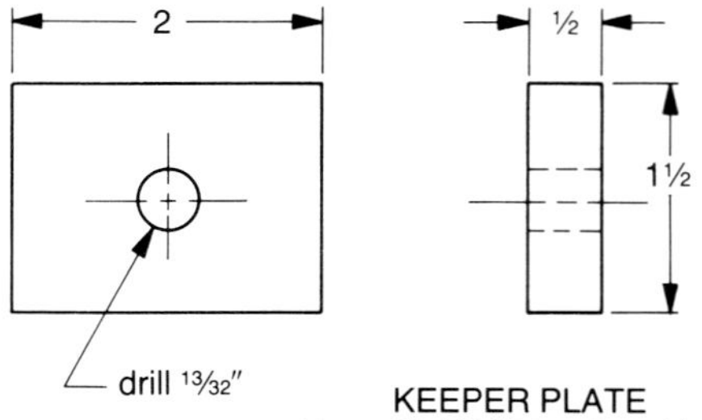

KEEPER PLATE
Use a 3/8-16NC hex head bolt.
Use cold rolled or hot rolled
mild steel for the plate.

Figure 10

By this time, you've tried bolting the adapter in position and made the necessary minor adjustments to make it secure.

The compound idler shaft "A" is shown in Figure 11. This must fit the compound bushing (Boston GAB20). The 7/16-20 NF thread should be cut on the lathe with a 60° threading tool bit for concentricity with the rest of the shaft. Do not use a die. This threading operation is called "single pointing" in the commercial shops. The thickness of the washer may have to be changed to fit your machine so the 120-tooth idler will clear the wheel shaft bearing housing (Photos 12 and 14).

Note the oiling hole drilled from the center hole in the head and the hole drilled across the shaft. This is for oiling with a pump-type oil can. A 7/16-20 NF hex nut is used on the end of the thread on the other side of the adapter as a locknut. Refer to Photo 15.

The compound (coupling) sleeve or bushing and the two idler gears are shown separately in Photo 16 and assembled in Photo 17.

The two half-round notches shown on the head are made with a 1/8" end mill, and are used to turn the shaft into the adapter and to hold the shaft when the locknut is tightened. Use round nose pliers to hold the head by the notches.

We will make the bronze sleeve and then the shaft "B" next, as shown in Photos 18 and 20 and in Figures 21 and 22.

You can use 3/4" bronze bearing stock if the bronze is a good fit in the pulley. The bronze will have to be held in a 3/4" collet or a four-jaw chuck and centered with a dial indicator. I used a piece of 7/8" stock, and held it in the three-jaw chuck because the bore of my pulley was slightly oversize. I turned the bar until the pulley would "just" slip on easily.

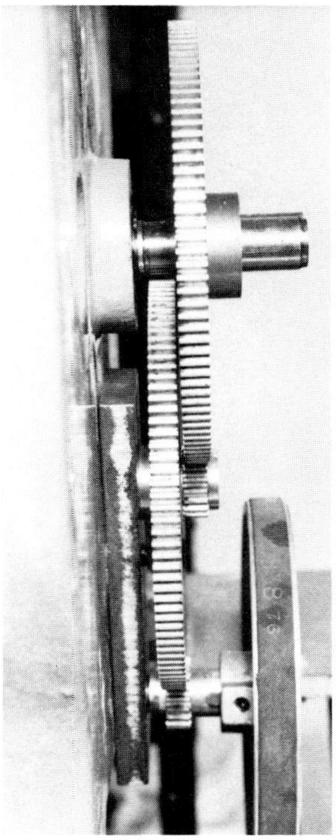

12 ***The slow speed attachment engaged for sawing metal. The pulley has been transferred to the lower, or drive, shaft.***

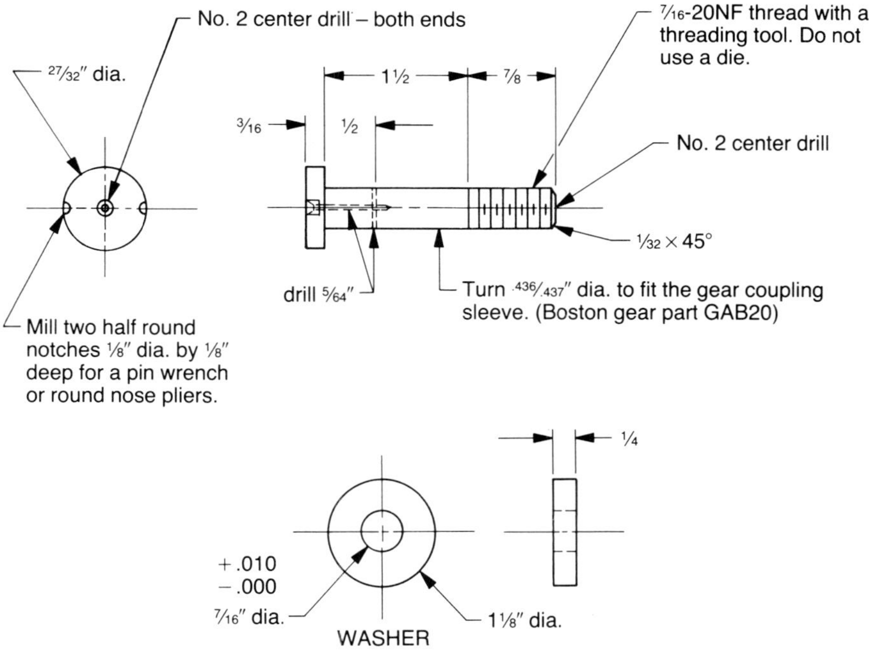

INTERMEDIATE SHAFT "A" FOR THE COMPOUND IDLER GEARS
bearing bronze

Figure 11

13 ***A second view showing the attachment engaged for cutting metal.***

14 ***The intermediate shaft for the compound idler gears.***

Next, I turned a 5/8" length to .001" larger than the bore of the 20-tooth gear (.626" diameter) to have .001" interference for a press fit. Drill and ream to 7/16" diameter, and then part off. Turn end for end for facing to 1-1/4" long.

The next operation will be the sunk key seat, more correctly called a Pratt & Whitney key seat (refer to Figure 21). You will need to use a two-flute end mill, called a slot drill. Mark the ends of the key seat with a rule and a scriber. Do not bother to mark a center line. Hold the sleeve in the milling machine vise by the ends – like an egg. If you hold the sleeve by the sides, the sleeve will squash out of round and you will not be able to center the cutter over the sleeve easily.

To cut the key seat, follow this procedure: Bring the cutter down and move the table until the cutter touches the side of the sleeve. Reset the dial to zero, or note the reading. Raise the cutter to clear the sleeve and move the table over until the cutter is centered over the sleeve. This is the "setover" or "offset." This distance is calculated by adding the diameter of the work to the diameter of the cutter and dividing by two.

My sleeve measured .626" diameter, and I considered the 3/16" end mill to be .188" because there will be a slight runout. Therefore, .626 + .188" = .814 − 2 = .407" offset.

The 3/16" end mill should turn at least 2,000 rpm. Start the machine and bring the cutter down slowly until the cutter is cutting its full width. Reset the depth dial to zero or note the reading. Now feed the cutter down slowly to a depth of .020", then feed the cutter back and forth to touch the lines indicating the ends of the key seat. Repeat four more times until you reach a depth of .100".

You will have to make a key to fit this slot, and it will have to be a light press fit so the key will stay in place when the pulley is transferred to the wheel shaft for cutting wood. Cut a piece of 3/16" square CRS slightly longer than the key seat (key stock). Place the key vertically in the bench vise with about 5/16" projecting above the vise jaws. File the end flat, then file two chamfers on two corners at 45°.

Then, using a rotary motion on the forward stoke, file one corner 1/4 round. Repeat on the second corner. Now make a 180° rotation on the forward strokes to finish the key 1/2 round.

15 ***A view of the slow speed attachment resting on the workbench. The keeper plate is not shown on the clamping bolt.***

This should not take more than two minutes for one end. File the other end the same way, then try the key in the slot. Continue filing until the key will almost enter the slot. Press in place using an arbor press or the bench vise. Note: a piece of 7/16" CRS or drill rod placed in the sleeve will prevent the sleeve from being squeezed out of round.

Now the 20-tooth gear will have to be pressed on the sleeve, or use *Loctite*, if you cannot turn well enough for a press fit. The gear could also be silver soldered in place. There is very little torque on this joint because the gear reduction is 1:36.

Pass a 7/16" reamer through the bore in case the sleeve "closed in" when the gear was pressed on.

You will need to locate accurately the position of the 7/16-20 NF hole for the lower shaft. A commercial transfer punch will be suitable if the punch is a close fit in the bore of the sleeve. If you have a piece of 7/16" drill rod that fits the sleeve, use this to make a punch (Figure 23). Cut off a 3" length and hold it in the collet or four-jaw chuck. Face and turn the cone as shown in the drawing.

It will not be necessary to harden the punch, as it will be used only once. A piece of 1/2" tool steel could be turned to size instead of using drill rod. I located the center point for the shaft before I pressed the gear onto the sleeve by making a punch 5/8" diameter to fit the gear. Either way will work.

Now bring the 20-tooth gear into mesh with the 120-tooth idler gear. There should be a slight clearance between the teeth. This can be easily achieved if you place a piece of thin paper, such as newspaper or cigarette paper, between the gear teeth. Locate the gear visually over the center line, then place the transfer punch in the bore and strike the punch once with a hammer.

Set the gear bracket in the vertical milling machine and locate the center mark with a wiggler. Drill and tap 7/16-20 NF. The hole must be deburred by filing so the surface will be flat.

The lower drive shaft "B" is made from 5/8" hexagon screw stock C12L14, or from cold rolled steel. The screw stock will give you a much better finish.

16 ***The 120-teeth gears and 20-teeth gears with the coupling sleeve shown separately.***

17 ***The compound idler gears, joined with the coupling sleeve.***

18 ***The band saw set for sawing wood.***

19 ***A second view showing the band saw set for sawing wood. The vee pulley is on the lower wheel shaft. The slow speed attachment is disengaged.***

20 ***The slow speed attachment, again shown without the vee pulley.***

Making the shaft is more involved than Figure 21 indicates at the first glance. Start with a piece of stock about 6" long. Face and center drill one end. Support this end on the tailstock center and turn to fit the sleeve for a length of 1-3/4" for now.

The head end will have to be faced back to fit the sleeve and the washer when the shaft has been turned into the gear bracket so the end of the thread jams into it.

Single point the thread (7/16-20). Be sure the thread is a snug fit in the tapped hole in the gear bracket. There is no excuse for sloppy work here.

When facing the shoulder or head to length, screw the shaft into the gear bracket with a wrench on the hexagon until the shaft is tight in the gear bracket. Now mark the hexagon where the sleeve and the washer will terminate. Remove the shaft for turning and facing. I used a 7/16" collet to hold the shaft accurately in the lathe, and a left-hand tool was used to turn the shaft up to the mark. Next, I parted off the excess, faced the head and drilled the center hole and oil hole. The cross hole can be drilled on a drill press or a vertical mill.

If a 7/16" collet is not available, use a four-jaw chuck and a dial indicator to set the work true or use a three-jaw chuck and check the concentricity with a dial indicator. A piece of paper may be placed between the work and one of the jaws, if necessary.

Now comes the final assembly (Figure 24). Make sure the gears turn freely on the shafts, and the large idler gear does not touch the ball bearing housing of the wheel shaft. Slide the adapter up until the idler gear meshes with the driven gear on the wheel shaft (Photos 12 and 13). Continue as previously described.

I recommend DoAll *Dart* blades for cutting metal. For sheet metal, I use a blade 1/4" wide, 32 teeth per inch, wavy set. These will be cut to length and welded by your local DoAll distributor, and the weld is guaranteed. For other metal cutting, use a blade with 14 teeth per inch and 1/2" wide.

DoAll has distributors in all the major cities in North America. If you cannot find one in your telephone book, the head office can help. Write to DoAll Company, 254 North Laurel Avenue, Des Plaines, IL 60016, or call (312) 824-1122. In Canada: DoAll Canada Ltd., 10

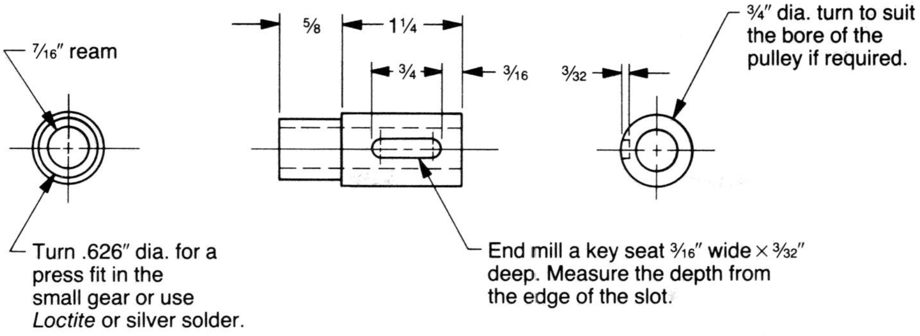

BRONZE SLEEVE TO FIT THE LOWER SHAFT "B"
Press fit the gear. The pulley is a slip fit.

BRONZE SLEEVE
to fit the lower shaft "B"
Press fit the gear. The pulley is a slip fit.
bearing bronze

Figure 21

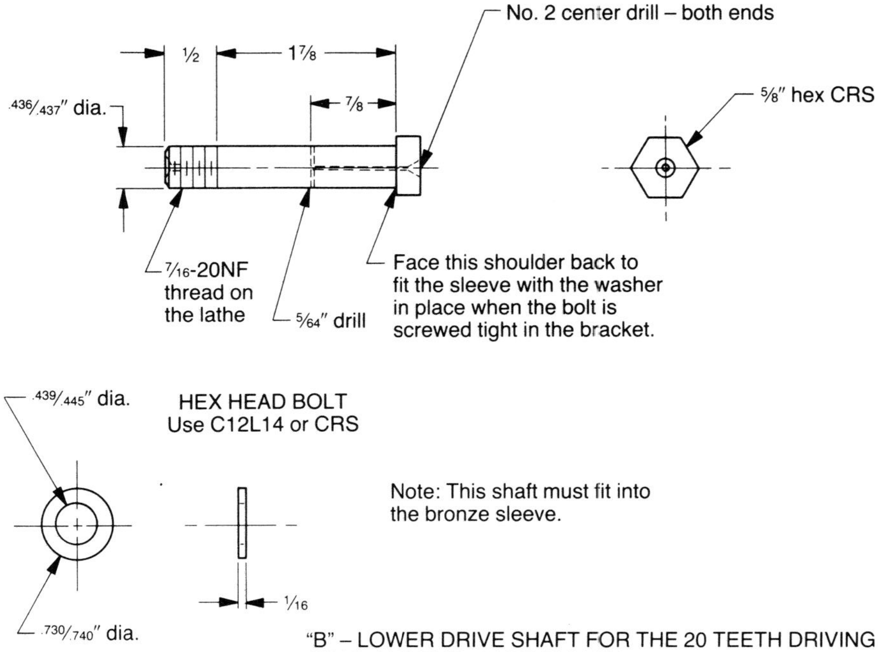

WASHER
CRS

"B" – LOWER DRIVE SHAFT FOR THE 20 TEETH DRIVING GEAR AND THE 6″ VEE PULLEY

Figure 22

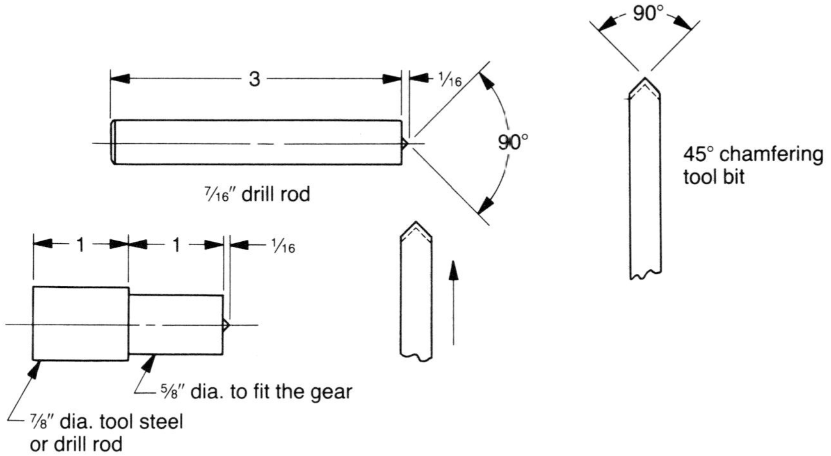

ALTERNATE TRANSFER PUNCHES

Figure 23

Meridian Road, Rexdale, Ontario M9W 4Z6. Phone 1 (800) 268-0938.

Boston Gear products are widely distributed. If you cannot find a local dealer, contact their head office: Boston Gear Incom International, Inc., 14 Hayward Street, Quincy, MA 02171, or call 1 (800) 343-3353. In Massachusetts, call (617) 328-5690. The gears required are shown on page A26 of their catalogue.

The gears are No. 20 diametral pitch, 14-1/2° pressure angle, 3/8" face width and 5/8" bore. You will need two each of the gears and one compound sleeve.

20 teeth	GA20
120 teeth	GA120B
Sleeve	GAB20

Some Canadian readers may not have local distributors, so I can recommend the

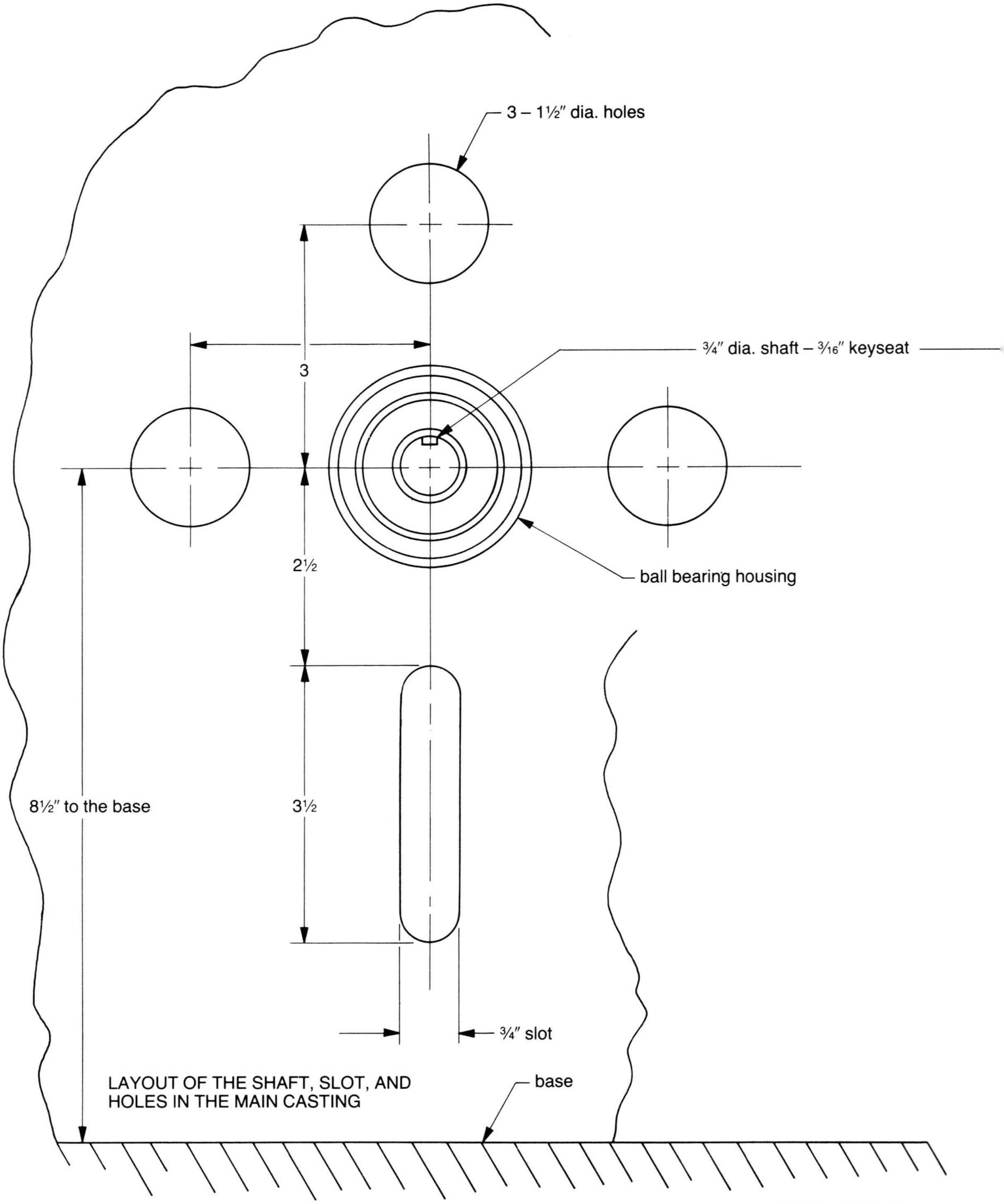

GEAR LAYOUT

companies I deal with. The Boston Gears and other bearing materials can be obtained from RCL Industrial Equipment, 112 Isabella Street, Ottawa, Ontario K1S 1V5. Phone (613) 233-1087. The owner is Robert Laflamme.

My band saw came from Mark's Machinery, 111 Greystone Drive, Carp, Ontario K0A 1L0. Phone (613) 831-3317. The owner is Mark Allen.

I hope you will find your enhanced band saw as useful and pleasant to use as I have found mine.

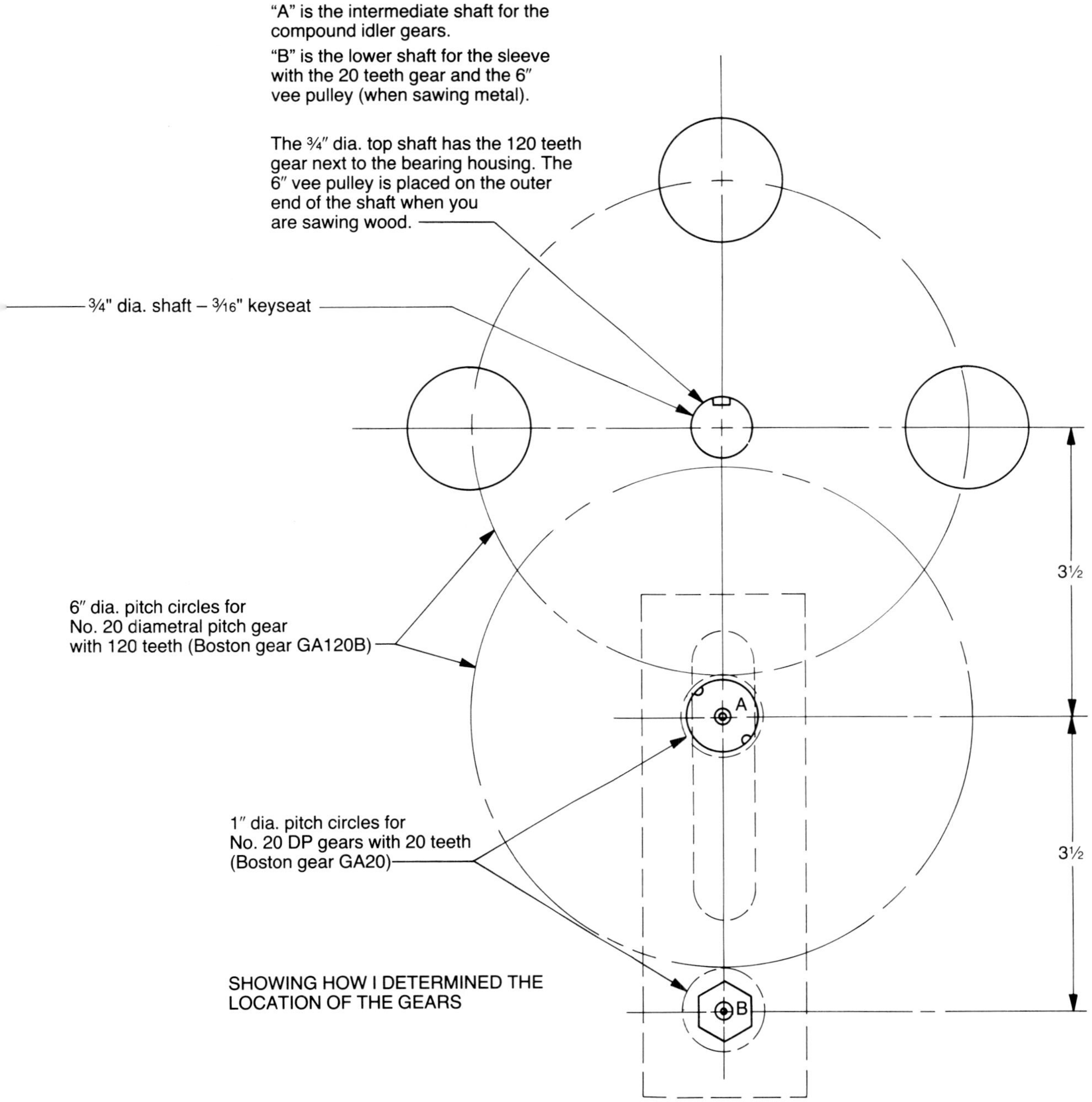

Figure 24

A New Centering Gage

Recently, I was making a new arbor for a shell end mill. The arbor has a boss 3/4" in diameter to locate the shell end mill, and two driving keys 5/16" wide. Milling one 5/16" keyway was no problem, but locating and milling the other keyway exactly opposite was not going to be easy without using a dividing head.

I had thought of silver soldering a blade in a round shank, but I could see problems making the gage with sufficient accuracy. While operating the lathe, I thought of the simple gage shown in the drawings.

I made one of brass, because I like working with brass, and another from CRS to use at work. To make the gage, follow the instructions in the drawings.

To use the gage, the work must be faced flat, and have a center line scribed accurately. Use a vernier height gage, if you have one, or a surface gage. I use a plate placed on the lathe bed as a surface plate. Some lathes have the cross slide machined flat on top, and this can be used as a surface plate. If you are in a hurry, and the lathe tool bit is set at the center height, scribe a line across the work with the tip of the tool bit. A shallow scratch is sufficient; you don't have to plow a furrow.

To use the centering gage, place the gage in a collet or a drill chuck, and bring the flat side to touch the end of the work. Rotate the work until the edge of the gage is parallel with the scribed line. Tighten the vise jaw, and check to be sure the line is still parallel with the edge of the gage. Now move the work over until the edge of the gage covers half of the line. Use a magnifying glass. Now move the work over half the diameter of the gage (.249" for my gage).

Note: when deburring the gage after milling the flat, deburr very carefully with a fine oilstone slip instead of a file to preserve the sharp edge.

The center line of the work is now under the spindle axis, and the flat on the gage is touching the end of the work, so use the table and cross slide dials to locate the work for machining. Repeat the process to align

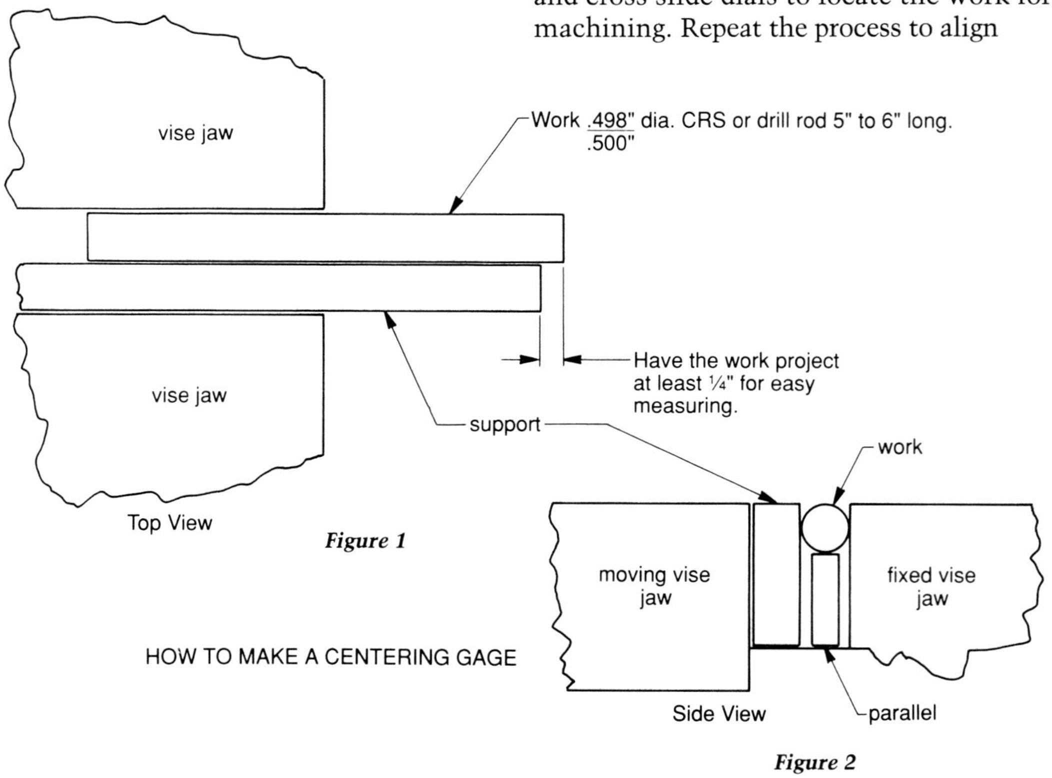

Figure 1

Figure 2

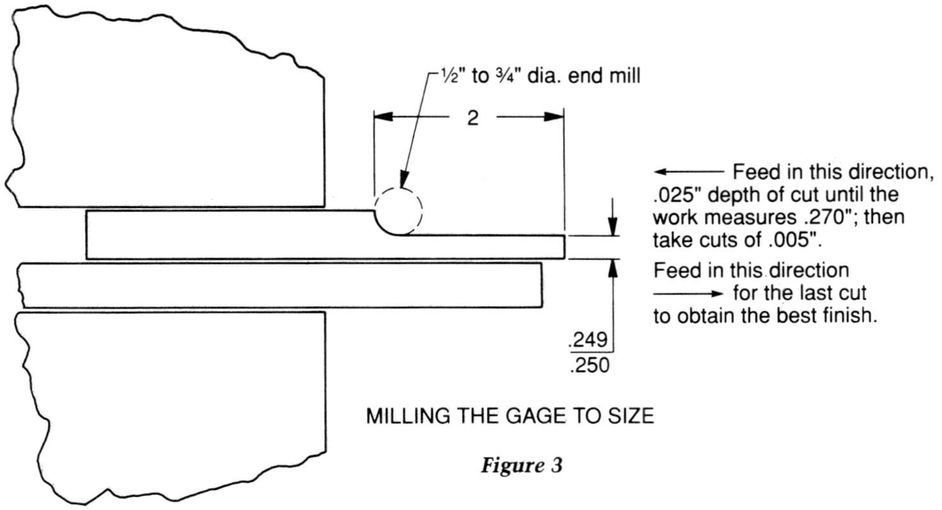

MILLING THE GAGE TO SIZE

Figure 3

the work to machine the other side.

You will find this gage is a great time saver. A precision square could be used to align the work, but the base at the vise is usually in the way, or the work will have to be extended way beyond the vise jaws.

The pointer of a wiggler can also be used, and when viewed from the end of the work with the pointer at the top, and then at the bottom of the line, the work can be located but not so accurately nor so quickly as when this centering gage is used. Try one. I think you'll like it.

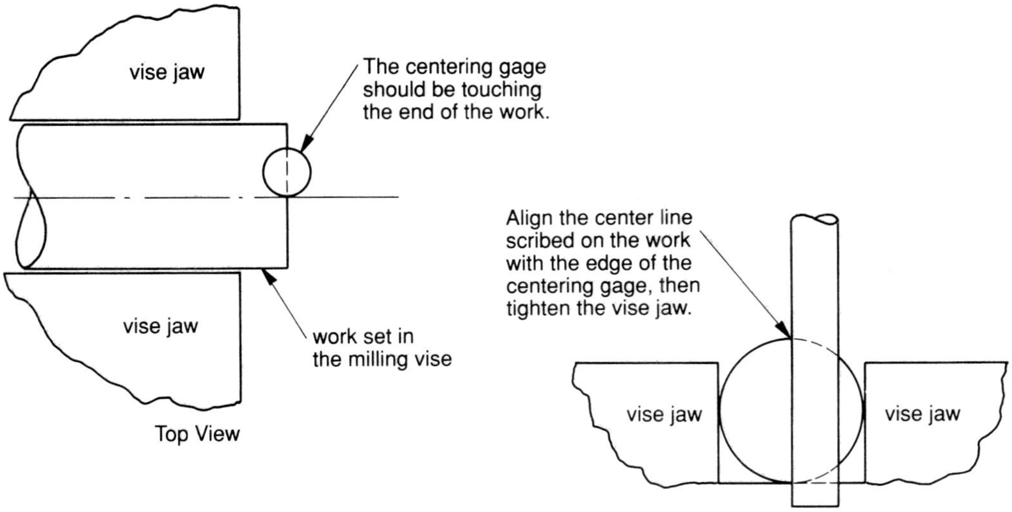

USING THE CENTERING GAGE

Figure 4

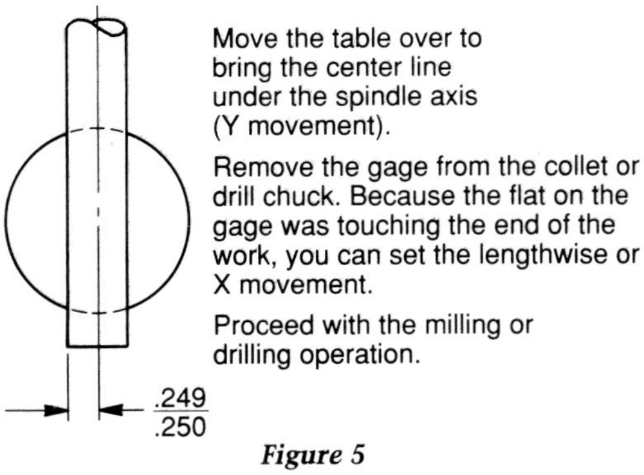

Figure 5

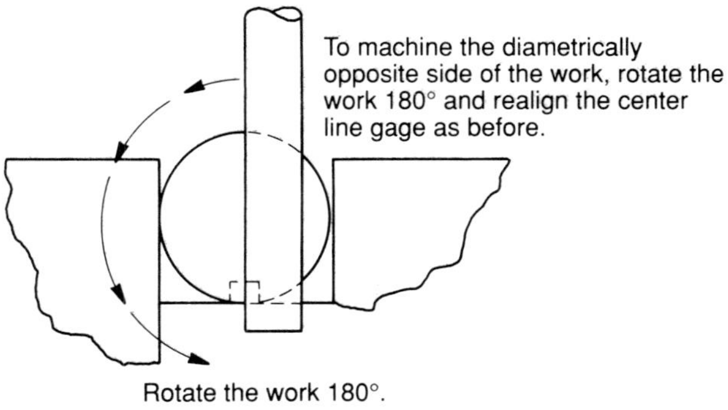

Figure 6

INDEX